PERIODS OF SUSCEPTIBILITY TO TERATOGENESIS

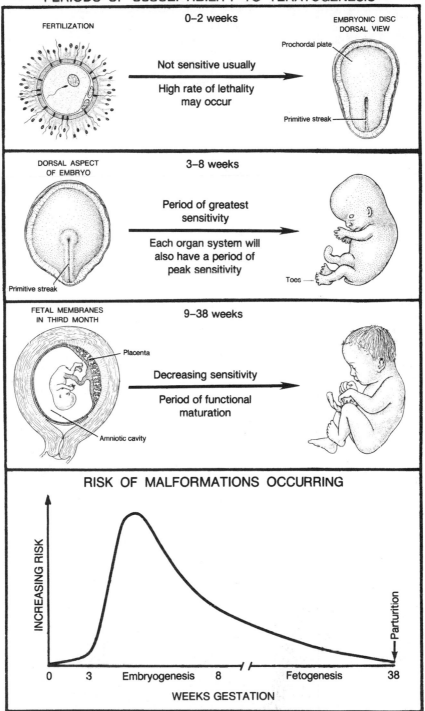

0–2 weeks

FERTILIZATION

EMBRYONIC DISC
DORSAL VIEW

Prochordal plate

Not sensitive usually

High rate of lethality
may occur

Primitive streak

3–8 weeks

DORSAL ASPECT
OF EMBRYO

Period of greatest
sensitivity

Each organ system will
also have a period of
peak sensitivity

Primitive streak

Toes

9–38 weeks

FETAL MEMBRANES
IN THIRD MONTH

Placenta

Decreasing sensitivity

Period of functional
maturation

Amniotic cavity

RISK OF MALFORMATIONS OCCURRING

INCREASING RISK

Parturition

0   3   Embryogenesis   8   Fetogenesis   38

WEEKS GESTATION

EMBRYONIC DEVELOPMENT IN DAYS

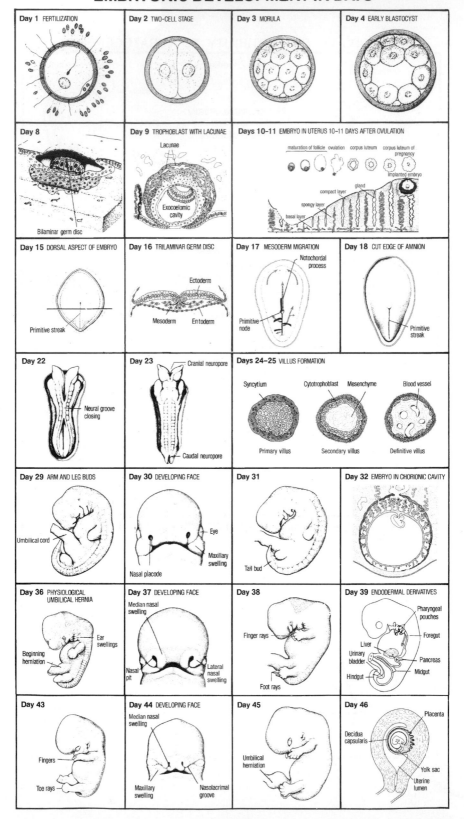

EMBRYONIC DEVELOPMENT IN DAYS

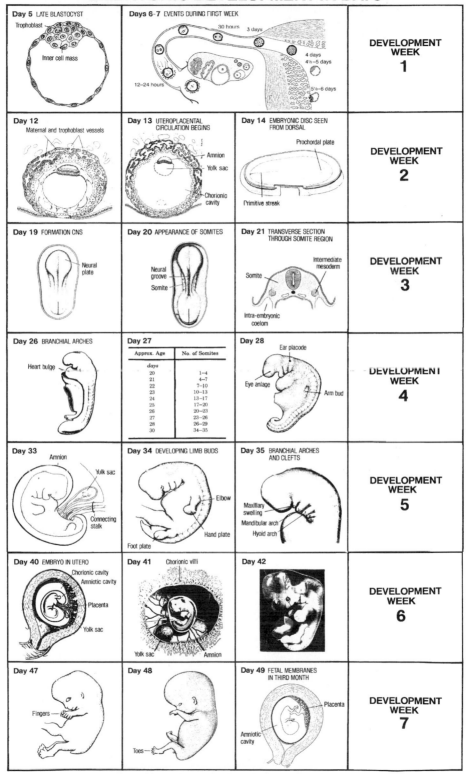

Day 5 LATE BLASTOCYST
- Trophoblast
- Inner cell mass

Days 6-7 EVENTS DURING FIRST WEEK
- 30 hours
- 3 days
- 4 days
- 4½–5 days
- 12–24 hours
- 5½–6 days

DEVELOPMENT WEEK **1**

Day 12
Maternal and trophoblast vessels

Day 13 UTEROPLACENTAL CIRCULATION BEGINS
- Amnion
- Yolk sac
- Chorionic cavity

Day 14 EMBRYONIC DISC SEEN FROM DORSAL
- Prochordal plate
- Primitive streak

DEVELOPMENT WEEK **2**

Day 19 FORMATION CNS
- Neural plate

Day 20 APPEARANCE OF SOMITES
- Neural groove
- Somite

Day 21 TRANSVERSE SECTION THROUGH SOMITE REGION
- Intermediate mesoderm
- Somite
- Intra-embryonic coelom

DEVELOPMENT WEEK **3**

Day 26 BRANCHIAL ARCHES
- Heart bulge

Day 27

Approx. Age	No. of Somites
days	
20	1–4
21	4–7
22	7–10
23	10–13
24	13–17
25	17–20
26	20–23
27	23–26
28	26–29
30	34–35

Day 28
- Ear placode
- Eye anlage
- Arm bud

DEVELOPMENT WEEK **4**

Day 33
- Amnion
- Yolk sac
- Connecting stalk

Day 34 DEVELOPING LIMB BUDS
- Elbow
- Hand plate
- Foot plate

Day 35 BRANCHIAL ARCHES AND CLEFTS
- Maxillary swelling
- Mandibular arch
- Hyoid arch

DEVELOPMENT WEEK **5**

Day 40 EMBRYO IN UTERO
- Chorionic cavity
- Amniotic cavity
- Placenta
- Yolk sac

Day 41
- Chorionic villi
- Yolk sac
- Amnion

Day 42

DEVELOPMENT WEEK **6**

Day 47
- Fingers

Day 48
- Toes

Day 49 FETAL MEMBRANES IN THIRD MONTH
- Placenta
- Amniotic cavity

DEVELOPMENT WEEK **7**

Langman's
Medical Embryology

NINTH EDITION

Langman's
Medical
Embryology

NINE EDITION

T. W. Sadler, Ph.D.
Consultant, Birth Defects Prevention
Twin Bridges
Madison County, Montana

Original Illustrations by **Jill Leland**

Computer Illustrations by **Susan L. Sadler-Redmond**

Scanning Electron Micrographs by
Kathleen K. Sulik and Jennifer Burgoon

Ultrasound Images by **Nancy Chescheir and Hytham Imseis**

LIPPINCOTT WILLIAMS & WILKINS
A Wolters Kluwer Company

Philadelphia • Baltimore • New York • London
Buenos Aires • Hong Kong • Sydney • Tokyo

Editor: Betty Sun
Managing Editor: Rebecca Kerins
Marketing Manager: Aimee Sirmon
Project Editor: Caroline Define
Designer: Doug Smock
Compositor: TechBooks
Printer: R.R. Donnelley & Sons

351 West Camden Street
Baltimore, Maryland 21201-2436 USA

530 Walnut Street
Philadelphia, Pennsylvania 19106-3621 USA

The publisher is not responsible (as a matter of product liability, negligence, or otherwise) for any injury resulting from any material contained herein. This publication contains information relating to general principles of medical care which should not be construed as specific instructions for individual patients. Manufacturers' product information and package inserts should be reviewed for current information, including contraindications, dosages, and precautions.

Printed in the United States of America

First edition, 1963	Fourth edition, 1981	Seventh edition, 1995
Second edition, 1969	Fifth edition, 1985	Eighth edition, 2000
Third edition, 1975	Sixth edition, 1990	

Library of Congress Cataloging-in-Publication Data

Sadler, T. W. (Thomas W.)
  Langman's medical embryology.—9th ed. / T. W. Sadler ; original illustrations by Jill Leland ; computer illustrations by Susan L. Sadler-Redmond ; scanning electron micrographs by Kathleen K. Sulik and Jennifer Burgoon ; ultrasound images by Nancy Chescheir and Hytham Imseis.
    p. ; cm.
  Includes bibliographical references and index.
  ISBN 0-7817-4310-9
  1. Embryology, Human. 2. Abnormalities, Human. I. Title: Medical embryology. II. Langman, Jan. III. Title.
  [DNLM: 1. Embryology. 2. Abnormalities. QS 604 S126m 2003]
  QM601 .L35 2003
  612.6'4—dc21

                                                                        2002043361

The publishers have made every effort to trace the copyright holders for borrowed material. If they have inadvertently overlooked any, they will be pleased to make the necessary arrangements at the first opportunity.

To purchase additional copies of this book call our customer service department at **(800) 638-3030** or fax orders to **(301) 824-7390**. International customers should call **(301) 714-2324**.

Visit Lippincott Williams & Wilkins on the Internet: http://www.lww.com. Lippincott Williams & Wilkins customer service representatives are available from 8:30 am to 6:00 pm, EST, Monday through Friday, for telephone access.

03 04 05
1 2 3 4 5 6 7 8 9 10

For each and every child and for Jaina and our friends, Parker, Ellie Mae, Colonel, Jack, Peanut, and Sam I Am.

Preface

The ninth edition of *Langman's Medical Embryology* adheres to the tradition established by the original publication—it provides a concise but thorough description of embryology and its clinical significance, an awareness of which is essential in the diagnosis and prevention of birth defects. Recent advances in genetics, developmental biology, maternal-fetal medicine, and public health have significantly increased our knowledge of embryology and its relevance. Because birth defects are the leading cause of infant mortality and a major contributor to disabilities, and because new prevention strategies have been developed, understanding the principles of embryology is important for health care professionals.

To accomplish its goal, *Langman's Medical Embryology* retains its unique approach of combining an economy of text with excellent diagrams and scanning electron micrographs. It reinforces basic embryologic concepts by providing numerous clinical examples that result from abnormalities in developmental processes. The following pedagogic features and updates in the ninth edition help facilitate student learning:

Organization of Material: *Langman's Medical Embryology* is organized into two parts. The first provides an overview of early development from gametogenesis through the embryonic period; also included in this section are chapters on placental and fetal development and prenatal diagnosis and birth defects. The second part of the text provides a description of the fundamental processes of embryogenesis for each organ system.

Molecular Biology: New information is provided about the molecular basis of normal and abnormal development.

Extensive Art Program: This edition features almost 400 illustrations, including new 4-color line drawings, scanning electron micrographs, and ultrasound images.

Clinical Correlates: In addition to describing normal events, each chapter contains clinical correlates that appear in highlighted boxes. This material is designed to provide information about birth defects and other clinical entities that are directly related to embryologic concepts.

Summary: At the end of each chapter is a summary that serves as a concise review of the key points described in detail throughout the chapter.

Problems to Solve: These problems test a student's ability to apply the information covered in a particular chapter. Detailed answers are provided in an appendix in the back of the book.

Simbryo: New to this edition, *Simbryo,* located in the back of the book, is an interactive CD-ROM that demonstrates normal embryologic events and the origins of some birth defects. This unique educational tool offers six original vector art animation modules to illustrate the complex, three-dimensional aspects of embryology. Modules include normal early development as well as head and neck, cardiovascular, gastrointestinal, genitourinary, and pulmonary system development.

Connection Web Site: This student and instructor site (http://connection. LWW.com/go/sadler) provides updates on new advances in the field and a syllabus designed for use with the book. The syllabus contains objectives and definitions of key terms organized by chapters and the "bottom line," which provides a synopsis of the most basic information that students should have mastered from their studies.

I hope you find this edition of *Langman's Medical Embryology* to be an excellent resource. Together, the textbook, CD, and connection site provide a user-friendly and innovative approach to learning embryology and its clinical relevance.

T. W. Sadler
Twin Bridges, Montana

Contents

part one

General Embryology

chapter 1

Gametogenesis: Conversion of Germ Cells Into Male and Female Gametes

Primordial Germ Cells

Development begins with fertilization, the process by which the male gamete, the **sperm,** and the female gamete, the **oocyte,** unite to give rise to a **zygote.** Gametes are derived from **primordial germ cells (PGCs** that are formed in the epiblast during the second week and that move to the wall of the yolk sac (Fig. 1.1). During the fourth week these cells begin to migrate from the yolk sac toward the developing gonads, where they arrive by the end of the fifth week. Mitotic divisions increase their number during their migration and also when they arrive in the gonad. In preparation for fertilization, germ cells undergo **gametogenesis,** which includes meiosis, to reduce the number of chromosomes and **cytodifferentiation** to complete their maturation.

CLINICAL CORRELATE

Primordial Germ Cells (PGCs) and Teratomas

Teratomas are tumors of disputed origin that often contain a variety of tissues, such as bone, hair, muscle, gut epithelia, and others. It is thought that these tumors arise from a pluripotent stem cell that can differentiate into any of the three germ layers or their derivatives.

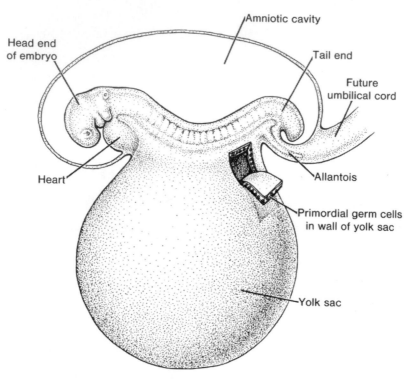

Figure 1.1 An embryo at the end of the third week, showing the position of primordial germ cells in the wall of the yolk sac, close to the attachment of the future umbilical cord. From this location, these cells migrate to the developing gonad.

Some evidence suggests that **PGCs** that have strayed from their normal migratory paths could be responsible for some of these tumors. Another source is epiblast cells migrating through the primitive streak during gastrulation (see page 80).

The Chromosome Theory of Inheritance

Traits of a new individual are determined by specific genes on chromosomes inherited from the father and the mother. Humans have approximately 35,000 genes on 46 chromosomes. Genes on the same chromosome tend to be inherited together and so are known as linked genes. In somatic cells, chromosomes appear as 23 **homologous** pairs to form the **diploid** number of 46. There are 22 pairs of matching chromosomes, the **autosomes,** and one pair of **sex chromosomes.** If the sex pair is XX, the individual is genetically female; if the pair is XY, the individual is genetically male. One chromosome of each pair is derived from the maternal gamete, the **oocyte,** and one from the paternal gamete, the

sperm. Thus each gamete contains a **haploid** number of 23 chromosomes, and the union of the gametes at **fertilization** restores the diploid number of 46.

MITOSIS

Mitosis is the process whereby one cell divides, giving rise to two daughter cells that are genetically identical to the parent cell (Fig. 1.2). Each daughter cell receives the complete complement of 46 chromosomes. Before a cell enters mitosis, each chromosome replicates its deoxyribonucleic acid (DNA). During this replication phase the chromosomes are extremely long, they are spread diffusely through the nucleus, and they cannot be recognized with the light microscope. With the onset of mitosis the chromosomes begin to coil, contract, and condense; these events mark the beginning of prophase. Each chromosome now consists of two parallel subunits, **chromatids,** that are joined at a narrow region common to both called the **centromere.** Throughout prophase the chromosomes continue to condense, shorten, and thicken (Fig. 1.2*A*), but only at prometaphase do the chromatids become distinguishable (Fig. 1.2*B*). During metaphase the chromosomes line up in the equatorial plane,

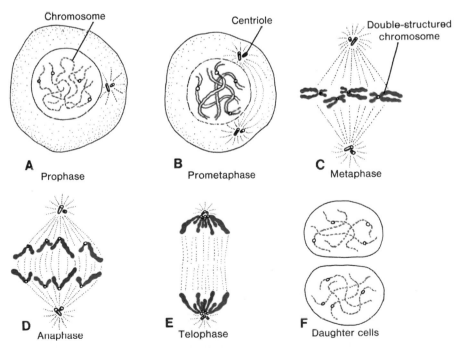

Figure 1.2 Various stages of mitosis. In prophase, chromosomes are visible as slender threads. Doubled chromatids become clearly visible as individual units during metaphase. At no time during division do members of a chromosome pair unite. *Blue,* paternal chromosomes; *red,* maternal chromosomes.

and their doubled structure is clearly visible (Fig. 1.2C). Each is attached by **microtubules** extending from the centromere to the centriole, forming the **mitotic spindle.** Soon the centromere of each chromosome divides, marking the beginning of anaphase, followed by migration of chromatids to opposite poles of the spindle. Finally, during telophase, chromosomes uncoil and lengthen, the nuclear envelope reforms, and the cytoplasm divides (Fig. 1.2, *D* and *E*). Each daughter cell receives half of all doubled chromosome material and thus maintains the same number of chromosomes as the mother cell.

MEIOSIS

Meiosis is the cell division that takes place in the **germ cells** to generate male and female gametes, sperm and egg cells, respectively. Meiosis requires two cell divisions, **meiosis I** and **meiosis II,** to reduce the number of chromosomes to the haploid number of 23 (Fig. 1.3). As in mitosis, male and female germ cells (**spermatocytes** and **primary oocytes**) at the beginning of meiosis I replicate their DNA so that each of the 46 chromosomes is duplicated into sister chromatids. In contrast to mitosis, however, **homologous chromosomes** then align themselves in **pairs,** a process called **synapsis.** The pairing is exact and point for point except for the XY combination. Homologous pairs then separate into two daughter cells. Shortly thereafter meiosis II separates sister chromatids. Each gamete then contains 23 chromosomes.

Crossover

Crossovers, critical events in meiosis I, are the **interchange of chromatid segments** between paired homologous chromosomes (Fig. 1.3C). Segments of chromatids break and are exchanged as homologous chromosomes separate. As separation occurs, points of interchange are temporarily united and form an X-like structure, a **chiasma** (Fig. 1.3C). The approximately 30 to 40 crossovers (one or two per chromosome) with each meiotic I division are most frequent between genes that are far apart on a chromosome.

As a result of meiotic divisions, (*a*) genetic variability is enhanced through crossover, which redistributes genetic material, and through random distribution of homologous chromosomes to the daughter cells; and (*b*) each germ cell contains a haploid number of chromosomes, so that at fertilization the diploid number of 46 is restored.

Polar Bodies

Also during meiosis one primary oocyte gives rise to four daughter cells, each with 22 plus 1 X chromosomes (Fig. 1.4A). However, only one of these develops into a mature gamete, the oocyte; the other three, the **polar bodies,** receive little cytoplasm and degenerate during subsequent development. Similarly, one primary spermatocyte gives rise to four daughter cells, two with 22 plus 1

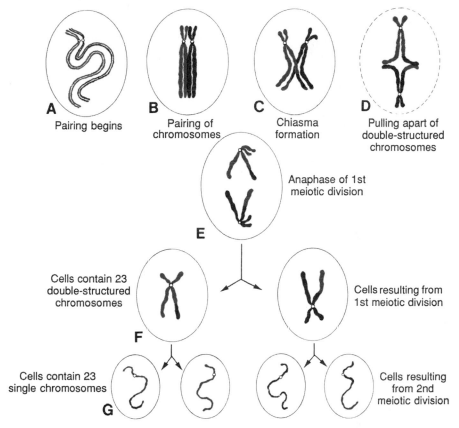

Figure 1.3 First and second meiotic divisions. **A.** Homologous chromosomes approach each other. **B.** Homologous chromosomes pair, and each member of the pair consists of two chromatids. **C.** Intimately paired homologous chromosomes interchange chromatid fragments (crossover). Note the chiasma. **D.** Double-structured chromosomes pull apart. **E.** Anaphase of the first meiotic division. **F** and **G.** During the second meiotic division, the double-structured chromosomes split at the centromere. At completion of division, chromosomes in each of the four daughter cells are different from each other.

X chromosomes and two with 22 plus 1 Y chromosomes (Fig. 1.4*B*). However, in contrast to oocyte formation, all four develop into mature gametes.

CLINICAL CORRELATES

Birth Defects and Spontaneous Abortions: Chromosomal and Genetic Factors

Chromosomal abnormalities, which may be **numerical** or **structural,** are important causes of birth defects and spontaneous abortions. It is estimated that 50% of conceptions end in spontaneous abortion and that 50% of these

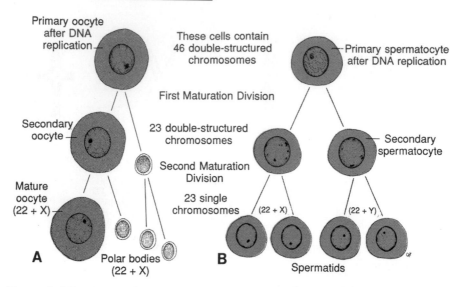

Figure 1.4 Events occurring during the first and second maturation divisions. **A.** The primitive female germ cell (primary oocyte) produces only one mature gamete, the mature oocyte. **B.** The primitive male germ cell (primary spermatocyte) produces four spermatids, all of which develop into spermatozoa.

abortuses have major chromosomal abnormalities. Thus approximately 25 % of conceptuses have a major chromosomal defect. The most common chromosomal abnormalities in abortuses are 45,X (Turner syndrome), triploidy, and trisomy 16. Chromosomal abnormalities account for 7 % of major birth defects, and **gene mutations** account for an additional 8 %.

Numerical Abnormalities

The normal human somatic cell contains 46 chromosomes; the normal gamete contains 23. Normal somatic cells are **diploid**, or $2n$; normal gametes are **haploid**, or n. **Euploid** refers to any exact multiple of n, e.g., diploid or triploid. **Aneuploid** refers to any chromosome number that is not euploid; it is usually applied when an extra chromosome is present (**trisomy**) or when one is missing (**monosomy**). Abnormalities in chromosome number may originate during meiotic or mitotic divisions. In **meiosis**, two members of a pair of homologous chromosomes normally separate during the first meiotic division so that each daughter cell receives one member of each pair (Fig. 1.5*A*). Sometimes, however, separation does not occur (**nondisjunction**), and both members of a pair move into one cell (Fig. 1.5, *B* and *C*). As a result of nondisjunction of the chromosomes, one cell receives 24 chromosomes, and the other receives 22 instead of the normal 23. When, at fertilization, a gamete having 23 chromosomes fuses with a gamete having 24 or

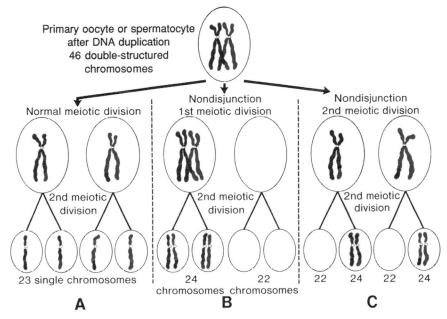

Primary oocyte or spermatocyte
after DNA duplication
46 double-structured
chromosomes

Normal meiotic division

Nondisjunction
1st meiotic division

Nondisjunction
2nd meiotic division

2nd meiotic
division

2nd meiotic
division

2nd meiotic
division

23 single chromosomes

24
chromosomes

22
chromosomes

22   24   22   24

A   **B**   **C**

Figure 1.5 A. Normal maturation divisions. **B.** Nondisjunction in the first meiotic division. **C.** Nondisjunction in the second meiotic division.

22 chromosomes, the result is an individual with either 47 chromosomes (trisomy) or 45 chromosomes (monosomy). Nondisjunction, which occurs during either the first or the second meiotic division of the germ cells, may involve the autosomes or sex chromosomes. In women, the incidence of chromosomal abnormalities, including nondisjunction, increases with age, especially at 35 years and older.

Occasionally nondisjunction occurs during mitosis (**mitotic nondisjunction**) in an embryonic cell during the earliest cell divisions. Such conditions produce **mosaicism,** with some cells having an abnormal chromosome number and others being normal. Affected individuals may exhibit few or many of the characteristics of a particular syndrome, depending on the number of cells involved and their distribution.

Sometimes chromosomes break, and pieces of one chromosome attach to another. Such **translocations** may be **balanced,** in which case breakage and reunion occur between two chromosomes but no critical genetic material is lost and individuals are normal; or they may be **unbalanced,** in which case part of one chromosome is lost and an altered phenotype is produced. For example, unbalanced translocations between the long arms of chromosomes 14 and 21 during meiosis I or II produce gametes with an extra copy of chromosome 21, one of the causes of Down syndrome (Fig. 1.6). Translocations

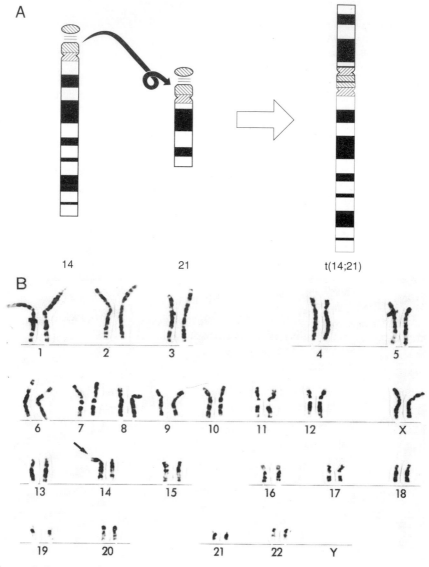

Figure 1.6 A. Translocation of the long arms of chromosomes 14 and 21 at the centromere. Loss of the short arms is not clinically significant, and these individuals are clinically normal, although they are at risk for producing offspring with unbalanced translocations. **B.** Karyotype of translocation of chromosome 21 onto 14, resulting in Down syndrome.

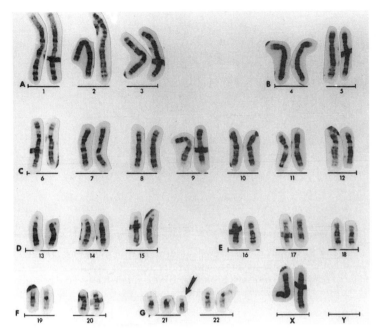

Figure 1.7 Karyotype of trisomy 21 (*arrow*), Down syndrome.

are particularly common between chromosomes 13, 14, 15, 21, and 22 because they cluster during meiosis.

TRISOMY 21 (DOWN SYNDROME)

Down syndrome is usually caused by an extra copy of **chromosome 21 (trisomy 21,** Fig. 1.7). Features of children with Down syndrome include growth retardation; varying degrees of mental retardation; craniofacial abnormalities, including upward slanting eyes, epicanthal folds (extra skin folds at the medial corners of the eyes), flat facies, and small ears; cardiac defects; and hypotonia (Fig. 1.8). These individuals also have relatively high incidences of leukemia, infections, thyroid dysfunction, and premature aging. Furthermore, nearly all develop signs of Alzheimer's disease after age 35. In 95% of cases, the syndrome is caused by trisomy 21 resulting from meiotic nondisjunction, and in 75% of these instances, nondisjunction occurs during oocyte formation. The incidence of Down syndrome is approximately 1 in 2000 conceptuses for women under age 25. This risk increases with maternal age to 1 in 300 at age 35 and 1 in 100 at age 40.

In approximately 4% of cases of Down syndrome, there is an unbalanced translocation between chromosome 21 and chromosome 13, 14, or 15 (Fig. 1.6). The final 1% are caused by mosaicism resulting from mitotic

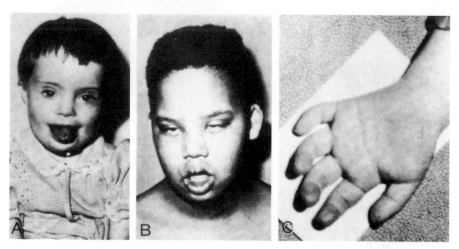

Figure 1.8 A and **B.** Children with Down syndrome, which is characterized by a flat, broad face, oblique palpebral fissures, epicanthus, and furrowed lower lip. **C.** Another characteristic of Down syndrome is a broad hand with single transverse or simian crease. Many children with Down syndrome are mentally retarded and have congenital heart abnormalities.

nondisjunction. These individuals have some cells with a normal chromosome number and some that are aneuploid. They may exhibit few or many of the characteristics of Down syndrome.

TRISOMY 18

Patients with **trisomy 18** show the following features: mental retardation, congenital heart defects, low-set ears, and flexion of fingers and hands (Fig. 1.9). In addition, patients frequently show micrognathia, renal anomalies, syndactyly, and malformations of the skeletal system. The incidence of this condition is approximately 1 in 5000 newborns. Eighty-five percent are lost between 10 weeks of gestation and term, whereas those born alive usually die by age 2 months.

TRISOMY 13

The main abnormalities of **trisomy 13** are mental retardation, holoprosencephaly, congenital heart defects, deafness, cleft lip and palate, and eye defects, such as microphthalmia, anophthalmia, and coloboma (Fig. 1.10). The incidence of this abnormality is approximately 1 in 20,000 live births, and over 90% of the infants die in the first month after birth.

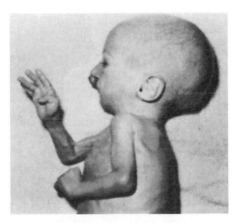

Figure 1.9 Photograph of child with trisomy 18. Note the prominent occiput, cleft lip, micrognathia, low-set ears, and one or more flexed fingers.

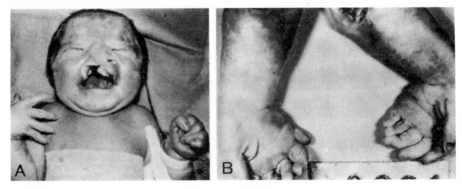

Figure 1.10 A. Child with trisomy 13. Note the cleft lip and palate, the sloping forehead, and microphthalmia. **B.** The syndrome is commonly accompanied by polydactyly.

KLINEFELTER SYNDROME

The clinical features of **Klinefelter syndrome,** found only in males and usually detected at puberty, are sterility, testicular atrophy, hyalinization of the seminiferous tubules, and usually gynecomastia. The cells have 47 chromosomes with a sex chromosomal complement of the XXY type, and a **sex chromatin body (Barr body:** formed by condensation of an inactivated sex chromosome; a Barr body is also present in normal females) is found in 80% of cases (Fig. 1.11). The incidence is approximately 1 in 500 males. Nondisjunction of the XX homologues is the most common causative event. Occasionally, patients with Klinefelter syndrome have 48 chromosomes: 44 autosomes and four sex chromosomes (XXXY). Although mental retardation is not generally

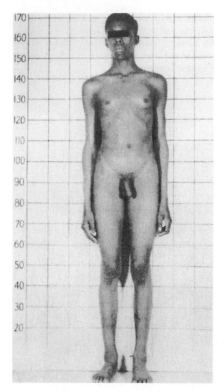

Figure 1.11 Patient with Klinefelter syndrome showing normal phallus development but gynecomastia (enlarged breasts).

part of the syndrome, the more X chromosomes there are, the more likely there will be some degree of mental impairment.

TURNER SYNDROME

Turner syndrome, with a 45,X karyotype, is the only monosomy compatible with life. Even then, 98% of all fetuses with the syndrome are spontaneously aborted. The few that survive are unmistakably female in appearance (Fig. 1.12) and are characterized by the absence of ovaries (**gonadal dysgenesis**) and short stature. Other common associated abnormalities are webbed neck, lymphedema of the extremities, skeletal deformities, and a broad chest with widely spaced nipples. Approximately 55% of affected women are monosomic for the X and chromatin body negative because of nondisjunction. In 80% of these women, nondisjunction in the male gamete is the cause. In the remainder of women, structural abnormalities of the X chromosome or mitotic nondisjunction resulting in mosaicism are the cause.

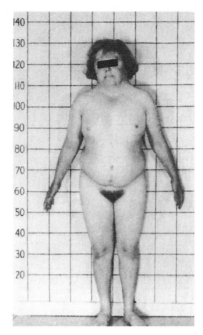

Figure 1.12 Patient with Turner syndrome. The main characteristics are webbed neck, short stature, broad chest, and absence of sexual maturation.

TRIPLE X SYNDROME

Patients with **triple X syndrome** are infantite, with scanty menses and some degree of mental retardation. They have two sex chromatin bodies in their cells.

Structural Abnormalities

Structural chromosome abnormalities, which involve one or more chromosomes, usually result from chromosome breakage. Breaks are caused by environmental factors, such as viruses, radiation, and drugs. The result of breakage depends on what happens to the broken pieces. In some cases, the broken piece of a chromosome is lost, and the infant with partial **deletion** of a chromosome is abnormal. A well-known syndrome, caused by partial deletion of the short arm of chromosome 5, is the **cri-du-chat syndrome.** Such children have a catlike cry, microcephaly, mental retardation, and congenital heart disease. Many other relatively rare syndromes are known to result from a partial chromosome loss.

   Microdeletions, spanning only a few **contiguous genes,** may result in **microdeletion syndrome** or **contiguous gene syndrome.** Sites where these deletions occur, called **contiguous gene complexes,** can be identified by **high-resolution chromosome banding.** An example of a microdeletion

Figure 1.13 Patient with Angelman syndrome resulting from a microdeletion on maternal chromosome 15. If the defect is inherited on the paternal chromosome, Prader-Willi syndrome occurs (Fig. 1.14).

occurs on the long arm of chromosome 15 (15q11–15q13). Inheriting the deletion on the maternal chromosome results in **Angelman syndrome,** and the children are mentally retarded, cannot speak, exhibit poor motor development, and are prone to unprovoked and prolonged periods of laughter (Fig. 1.13). If the defect is inherited on the paternal chromosome, **Prader-Willi syndrome** is produced; affected individuals are characterized by hypotonia, obesity, mental retardation, hypogonadism, and cryptorchidism (Fig. 1.14). Characteristics that are differentially expressed depending upon whether the genetic material is inherited from the mother or the father are examples of **genomic imprinting.** Other contiguous gene syndromes may be inherited from either parent, including **Miller-Dieker syndrome** (lissencephaly, developmental delay, seizures, and cardiac and facial abnormalities resulting from a deletion at 17p13) and most cases of **velocardiofacial (Shprintzen) syndrome** (palatal defects, conotruncal heart defects, speech delay, learning disorders, and schizophrenia-like disorder resulting from a deletion in 22q11).

    Fragile sites are regions of chromosomes that demonstrate a propensity to separate or break under certain cell manipulations. For example, fragile sites can be revealed by culturing lymphocytes in folate-deficient medium. Although numerous fragile sites have been defined and consist of **CGG repeats,** only the site on the long arm of the X chromosome (Xq27) has been

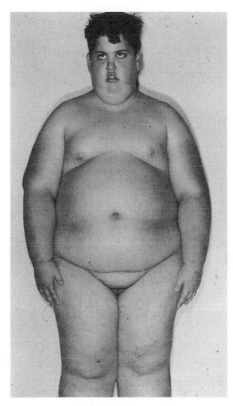

Figure 1.14 Patient with Prader-Willi syndrome resulting from a microdeletion on pater-nal chromosome 15. If the defect is inherited on the maternal chromosome, Angelman syndrome occurs (Fig. 1.13).

correlated with an altered phenotype and is called the **fragile X syndrome.** Fragile X syndrome is characterized by mental retardation, large ears, promi-nent jaw, and pale blue irides. Males are affected more often than females (1/1000 versus 1/2000), which may account for the preponderance of males among the mentally retarded. Fragile X syndrome is second only to Down syndrome as a cause of mental retardation because of chromosomal abnor-malities.

Gene Mutations

Many congenital formations in humans are inherited, and some show a clear mendelian pattern of inheritance. Many birth defects are directly attributable to a change in the structure or function of a single gene, hence the name **single gene mutation.** This type of defect is estimated to account for approximately 8% of all human malformations.

With the exception of the X and Y chromosomes in the male, genes exist as pairs, or **alleles,** so that there are two doses for each genetic determinant, one from the mother and one from the father. If a mutant gene produces an abnormality in a single dose, despite the presence of a normal allele, it is a **dominant mutation.** If both alleles must be abnormal (double dose) or if the mutation is X-linked in the male, it is a **recessive mutation.** Gradations in the effects of mutant genes may be a result of modifying factors.

The application of molecular biological techniques has increased our knowledge of genes responsible for normal development. In turn, genetic analysis of human syndromes has shown that mutations in many of these same genes are responsible for some congenital abnormalities and childhood diseases. Thus, the link between key genes in development and their role in clinical syndromes is becoming clearer.

In addition to causing congenital malformations, mutations can result in **inborn errors of metabolism.** These diseases, among which phenylketonuria, homocystinuria, and galactosemia are the best known, are frequently accompanied by or cause various degrees of mental retardation.

Diagnostic Techniques for Identifying Genetic Abnormalities

Cytogenetic analysis is used to assess chromosome number and integrity. The technique requires dividing cells, which usually means establishing cell cultures that are arrested in metaphase by chemical treatment. Chromosomes are stained with **Giemsa stain** to reveal light and dark banding patterns (G-bands; Fig. 1.6) unique for each chromosome. Each band represents 5 to 10×10^6 base pairs of DNA, which may include a few to several hundred genes. Recently, **high resolution metaphase banding techniques** have been developed that demonstrate greater numbers of bands representing even smaller pieces of DNA, thereby facilitating diagnosis of small deletions.

New molecular techniques, such as **fluorescence in situ hybridization** (FISH), use specific DNA probes to identify ploidy for a few selected chromosomes. Fluorescent probes are hybridized to chromosomes or genetic loci using cells on a slide, and the results are visualized with a fluorescence microscope (Fig.1.15). **Spectral karyotype analysis** is a technique in which every chromosome is hybridized to a unique fluorescent probe of a different color. Results are then analyzed by a computer.

Morphological Changes During Maturation of the Gametes

OOGENESIS

Maturation of Oocytes Begins Before Birth

Once primordial germ cells have arrived in the gonad of a genetic female, they differentiate into **oogonia** (Fig. 1.16, *A* and *B*). These cells undergo a number

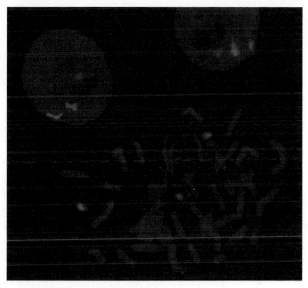

Figure 1.15 Fluorescence in situ hybridization (FISH) using a probe for chromosome 21. Two interphase cells and a metaphase spread of chromosomes are shown; each has three domains, indicated by the probe, characteristic of trisomy 21 (Down syndrome).

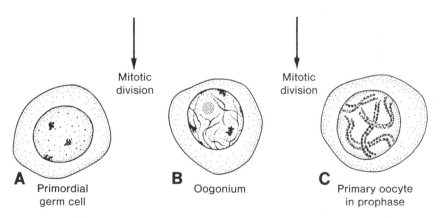

Figure 1.16 Differentiation of primordial germ cells into oogonia begins shortly after their arrival in the ovary. By the third month of development, some oogonia give rise to primary oocytes that enter prophase of the first meiotic division. This prophase may last 40 or more years and finishes only when the cell begins its final maturation. During this period it carries 46 double-structured chromosomes.

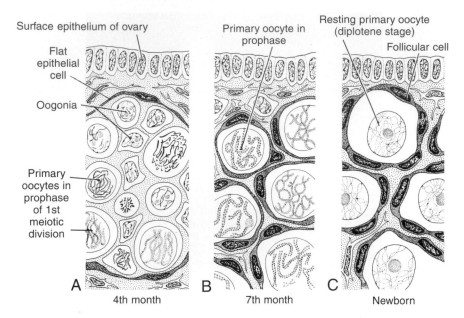

Surface epithelium of ovary

Flat
epithelial
cell

Oogonia

Primary
oocytes in
prophase
of 1st
meiotic
division

A   4th month

Primary oocyte in
prophase

B   7th month

Resting primary oocyte
(diplotene stage)

Follicular cell

C   Newborn

Figure 1.17 Segment of the ovary at different stages of development. **A.** Oogonia are grouped in clusters in the cortical part of the ovary. Some show mitosis; others have differentiated into primary oocytes and entered prophase of the first meiotic division. **B.** Almost all oogonia are transformed into primary oocytes in prophase of the first meiotic division. **C.** There are no oogonia. Each primary oocyte is surrounded by a single layer of follicular cells, forming the primordial follicle. Oocytes have entered the diplotene stage of prophase, in which they remain until just before ovulation. Only then do they enter metaphase of the first meiotic division.

of mitotic divisions and, by the end of the third month, are arranged in clusters surrounded by a layer of flat epithelial cells (Fig. 1.17 and 1.18). Whereas all of the oogonia in one cluster are probably derived from a single cell, the flat epithelial cells, known as **follicular cells,** originate from surface epithelium covering the ovary.

The majority of oogonia continue to divide by mitosis, but some of them arrest their cell division in prophase of meiosis I and form **primary oocytes** (Figs. 1.16*C* and 1.17*A*). During the next few months, oogonia increase rapidly in number, and by the fifth month of prenatal development, the total number of germ cells in the ovary reaches its maximum, estimated at 7 million. At this time, cell death begins, and many oogonia as well as primary oocytes become atretic. By the seventh month, the majority of oogonia have degenerated except for a few near the surface. All surviving primary oocytes have entered prophase of meiosis I, and most of them are individually surrounded by a layer of flat epithelial cells (Fig. 1.17*B*). A primary oocyte, together with its surrounding flat epithelial cells, is known as a **primordial follicle** (Fig. 1.19*A*).

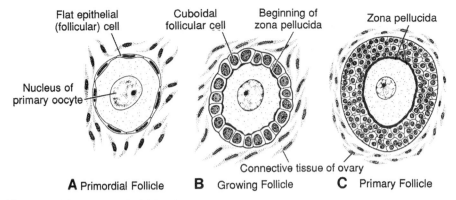

A Primordial Follicle   **B** Growing Follicle   **C** Primary Follicle

Figure 1.18 A. Primordial follicle consisting of a primary oocyte surrounded by a layer of flattened epithelial cells. **B.** Early primary or preantral stage follicle recruited from the pool of primordial follicles. As the follicle grows, follicular cells become cuboidal and begin to secrete the zona pellucida, which is visible in irregular patches on the surface of the oocyte. **C.** Mature primary (preantral) follicle with follicular cells forming a stratified layer of granulosa cells around the oocyte and the presence of a well-defined zona pellucida.

Maturation of Oocytes Continues at Puberty

Near the time of birth, all primary oocytes have started prophase of meiosis I, but instead of proceeding into metaphase, they enter the **diplotene stage,** a resting stage during prophase that is characterized by a lacy network of chromatin (Fig. 1.17*C*). **Primary oocytes remain in prophase** and **do not finish their first meiotic division before puberty is reached,** apparently because of **oocyte maturation inhibitor** (OMI), a substance secreted by follicular cells. The total number of primary oocytes at birth is estimated to vary from 700,000 to 2 million. During childhood most oocytes become atretic; only approximately 400,000 are present by the beginning of puberty, and fewer than 500 will be ovulated. Some oocytes that reach maturity late in life have been dormant in the diplotene stage of the first meiotic division for 40 years or more before ovulation. Whether the diplotene stage is the most suitable phase to protect the oocyte against environmental influences is unknown. The fact that the risk of having children with chromosomal abnormalities increases with maternal age indicates that primary oocytes are vulnerable to damage as they age.

At puberty, a pool of growing follicles is established and continuously maintained from the supply of primordial follicles. Each month, 15 to 20 follicles selected from this pool begin to mature, passing through three stages: 1) **primary** or **preantral;** 2) **secondary** or **antral** (also called **vesicular** or **Graafian);** and 3) **preovulatory.** The antral stage is the longest, whereas the preovulatory stage encompasses approximately 37 hours before ovulation. As the primary oocyte begins to grow, surrounding follicular cells change from flat to cuboidal and proliferate to produce a stratified epithelium of **granulosa cells,** and the unit

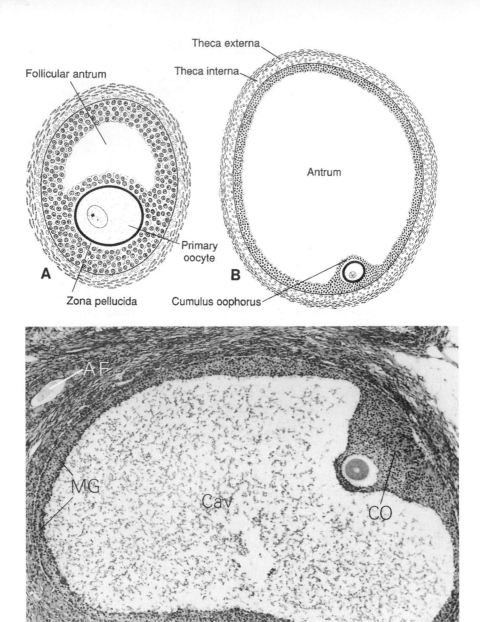

Figure 1.19 A. Secondary (antral) stage follicle. The oocyte, surrounded by the zona pellucida, is off-center; the antrum has developed by fluid accumulation between intercellular spaces. Note the arrangement of cells of the theca interna and the theca externa. **B.** Mature secondary (graafian) follicle. The antrum has enlarged considerably, is filled with follicular fluid, and is surrounded by a stratified layer of granulosa cells. The oocyte is embedded in a mound of granulosa cells, the cumulus oophorus. **C.** Photomicrograph of a mature secondary follicle with an enlarged fluid-filled antrum (cavity, *Cav*) and a diameter of 20 mm (×65). *CO,* cumulus oophorus; *MG,* granulosa cells; *AF,* atretic follicle.

is called a **primary follicle** (Fig. 1.18, *B* and *C*). Granulosa cells rest on a basement membrane separating them from surrounding stromal cells that form the **theca folliculi.** Also, granulosa cells and the oocyte secrete a layer of glycoproteins on the surface of the oocyte, forming the **zona pellucida** (Fig. 1.18*C*). As follicles continue to grow, cells of the theca folliculi organize into an inner layer of secretory cells, the **theca interna,** and an outer fibrous capsule, the **theca externa.** Also, small, finger-like processes of the follicular cells extend across the zona pellucida and interdigitate with microvilli of the plasma membrane of the oocyte. These processes are important for transport of materials from follicular cells to the oocyte.

As development continues, fluid-filled spaces appear between granulosa cells. Coalescence of these spaces forms the **antrum,** and the follicle is termed a **secondary (vesicular, Graafian) follicle.** Initially, the antrum is crescent shaped, but with time, it enlarges (Fig. 1.19). Granulosa cells surrounding the oocyte remain intact and form the **cumulus oophorus.** At maturity, the **secondary follicle** may be 25 mm or more in diameter. It is surrounded by the theca interna, which is composed of cells having characteristics of steroid secretion, rich in blood vessels, and the theca externa, which gradually merges with the ovarian stroma (Fig. 1.19).

With each ovarian cycle, a number of follicles begin to develop, but usually only one reaches full maturity. The others degenerate and become atretic (Fig. 1.19*C*). When the secondary follicle is mature, a surge in **luteinizing hormone (LH)** induces the preovulatory growth phase. Meiosis I is completed, resulting in formation of two daughter cells of unequal size, each with 23 double-structured chromosomes (Fig. 1.20, *A* and *B*). One cell, the **secondary oocyte,** receives most of the cytoplasm; the other, the **first polar** body, receives practically none. The first polar body lies between the zona pellucida and the cell

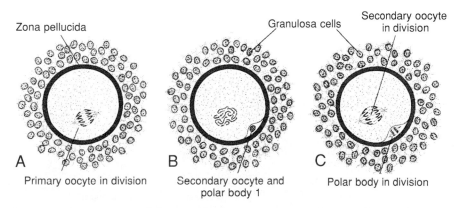

Zona pellucida · Granulosa cells · Secondary oocyte in division

A Primary oocyte in division · **B** Secondary oocyte and polar body 1 · **C** Polar body in division

Figure 1.20 Maturation of the oocyte. **A.** Primary oocyte showing the spindle of the first meiotic division. **B.** Secondary oocyte and first polar body. The nuclear membrane is absent. **C.** Secondary oocyte showing the spindle of the second meiotic division. The first polar body is also dividing.

membrane of the secondary oocyte in the perivitelline space (Fig. 1.20*B*). The cell then enters meiosis II but arrests in metaphase approximately 3 hours before ovulation. Meiosis II is completed only if the oocyte is fertilized; otherwise, the cell degenerates approximately 24 hours after ovulation. The first polar body also undergoes a second division (Fig. 1.20C).

SPERMATOGENESIS

Maturation of Sperm Begins at Puberty

Spermatogenesis, which begins at puberty, includes all of the events by which **spermatogonia** are transformed into **spermatozoa.** At birth, germ cells in the male can be recognized in the sex cords of the testis as large, pale cells surrounded by supporting cells (Fig. 1.21*A*). Supporting cells, which are derived from the surface epithelium of the gland in the same manner as follicular cells, become **sustentacular cells,** or **Sertoli cells** (Fig. 1.21*C*).

Shortly before puberty, the sex cords acquire a lumen and become the **seminiferous tubules.** At about the same time, primordial germ cells give rise to spermatogonial stem cells. At regular intervals, cells emerge from this stem cell population to form **type A spermatogonia,** and their production marks the initiation of spermatogenesis. Type A cells undergo a limited number of mitotic divisions to form a clone of cells. The last cell division produces **type B spermatogonia,** which then divide to form **primary spermatocytes** (Figs. 1.21 and 1.22). Primary spermatocytes then enter a prolonged

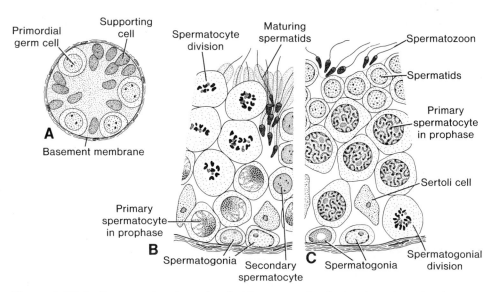

Figure 1.21 A. Cross section through primitive sex cords of a newborn boy showing primordial germ cells and supporting cells. **B** and **C.** Two segments of a seminiferous tubule in transverse section. Note the different stages of spermatogenesis.

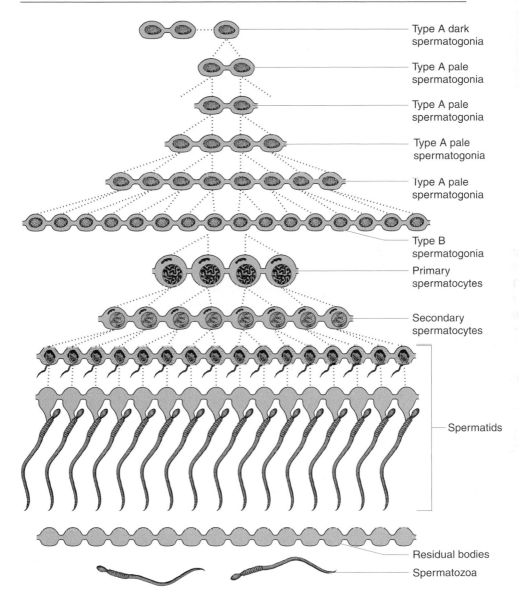

Type A dark spermatogonia

Type A pale spermatogonia

Type A pale spermatogonia

Type A pale spermatogonia

Type A pale spermatogonia

Type B spermatogonia

Primary spermatocytes

Secondary spermatocytes

Spermatids

Residual bodies

Spermatozoa

Figure 1.22 Type A spermatogonia, derived from the spermatogonial stem cell population, represent the first cells in the process of spermatogenesis. Clones of cells are established and cytoplasmic bridges join cells in each succeeding division until individual sperm are separated from residual bodies. In fact, the number of individual interconnected cells is considerably greater than depicted in this figure.

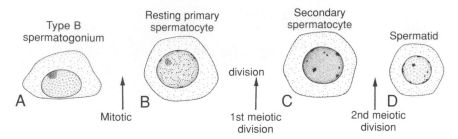

Figure 1.23 The products of meiosis during spermatogenesis in humans.

prophase (22 days) followed by rapid completion of meiosis I and formation of **secondary spermatocytes.** During the second meiotic division, these cells immediately begin to form haploid **spermatids** (Figs. 1.21–1.23). Throughout this series of events, from the time type A cells leave the stem cell population to formation of spermatids, cytokinesis is incomplete, so that successive cell generations are joined by cytoplasmic bridges. Thus, the progeny of a single type A spermatogonium form a clone of germ cells that maintain contact throughout differentiation (Fig. 1.22). Furthermore, spermatogonia and spermatids remain embedded in deep recesses of Sertoli cells throughout their development (Fig. 1.24). In this manner, Sertoli cells support and protect the germ cells, participate in their nutrition, and assist in the release of mature spermatozoa.

Spermatogenesis is regulated by **luteinizing hormone (LH)** production by the pituitary. LH binds to receptors on Leydig cells and stimulates testosterone production, which in turn binds to Sertoli cells to promote spermatogenesis. **Follicle stimulating hormone (FSH)** is also essential because its binding to Sertoli cells stimulates testicular fluid production and synthesis of intracellular androgen receptor proteins.

Spermiogenesis

The series of changes resulting in the transformation of spermatids into spermatozoa is **spermiogenesis.** These changes include (*a*) formation of the **acrosome,** which covers half of the nuclear surface and contains enzymes to assist in penetration of the egg and its surrounding layers during fertilization (Fig. 1.25); (*b*) condensation of the nucleus; (*c*) formation of neck, middle piece, and tail; and (*d*) shedding of most of the cytoplasm. In humans, the time required for a spermatogonium to develop into a mature spermatozoon is approximately 64 days.

When fully formed, spermatozoa enter the lumen of seminiferous tubules. From there, they are pushed toward the epididymis by contractile elements in the wall of the seminiferous tubules. Although initially only slightly motile, spermatozoa obtain full motility in the epididymis.

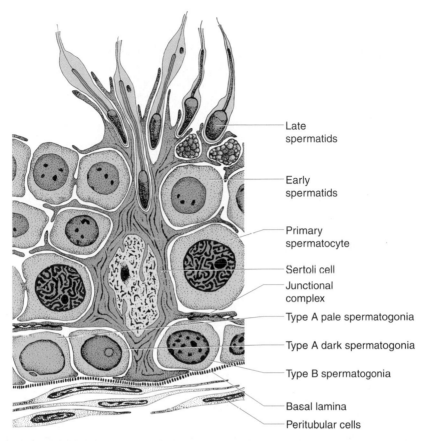

Late
spermatids

Early
spermatids

Primary
spermatocyte

Sertoli cell

Junctional
complex

Type A pale spermatogonia

Type A dark spermatogonia

Type B spermatogonia

Basal lamina

Peritubular cells

Figure 1.24 Sertoli cells and maturing spermatocytes. Spermatogonia, spermatocytes, and early spermatids occupy depressions in basal aspects of the cell; late spermatids are in deep recesses near the apex.

CLINICAL CORRELATES

Abnormal Gametes

In humans and in most mammals, one ovarian follicle occasionally contains two or three clearly distinguishable primary oocytes (Fig. 1.26*A*). Although these oocytes may give rise to twins or triplets, they usually degenerate before reaching maturity. In rare cases, one primary oocyte contains two or even three nuclei (Fig. 1.26*B*). Such binucleated or trinucleated oocytes die before reaching maturity.

In contrast to atypical oocytes, abnormal spermatozoa are seen frequently, and up to 10% of all spermatozoa have observable defects. The head or the tail may be abnormal; spermatozoa may be giants or dwarfs; and sometimes they are joined (Fig. 1.26*C*). Sperm with morphologic abnormalities lack normal motility and probably do not fertilize oocytes.

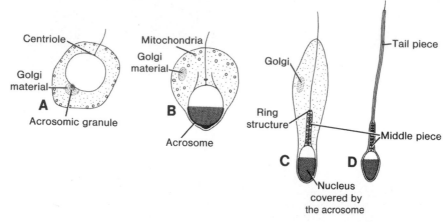

Figure 1.25 Important stages in transformation of the human spermatid into the spermatozoon.

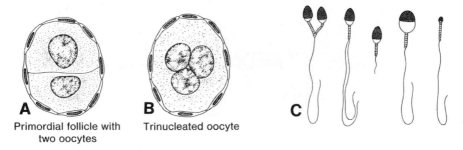

Figure 1.26 Abnormal germ cells. **A.** Primordial follicle with two oocytes. **B.** Trinucleated oocyte. **C.** Various types of abnormal spermatozoa.

Summary

Primordial germ cells appear in the wall of the yolk sac in the fourth week and migrate to the indifferent gonad (Fig. 1.1), where they arrive at the end of the fifth week. In preparation for fertilization, both male and female germ cells undergo **gametogenesis,** which includes **meiosis** and **cytodifferentiation.** During meiosis I, **homologous chromosomes pair** and **exchange genetic material;** during meiosis II, cells fail to replicate DNA, and each cell is thus provided with a **haploid** number of chromosomes and half the amount of DNA of a normal somatic cell (Fig. 1.3). Hence, mature male and female gametes have, respectively, 22 plus X or 22 plus Y chromosomes.

Birth defects may arise through abnormalities in **chromosome number** or **structure** and from **single gene mutations.** Approximately 7% of major birth defects are a result of chromosome abnormalities, and 8%, are a result of gene mutations. **Trisomies** (an extra chromosome) and **monosomies** (loss of a chromosome) arise during mitosis or meiosis. During meiosis, homologous chromosomes normally pair and then separate. However, if separation fails (**nondisjunction**), one cell receives too many chromosomes and one receives too few (Fig. 1.5). The incidence of abnormalities of chromosome number increases with age of the mother, particularly with mothers aged 35 years and older. Structural abnormalities of chromosomes include large **deletions (cri-du-chat syndrome)** and **microdeletions.** Microdeletions involve **contiguous genes** that may result in defects such as **Angelman syndrome** (maternal deletion, chromosome 15q11–15q13) or **Prader-Willi syndrome** (paternal deletion, 15q11–15q13). Because these syndromes depend on whether the affected genetic material is inherited from the mother or the father, they also are an example of **imprinting.** Gene mutations may be **dominant** (only one gene of an allelic pair has to be affected to produce an alteration) or **recessive** (both allelic gene pairs must be mutated). Mutations responsible for many birth defects affect genes involved in normal embryological development.

In the female, maturation from primitive germ cell to mature gamete, which is called **oogenesis, begins before birth;** in the male, it is called **spermatogenesis,** and it **begins at puberty.** In the female, primordial germ cells form **oogonia.** After repeated mitotic divisions, some of these arrest in prophase of meiosis I to form **primary oocytes.** By the seventh month, nearly all oogonia have become atretic, and only primary oocytes remain surrounded by a layer of **follicular cells** derived from the surface epithelium of the ovary (Fig. 1.17). Together, they form the **primordial follicle.** At puberty, a pool of growing follicles is recruited and maintained from the finite supply of primordial follicles. Thus, everyday 15 to 20 follicles begin to grow, and as they mature, they pass through three stages: 1) **primary** or **preantral;** 2) **secondary** or **antral (vesicular, Graafian);** and 3) **preovulatory.** The primary oocyte remains in prophase of the first meiotic division until the secondary follicle is mature. At this point, a surge in **luteinizing hormone (LH)** stimulates preovulatory growth: meiosis I is completed and a secondary oocyte and polar body are formed. Then, the secondary oocyte is arrested in metaphase of meiosis II approximately 3 hours before ovulation and will not complete this cell division until fertilization. In the male, primordial cells remain dormant until puberty, and only then do they differentiate into spermatogonia. These stem cells give rise to primary spermatocytes, which through two successive meiotic divisions produce four **spermatids** (Fig. 1.4). Spermatids go through a series of changes (**spermiogenesis**) (Fig. 1.25) including (*a*) formation of the acrosome, (*b*) condensation of the nucleus, (*c*) formation of neck, middle piece, and tail, and (*d*) shedding of most of the cytoplasm. The time required for a spermatogonium to become a mature spermatozoon is approximately 64 days.

Problems to Solve

1. *What is the most common cause of abnormal chromosome number? Give an example of a clinical syndrome involving abnormal numbers of chromosomes.*

2. *In addition to numerical abnormalities, what types of chromosomal alterations occur?*

3. *What is mosaicism, and how does it occur?*

SUGGESTED READING

Chandley AC: Meiosis in man. *Trends Genet* 4:79, 1988.

Clermont Y: Kinetics of spermatogenesis in mammals: seminiferous epithelium cycle and spermatogonial renewal. *Physiol Rev* 52:198, 1972.

Eddy EM, Clark JM, Gong D, Fenderson BA: Origin and migration of primordial germ cells in mammals. *Gamete Res* 4:333, 1981.

Gelchrter TD, Collins FS: *Principles of Medical Genetics.* Baltimore, Williams & Wilkins, 1990.

Gorlin RJ, Cohen MM, Levin LS (eds): *Syndromes of the Head and Neck.* 3rd ed. New York, Oxford University, 1990.

Heller CG, Clermont Y: Kinetics of the germinal epithelium in man. *Recent Prog Horm Res* 20:545, 1964.

Johnson MH, Everett BJ: *Essential Reproduction.* 5th ed. London, Blackwell Science Limited, 2000.

Jones KL (ed): *Smith's Recognizable Patterns of Human Malformation.* 4th ed. Philadelphia, WB Saunders, 1988.

Larsen WJ, Wert SE: Roles of cell junctions in gametogenesis and early embryonic development. *Tissue Cell* 20:809, 1988.

Lenke RR, Levy HL: Maternal phenylketonuria and hyperphenylalaninemia: an international survey of untreated and treated pregnancies. *N Engl J Med* 303:1202, 1980.

Pelletier RA, We K, Balakier H: Development of membrane differentiations in the guinea pig spermatid during spermiogenesis. *Am J Anat* 167:119, 1983.

Russell LD: Sertoligerm cell interactions: a review. *Gamete Res* 3:179, 1980.

Stevenson RE, Hall JG, Goodman RM (eds): *Human Malformations and Related Anomalies.* Vol I, II. New York, Oxford University Press, 1993.

Thorogood P (ed): *Embryos, Genes, and Birth Defects.* New York, Wiley, 1997.

Witschj E: Migration of the germ cells of the human embryos from the yolk sac to the primitive gonadal folds. *Contrib Embryol* 36:67, 1948.

chapter 2

First Week of Development: Ovulation to Implantation

Ovarian Cycle

At puberty, the female begins to undergo regular monthly cycles. These **sexual cycles** are controlled by the hypothalamus. **Gonadotropin-releasing hormone (GnRH)** produced by the hypothalamus acts on cells of the anterior pituitary gland, which in turn secrete **gonadotropins.** These hormones, **follicle-stimulating hormone (FSH)** and **luteinizing hormone (LH),** stimulate and control cyclic changes in the ovary.

At the beginning of each ovarian cycle, 15 to 20 primary (preantral) stage follicles are stimulated to grow under the influence of FSH. (The hormone is not necessary to promote development of primordial follicles to the primary follicle stage, but without it, these primary follicles die and become atretic.) Thus, FSH rescues 15 to 20 of these cells from a pool of continuously forming primary follicles (Fig. 2.1). Under normal conditions, only one of these follicles reaches full maturity, and only one oocyte is discharged; the others degenerate and become atretic. In the next cycle, another group of primary follicles is recruited, and again, only one follicle reaches maturity. Consequently, most follicles degenerate without ever reaching full maturity. When a follicle becomes atretic, the oocyte and surrounding follicular cells degenerate and are replaced by connective tissue, forming a **corpus atreticum.** FSH also stimulates maturation of **follicular (granulosa)** cells surrounding the oocyte. In turn, proliferation of these cells is mediated by growth differentiation

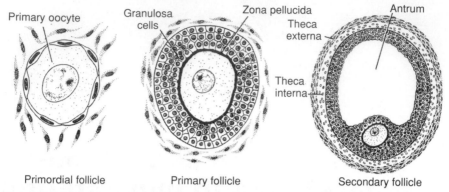

Figure 2.1 From the pool of primordial follicles, every day some begin to grow and develop into secondary (preantral) follicles, and this growth is independent of FSH. Then, as the cycle progresses, FSH secretion recruits primary follicles to begin development into secondary (antral, Graafian) follicles. During the last few days of maturation of secondary follicles, estrogens, produced by follicular and thecal cells, stimulate increased production of LH by the pituitary (Fig. 2.13), and this hormone causes the follicle to enter the preovulatory stage, to complete meiosis I, and to enter meiosis II where it arrests in metaphase approximately 3 hours before ovulation.

factor-9 (GDF-9), a member of the transforming growth factor-β (TGF-β) family. In cooperation, granulosa and thecal cells produce estrogens that (*a*) cause the uterine endometrium to enter the follicular or **proliferative phase;** (*b*) cause thinning of the cervical mucus to allow passage of sperm; and (*c*) stimulate the pituitary gland to secrete LH. At mid-cycle, there is an **LH surge** that (*a*) elevates concentrations of maturation-promoting factor, causing oocytes to complete meiosis I and initiate meiosis II; (*b*) stimulates production of progesterone by follicular stromal cells **(luteinization);** and (*c*) causes follicular rupture and ovulation.

OVULATION

In the days immediately preceding ovulation, under the influence of FSH and LH, the secondary follicle grows rapidly to a diameter of 25 mm. Coincident with final development of the secondary follicle, there is an abrupt increase in LH that causes the primary oocyte to complete meiosis I and the follicle to enter the preovulatory stage. Meiosis II is also initiated, but the oocyte is arrested in metaphase approximately 3 hours before ovulation. In the meantime, the surface of the ovary begins to bulge locally, and at the apex, an avascular spot, the **stigma,** appears. The high concentration of LH increases collagenase activity, resulting in digestion of collagen fibers surrounding the follicle. Prostaglandin levels also increase in response to the LH surge and cause local muscular contractions in the ovarian wall. Those contractions extrude the oocyte, which together with its surrounding granulosa cells from the region of the cumulus

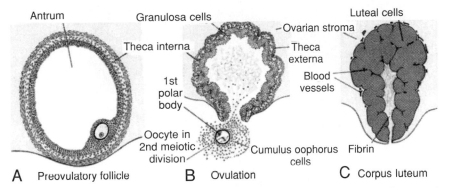

Figure 2.2 A. Preovulatory follicle bulging at the ovarian surface. **B.** Ovulation. The oocyte, in metaphase of meiosis II, is discharged from the ovary together with a large number of cumulus oophorus cells. Follicular cells remaining inside the collapsed follicle differentiate into lutean cells. **C.** Corpus luteum. Note the large size of the corpus luteum, caused by hypertrophy and accumulation of lipid in granulosa and theca interna cells. The remaining cavity of the follicle is filled with fibrin.

oophorus, breaks free (**ovulation**) and floats out of the ovary (Figs. 2.2 and 2.3). Some of the cumulus oophorus cells then rearrange themselves around the zona pellucida to form the **corona radiata** (Figs. 2.4–2.6).

CLINICAL CORRELATES

Ovulation

During ovulation, some women feel a slight pain, known as **middle pain** because it normally occurs near the middle of the menstrual cycle. Ovulation is also generally accompanied by a rise in basal temperature, which can be monitored to aid in determining when release of the oocyte occurs. Some women fail to ovulate because of a low concentration of gonadotropins. In these cases, administration of an agent to stimulate gonadotropin release and hence ovulation can be employed. Although such drugs are effective, they often produce multiple ovulations, so that the risk of multiple pregnancies is 10 times higher in these women than in the general population.

CORPUS LUTEUM

After ovulation, granulosa cells remaining in the wall of the ruptured follicle, together with cells from the theca interna, are vascularized by surrounding vessels. Under the influence of LH, these cells develop a yellowish pigment and change into **lutean cells,** which form the **corpus luteum** and secrete the hormone **progesterone** (Fig. 2.2C). Progesterone, together with estrogenic hormones, causes the uterine mucosa to enter the **progestational** or **secretory stage** in preparation for implantation of the embryo.

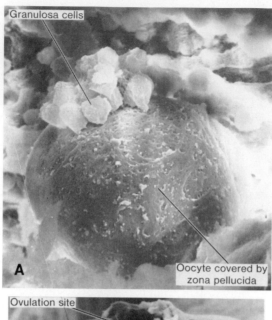

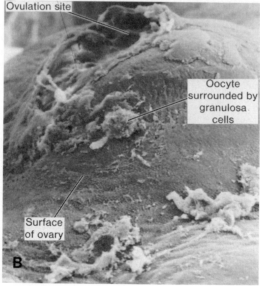

Figure 2.3 A. Scanning electron micrograph of ovulation in the mouse. The surface of the oocyte is covered by the zona pellucida. The cumulus oophorus is composed of granulosa cells. **B.** Scanning electron micrograph of a rabbit oocyte 1.5 hours after ovulation. The oocyte, which is surrounded by granulosa cells, lies on the surface of the ovary. Note the site of ovulation.

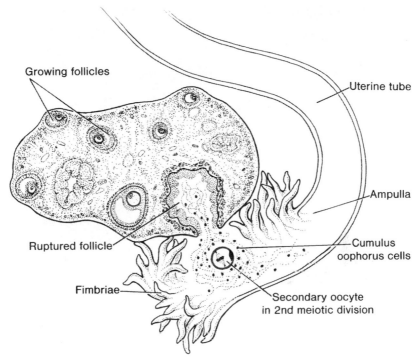

Figure 2.4 Relation of fimbriae and ovary. Fimbriae collect the oocyte and sweep it into the uterine tube.

OOCYTE TRANSPORT

Shortly before ovulation, fimbriae of the oviduct begin to sweep over the surface of the ovary, and the tube itself begins to contract rhythmically. It is thought that the oocyte surrounded by some granulosa cells (Figs. 2.3 and 2.4) is carried into the tube by these sweeping movements of the fimbriae and by motion of cilia on the epithelial lining. Once in the tube, cumulus cells withdraw their cytoplasmic processes from the zona pellucida and lose contact with the oocyte.

Once the oocyte is in the uterine tube, it is propelled by cilia with the rate of transport regulated by the endocrine status during and after ovulation. In humans, the fertilized oocyte reaches the uterine lumen in approximately 3 to 4 days.

CORPUS ALBICANS

If fertilization does not occur, the corpus luteum reaches maximum development approximately 9 days after ovulation. It can easily be recognized as a yellowish projection on the surface of the ovary. Subsequently, the corpus luteum shrinks because of degeneration of lutean cells and forms a mass of fibrotic

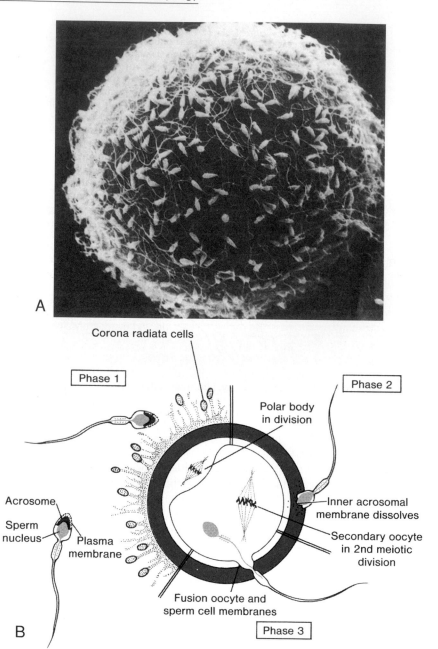

Corona radiata cells

Phase 1

Phase 2

Polar body
in division

Acrosome

Sperm
nucleus

Plasma
membrane

Inner acrosomal
membrane dissolves

Secondary oocyte
in 2nd meiotic
division

Fusion oocyte and
sperm cell membranes

B

Phase 3

Figure 2.5 A. Scanning electron micrograph of sperm binding to the zona pellucida. **B.** The three phases of oocyte penetration. In phase 1, spermatozoa pass through the corona radiata barrier; in phase 2, one or more spermatozoa penetrate the zona pellucida; in phase 3, one spermatozoon penetrates the oocyte membrane while losing its own plasma membrane. *Inset.* Normal spermatocyte with acrosomal head cap.

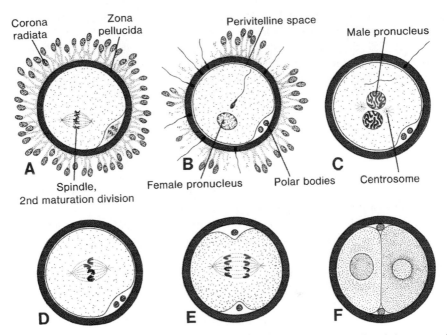

Figure 2.6 A. Oocyte immediately after ovulation, showing the spindle of the second meiotic division. **B.** A spermatozoon has penetrated the oocyte, which has finished its second meiotic division. Chromosomes of the oocyte are arranged in a vesicular nucleus, the female pronucleus. Heads of several sperm are stuck in the zona pellucida. **C.** Male and female pronuclei. **D** and **E.** Chromosomes become arranged on the spindle, split longitudinally, and move to opposite poles. **F.** Two-cell stage.

scar tissue, the **corpus albicans.** Simultaneously, progesterone production decreases, precipitating menstrual bleeding. If the oocyte is fertilized, degeneration of the corpus luteum is prevented by **human chorionic gonadotropin (hCG),** a hormone secreted by the syncytiotrophoblast of the developing embryo. The corpus luteum continues to grow and forms the **corpus luteum of pregnancy (corpus luteum graviditatis).** By the end of the third month, this structure may be one-third to one-half of the total size of the ovary. Yellowish luteal cells continue to secrete progesterone until the end of the fourth month; thereafter, they regress slowly as secretion of progesterone by the trophoblastic component of the placenta becomes adequate for maintenance of pregnancy. Removal of the corpus luteum of pregnancy before the fourth month usually leads to abortion.

Fertilization

Fertilization, the process by which male and female gametes fuse, occurs in the **ampullary region of the uterine tube.** This is the widest part of the tube and is close to the ovary (Fig. 2.4). Spermatozoa may remain viable in the female reproductive tract for several days.

Only 1 % of sperm deposited in the vagina enter the cervix, where they may survive for many hours. Movement of sperm from the cervix to the oviduct is accomplished primarily by their own propulsion, although they may be assisted by movements of fluids created by uterine cilia. The trip from cervix to oviduct requires a minimum of 2 to 7 hours, and after reaching the isthmus, sperm become less motile and cease their migration. At ovulation, sperm again become motile, perhaps because of chemoattractants produced by cumulus cells surrounding the egg, and swim to the ampulla where fertilization usually occurs. Spermatozoa are not able to fertilize the oocyte immediately upon arrival in the female genital tract but must undergo (*a*) **capacitation** and (*b*) the **acrosome reaction** to acquire this capability.

Capacitation is a period of conditioning in the female reproductive tract that in the human lasts approximately 7 hours. Much of this conditioning, which occurs in the uterine tube, entails epithelial interactions between the sperm and mucosal surface of the tube. During this time a glycoprotein coat and seminal plasma proteins are removed from the plasma membrane that overlies the acrosomal region of the spermatozoa. Only capacitated sperm can pass through the corona cells and undergo the acrosome reaction.

The **acrosome reaction,** which occurs after binding to the zona pellucida, is induced by zona proteins. This reaction culminates in the release of enzymes needed to penetrate the zona pellucida, including acrosin and trypsin-like substances (Fig. 2.5).

The phases of fertilization include phase 1, penetration of the corona radiata; phase 2, penetration of the zona pellucida; and phase 3, fusion of the oocyte and sperm cell membranes.

PHASE 1: PENETRATION OF THE CORONA RADIATA

Of the 200 to 300 million spermatozoa deposited in the female genital tract, only 300 to 500 reach the site of fertilization. Only one of these fertilizes the egg. It is thought that the others aid the fertilizing sperm in penetrating the barriers protecting the female gamete. Capacitated sperm pass freely through corona cells (Fig. 2.5).

PHASE 2: PENETRATION OF THE ZONA PELLUCIDA

The zona is a glycoprotein shell surrounding the egg that facilitates and maintains sperm binding and induces the acrosome reaction. Both binding and the acrosome reaction are mediated by the ligand ZP3, a zona protein. Release of acrosomal enzymes (acrosin) allows sperm to penetrate the zona, thereby coming in contact with the plasma membrane of the oocyte (Fig. 2.5). Permeability of the zona pellucida changes when the head of the sperm comes in contact with the oocyte surface. This contact results in release of lysosomal enzymes from cortical granules lining the plasma membrane of the oocyte. In turn, these enzymes alter properties of the zona pellucida **(zona reaction)** to prevent sperm penetration and inactivate species-specific receptor sites for spermatozoa on the zona surface. Other spermatozoa have been found embedded in the zona pellucida, but only one seems to be able to penetrate the oocyte (Fig. 2.6).

PHASE 3: FUSION OF THE OOCYTE AND SPERM CELL MEMBRANES

The initial adhesion of sperm to the oocyte is mediated in part by the interaction of integrins on the oocyte and their ligands, disintegrins, on sperm. After adhesion, the plasma membranes of the sperm and egg fuse (Fig. 2.5). Because the plasma membrane covering the acrosomal head cap disappears during the acrosome reaction, actual fusion is accomplished between the oocyte membrane and the membrane that covers the posterior region of the sperm head (Fig. 2.5). In the human, both the head and tail of the spermatozoon enter the cytoplasm of the oocyte, but the plasma membrane is left behind on the oocyte surface. As soon as the spermatozoon has entered the oocyte, the egg responds in three ways:

1. **Cortical and zona reactions.** As a result of the release of cortical oocyte granules, which contain lysosomal enzymes, (*a*) the oocyte membrane becomes impenetrable to other spermatozoa, and (*b*) the zona pellucida alters its structure and composition to prevent sperm binding and penetration. These reactions prevent polyspermy (penetration of more than one spermatozoon into the oocyte).

2. **Resumption of the second meiotic division.** The oocyte finishes its second meiotic division immediately after entry of the spermatozoon. One of the daughter cells, which receives hardly any cytoplasm, is known as the **second polar body;** the other daughter cell is the **definitive oocyte.** Its chromosomes $(22 + X)$ arrange themselves in a vesicular nucleus known as the **female pronucleus** (Figs. 2.6 and 2.7).

3. **Metabolic activation of the egg.** The activating factor is probably carried by the spermatozoon. Postfusion activation may be considered to encompass the initial cellular and molecular events associated with early embryogenesis.

The spermatozoon, meanwhile, moves forward until it lies close to the female pronucleus. Its nucleus becomes swollen and forms the **male pronucleus** (Fig. 2.6); the tail detaches and degenerates. Morphologically, the male and female pronuclei are indistinguishable, and eventually, they come into close contact and lose their nuclear envelopes (Fig. 2.7*A*). During growth of male and female pronuclei (both haploid), each pronucleus must replicate its DNA. If it does not, each cell of the two-cell zygote has only half of the normal amount of DNA. Immediately after DNA synthesis, chromosomes organize on

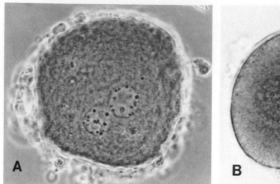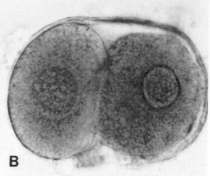

Figure 2.7 A. Phase contrast view of the pronuclear stage of a fertilized human oocyte with male and female pronuclei. **B.** Two-cell stage of human zygote.

the spindle in preparation for a normal mitotic division. The 23 maternal and 23 paternal (double) chromosomes split longitudinally at the centromere, and sister chromatids move to opposite poles, providing each cell of the zygote with the normal diploid number of chromosomes and DNA (Fig. 2.6, *D* and *E*). As sister chromatids move to opposite poles, a deep furrow appears on the surface of the cell, gradually dividing the cytoplasm into two parts (Figs. 2.6*F* and 2.7*B*).

The main results of fertilization are as follows:

- **Restoration of the diploid number of chromosomes,** half from the father and half from the mother. Hence, the zygote contains a new combination of chromosomes different from both parents.

- **Determination of the sex** of the new individual. An X-carrying sperm produces a female (XX) embryo, and a Y-carrying sperm produces a male (XY) embryo. Hence, the chromosomal sex of the embryo is determined at fertilization.

- **Initiation of cleavage.** Without fertilization, the oocyte usually degenerates 24 hours after ovulation.

CLINICAL CORRELATES

Contraceptive Methods

Barrier techniques of contraception include the male condom, made of latex and often containing chemical spermicides, which fits over the penis; and the female condom, made of polyurethane, which lines the vagina. Other barriers placed in the vagina include the diaphragm, the cervical cap, and the contraceptive sponge.

The **contraceptive pill** is a combination of estrogen and the progesterone analogue progestin, which together inhibit ovulation but permit menstruation.

Both hormones act at the level of FSH and LH, preventing their release from the pituitary. The pills are taken for 21 days and then stopped to allow menstruation, after which the cycle is repeated.

Depo-Provera is a **progestin** compound that can be implanted subdermally or injected intramuscularly to prevent ovulation for up to 5 years or 23 months, respectively.

A **male "pill"** has been developed and tested in clinical trials. It contains a synthetic androgen that prevents both LH and FSH secretion and either stops sperm production (70–90% of men) or reduces it to a level of infertility.

The **intrauterine device (IUD)** is placed in the uterine cavity. Its mechanism for preventing pregnancy is not clear but may entail direct effects on sperm and oocytes or inhibition of preimplantation stages of development.

The drug **RU-486 (mifepristone)** causes abortion if it is administered within 8 weeks of the previous menses. It initiates menstruation, possibly through its action as an antiprogesterone agent.

Vasectomy and **tubal ligation** are effective means of contraception, and both procedures are reversible, although not in every case.

Infertility

Infertility is a problem for 15% to 30% of couples. Male infertility may be a result of insufficient numbers of sperm and/or poor motility. Normally, the ejaculate has a volume of 3 to 4 ml, with approximately 100 million sperm per ml. Males with 20 million sperm per ml or 50 million sperm per total ejaculate are usually fertile. Infertility in a woman may be due to a number of causes, including occluded oviducts (most commonly caused by pelvic inflammatory disease), hostile cervical mucus, immunity to spermatozoa, absence of ovulation, and others.

In vitro fertilization (IVF) of human ova and embryo transfer is a frequent practice conducted by laboratories throughout the world. Follicle growth in the ovary is stimulated by administration of gonadotropins. Oocytes are recovered by laparoscopy from ovarian follicles with an aspirator just before ovulation when the oocyte is in the late stages of the first meiotic division. The egg is placed in a simple culture medium and sperm are added immediately. Fertilized eggs are monitored to the eight-cell stage and then placed in the uterus to develop to term. Fortunately, because preimplantation-stage embryos are resistant to teratogenic insult, the risk of producing malformed offspring by in vitro procedures is low.

A disadvantage of IVF is its low success rate; only 20% of fertilized ova implant and develop to term. Therefore, to increase chances of a successful pregnancy, four or five ova are collected, fertilized, and placed in the uterus. This approach sometimes leads to multiple births.

Another technique, **gamete intrafallopian transfer (GIFT),** introduces oocytes and sperm into the ampulla of the fallopian (uterine) tube, where fertilization takes place. Development then proceeds in a normal fashion. In a similar approach, **zygote intrafallopian transfer (ZIFT),** fertilized oocytes are placed in the ampullary region. Both of these methods require patent uterine tubes.

Severe male infertility, in which the ejaculate contains very few live sperm **(oligozoospermia)** or even no live sperm **(azoospermia),** can be overcome using **intracytoplasmic sperm injection (ICSI).** With this technique, a single sperm, which may be obtained from any point in the male reproductive tract, is injected into the cytoplasm of the egg to cause fertilization. This approach offers couples an alternative to using donor sperm for IVF. The technique carries an increased risk for fetuses to have Y chromosome deletions but no other chromosomal abnormalities.

Cleavage

Once the zygote has reached the two-cell stage, it undergoes a series of mitotic divisions, increasing the numbers of cells. These cells, which become smaller with each cleavage division, are known as **blastomeres** (Fig. 2.8). Until the eight-cell stage, they form a loosely arranged clump (Fig. 2.9*A*). However, after the third cleavage, blastomeres maximize their contact with each other, forming a compact ball of cells held together by tight junctions (Fig. 2.9*B*). This process, **compaction,** segregates inner cells, which communicate extensively by gap junctions, from outer cells. Approximately 3 days after fertilization, cells of the compacted embryo divide again to form a 16-cell **morula** (mulberry). Inner cells of the morula constitute the **inner cell mass,** and surrounding cells compose the **outer cell mass.** The inner cell mass gives rise to tissues of the

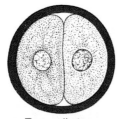

Two-cell stage Four-cell stage Morula

Figure 2.8 Development of the zygote from the two-cell stage to the late morula stage. The two-cell stage is reached approximately 30 hours after fertilization; the four-cell stage, at approximately 40 hours; the 12- to 16-cell stage, at approximately 3 days; and the late morula stage, at approximately 4 days. During this period, blastomeres are surrounded by the zona pellucida, which disappears at the end of the fourth day.

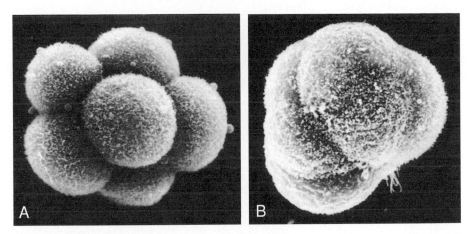

Figure 2.9 Scanning electron micrographs of uncompacted **(A)** and compacted **(B)** eight-cell mouse embryos. In the uncompacted state, outlines of each blastomere are distinct, whereas after compaction cell-cell contacts are maximized and cellular outlines are indistinct.

embryo proper, and the outer cell mass forms the **trophoblast,** which later contributes to the **placenta.**

Blastocyst Formation

About the time the morula enters the uterine cavity, fluid begins to penetrate through the zona pellucida into the intercellular spaces of the inner cell mass. Gradually the intercellular spaces become confluent, and finally a single cavity, the **blastocele,** forms (Fig. 2.10, *A* and *B*). At this time, the embryo is a **blastocyst.** Cells of the inner cell mass, now called the **embryoblast,** are at one pole, and those of the outer cell mass, or **trophoblast,** flatten and form the epithelial wall of the blastocyst (Fig. 2.10, *A* and *B*). The zona pellucida has disappeared, allowing implantation to begin.

In the human, trophoblastic cells over the embryoblast pole begin to penetrate between the epithelial cells of the uterine mucosa about the sixth day (Fig. 2.10*C*). Attachment and invasion of the trophoblast involve integrins, expressed by the trophoblast, and the extracellular matrix molecules laminin and fibronectin. Integrin receptors for laminin promote attachment, while those for fibronectin stimulate migration. These molecules also interact along signal transduction pathways to regulate trophoblast differentiation so that implantation is the result of mutual trophoblastic and endometrial action. Hence, by the end of the first week of development, the human zygote has passed through the morula and blastocyst stages and has begun implantation in the uterine mucosa.

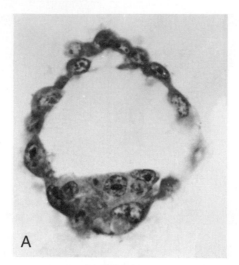

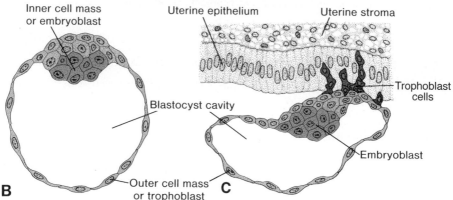

Figure 2.10 A. Section of a 107-cell human blastocyst showing inner cell mass and trophoblast cells. **B.** Schematic representation of a human blastocyst recovered from the uterine cavity at approximately 4.5 days. *Blue,* inner cell mass or embryoblast; *green,* trophoblast. **C.** Schematic representation of a blastocyst at the ninth day of development showing trophoblast cells at the embryonic pole of the blastocyst penetrating the uterine mucosa. The human blastocyst begins to penetrate the uterine mucosa by the sixth day of development.

CLINICAL CORRELATES

Abnormal Zygotes

The exact number of **abnormal zygotes** formed is unknown because they are usually lost within 2 to 3 weeks of fertilization, before the woman realizes she is pregnant, and therefore are not detected. Estimates are that as many as 50% of pregnancies end in spontaneous abortion and that half of these losses are a result of chromosomal abnormalities. These abortions are a natural means of screening embryos for defects, reducing the incidence of congenital malformations. Without this phenomenon, approximately 12% instead of 2% to 3% of infants would have birth defects.

With the use of a combination of IVF and **polymerase chain reaction (PCR),** molecular screening of embryos for genetic defects is being conducted. Single blastomeres from early-stage embryos can be removed and their DNA amplified for analysis. As the Human Genome Project provides more sequencing information and as specific genes are linked to various syndromes, such procedures will become more commonplace.

Uterus at Time of Implantation

The wall of the uterus consists of three layers: (*a*) **endometrium** or mucosa lining the inside wall; (*b*) **myometrium,** a thick layer of smooth muscle; and (*c*) **perimetrium,** the peritoneal covering lining the outside wall (Fig. 2.11). From puberty (11–13 years) until menopause (45–50 years), the endometrium undergoes changes in a cycle of approximately 28 days under hormonal control by the ovary. During this menstrual cycle, the uterine endometrium passes through three stages, the **follicular** or **proliferative phase,**

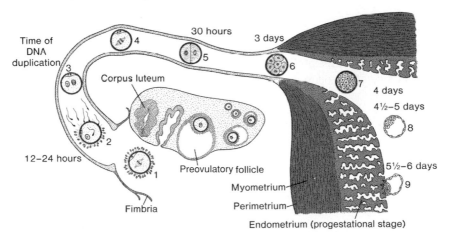

Figure 2.11 Events during the first week of human development. *1*, Oocyte immediately after ovulation. *2*, Fertilization, approximately 12 to 24 hours after ovulation. *3*, Stage of the male and female pronuclei. *4*, Spindle of the first mitotic division. *5*, Two-cell stage (approximately 30 hours of age). *6*, Morula containing 12 to 16 blastomeres (approximately 3 days of age). *7*, Advanced morula stage reaching the uterine lumen (approximately 4 days of age). *8*, Early blastocyst stage (approximately 4.5 days of age). The zona pellucida has disappeared. *9*, Early phase of implantation (blastocyst approximately 6 days of age). The ovary shows stages of transformation between a primary follicle and a preovulatory follicle as well as a corpus luteum. The uterine endometrium is shown in the progestational stage.

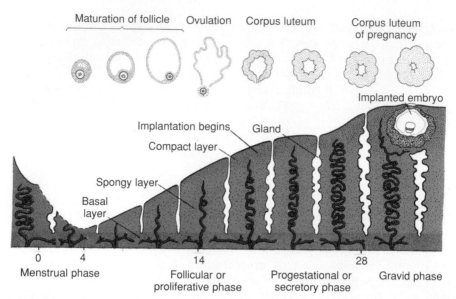

Maturation of follicle   Ovulation   Corpus luteum   Corpus luteum of pregnancy

Implanted embryo

Implantation begins   Gland

Compact layer

Spongy layer

Basal layer

0   4   14   28

Menstrual phase   Follicular or proliferative phase   Progestational or secretory phase   Gravid phase

Figure 2.12 Changes in the uterine mucosa correlated with those in the ovary. Implantation of the blastocyst has caused development of a large corpus luteum of pregnancy. Secretory activity of the endometrium increases gradually as a result of large amounts of progesterone produced by the corpus luteum of pregnancy.

the **secretory** or **progestational phase,** and the **menstrual phase** (Figs. 2.11– 2.13). The proliferative phase begins at the end of the menstrual phase, is under the influence of estrogen, and parallels growth of the ovarian follicles. The secretory phase begins approximately 2 to 3 days after ovulation in response to progesterone produced by the corpus luteum. If fertilization does not occur, shedding of the endometrium (compact and spongy layers) marks the beginning of the menstrual phase. If fertilization does occur, the endometrium assists in implantation and contributes to formation of the placenta.

At the time of implantation, the mucosa of the uterus is in the secretory phase (Figs. 2.11 and 2.12), during which time uterine glands and arteries become coiled and the tissue becomes succulent. As a result, three distinct layers can be recognized in the endometrium: a superficial **compact layer,** an intermediate **spongy layer,** and a thin **basal layer** (Fig. 2.12). Normally, the human blastocyst implants in the endometrium along the anterior or posterior wall of the body of the uterus, where it becomes embedded between the openings of the glands (Fig. 2.12).

If the oocyte is not fertilized, venules and sinusoidal spaces gradually become packed with blood cells, and an extensive diapedesis of blood into the tissue is seen. When the **menstrual phase** begins, blood escapes from superficial arteries, and small pieces of stroma and glands break away. During the following 3 or 4 days, the compact and spongy layers are expelled from the

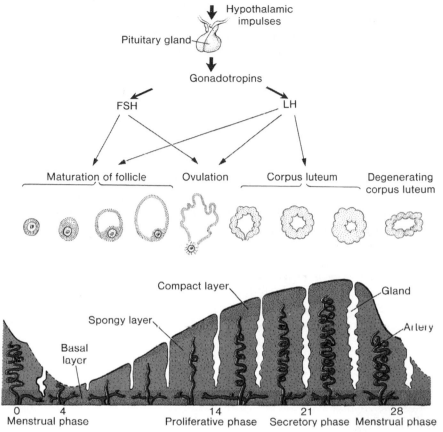

Figure 2.13 Changes in the uterine mucosa (endometrium) and corresponding changes in the ovary during a regular menstrual cycle without fertilization.

uterus, and the basal layer is the only part of the endometrium that is retained (Fig. 2.13). This layer, which is supplied by its own arteries, the **basal arteries,** functions as the regenerative layer in the rebuilding of glands and arteries in the **proliferative phase** (Fig. 2.13).

Summary

With each ovarian cycle, a number of primary follicles begin to grow, but usually only one reaches full maturity, and only one oocyte is discharged at **ovulation.** At ovulation, the oocyte is in metaphase of the **second meiotic division** and is surrounded by the zona pellucida and some granulosa cells (Fig. 2.4). Sweeping action of tubal fimbriae carries the oocyte into the uterine tube.

Before spermatozoa can fertilize the oocyte, they must undergo (*a*) **capacitation,** during which time a glycoprotein coat and seminal plasma proteins are removed from the spermatozoon head, and (*b*) the **acrosome reaction,** during which acrosin and trypsin-like substances are released to penetrate the zona pellucida. During fertilization, the spermatozoon must penetrate (*a*) the **corona radiata,** (*b*) the **zona pellucida,** and (*c*) the **oocyte cell membrane** (Fig. 2.5). As soon as the spermatocyte has entered the oocyte, (*a*) the oocyte finishes its second meiotic division and forms the **female pronucleus;** (*b*) the zona pellucida becomes impenetrable to other spermatozoa; and (*c*) the head of the sperm separates from the tail, swells, and forms the **male pronucleus** (Figs. 2.6 and 2.7). After both pronuclei have replicated their DNA, paternal and maternal chromosomes intermingle, split longitudinally, and go through a mitotic division, giving rise to the two-cell stage. The **results of fertilization** are (*a*) **restoration of the diploid number of chromosomes,** (*b*) **determination of chromosomal sex,** and (*c*) **initiation of cleavage.**

Cleavage is a series of mitotic divisions that results in an increase in cells, **blastomeres,** which become smaller with each division. After three divisions, blastomeres undergo **compaction** to become a tightly grouped ball of cells with inner and outer layers. Compacted blastomeres divide to form a 16-cell **morula.** As the morula enters the uterus on the third or fourth day after fertilization, a cavity begins to appear, and the **blastocyst** forms. The **inner cell mass,** which is formed at the time of compaction and will develop into the embryo proper, is at one pole of the blastocyst. The **outer cell mass,** which surrounds the inner cells and the blastocyst cavity, will form the trophoblast.

The uterus at the time of implantation is in the secretory phase, and the blastocyst implants in the endometrium along the anterior or posterior wall. If fertilization does not occur, then the menstrual phase begins and the spongy and compact endometrial layers are shed. The basal layer remains to regenerate the other layers during the next cycle.

Problems to Solve

1. *What are the primary causes of infertility in men and women?*
2. *A woman has had several bouts of pelvic inflammatory disease and now wants to have children. However, she has been having difficulty becoming pregnant. What is likely to be the problem, and what would you suggest?*

SUGGESTED READING

Allen CA, Green DPL: The mammalian acrosome reaction: gateway to sperm fusion with the oocyte? *Bioessays* 19:241, 1997.

Archer DF, Zeleznik AJ, Rockette HE: Ovarian follicular maturation in women: 2. Reversal of estrogen inhibited ovarian folliculogenesis by human gonadotropin. *Fertil Steril* 50:555, 1988.

Barratt CLR, Cooke ID: Sperm transport in the human female reproductive tract: a dynamic interaction. *Int J Androl* 14:394, 1991.

Boldt J, et al: Carbohydrate involvement in sperm-egg fusion in mice. *Biol Reprod* 40:887, 1989.

Burrows TD, King A, Loke YW: Expression of integrins by human trophoblast and differential adhesion to laminin or fibronectin. *Hum Reprod* 8:475, 1993.

Carr DH: Chromosome studies on selected spontaneous abortions: polyploidy in man. *J Med Genet* 8:164, 1971.

Chen CM, Sathananthan AH: Early penetration of human sperm through the vestments of human egg in vitro. *Arch Androl* 16:183, 1986.

Cowchock S: Autoantibodies and fetal wastage. *Am J Reprod Immunol* 26:38, 1991.

Edwards RG: A decade of in vitro fertilization. *Res Reprod* 22:1, 1990.

Edwards RG, Bavister BD, Steptoe PC: Early stages of fertilization in vitro of human oocytes matured in vitro. *Nature (Lond)* 221:632, 1969.

Egarter C: The complex nature of egg transport through the oviduct. *Am J Obstet Gynecol* 163:687, 1990.

Enders AC, Hendrickx AG, Schlake S: Implantation in the rhesus monkey: initial penetration of the endometrium. *Am J Anat* 167:275, 1983.

Gilbert SF: *Developmental Biology*. Sunderland, MA, Sinauer, 1991.

Handyside AH, Kontogianni EH, Hardy K, Winston RML: Pregnancies from biopsied human preimplantation embryos sexed by Y-specific DNA amplification. *Nature* 344:768, 1990.

Hertig AT, Rock J, Adams EC: A description of 34 human ova within the first 17 days of development. *Am J Anat* 98:435, 1956.

Johnson MH, Everitt BJ: *Essential Reproduction*. 5th ed. London, Blackwell Science Limited, 2000.

Liu J, et al: Analysis of 76 total fertilization failure cycles out of 2732 intracytoplasmic sperm injection cycles. *Hum Reprod* 10:2630, 1995.

Oura C, Toshimori K: Ultrasound studies on the fertilization of mammalian gametes. *Rev Cytol* 122:105, 1990.

Pedersen RA, We K, Balakier H: Origin of the inner cell mass in mouse embryos: cell lineage analysis by microinjection. *Dev Biol* 117:581, 1986.

Reproduction (entire issue). *J NIH Res* 9:1997.

Scott RT, Hodgen GD: The ovarian follicle: life cycle of a pelvic clock. *Clin Obstet Gynecol* 33:551, 1990.

Wasserman PM: Fertilization in mammals. *Sci Am* 259:78, 1988.

Wolf DP, Quigley MM (eds): *Human in Vitro Fertilization and Transfer*. New York, Plenum, 1984

Yen SC, Jaffe RB (eds): *Reproductive Endocrinology: Physiology, Pathophysiology, and Clinical Management*. 2nd ed. Philadelphia, WB Saunders, 1986.

chapter 3

Second Week of Development: Bilaminar Germ Disc

This chapter gives a day-by-day account of the major events of the second week of development. However, embryos of the same fertilization age do not necessarily develop at the same rate. Indeed, considerable differences in rate of growth have been found even at these early stages of development.

Day 8

At the eighth day of development, the blastocyst is partially embedded in the endometrial stroma. In the area over the embryoblast, the trophoblast has differentiated into two layers: (a) an inner layer of mononucleated cells, the **cytotrophoblast,** and (b) an outer multinucleated zone without distinct cell boundaries, the **syncytiotrophoblast** (Figs. 3.1 and 3.2). Mitotic figures are found in the cytotrophoblast but not in the syncytiotrophoblast. Thus, cells in the cytotrophoblast divide and migrate into the syncytiotrophoblast, where they fuse and lose their individual cell membranes.

Cells of the inner cell mass or embryoblast also differentiate into two layers: (a) a layer of small cuboidal cells adjacent to the blastocyst cavity, known as the **hypoblast layer,** and (b) a layer of high columnar cells adjacent to the amniotic cavity, the **epiblast layer** (Figs. 3.1 and 3.2). Together, the layers form a flat disc. At the same time, a small cavity appears within the epiblast. This cavity enlarges to become the

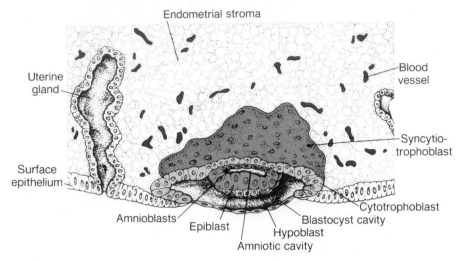

Figure 3.1 A 7.5-day human blastocyst, partially embedded in the endometrial stroma. The trophoblast consists of an inner layer with mononuclear cells, the cytotrophoblast, and an outer layer without distinct cell boundaries, the syncytiotrophoblast. The embryoblast is formed by the epiblast and hypoblast layers. The amniotic cavity appears as a small cleft.

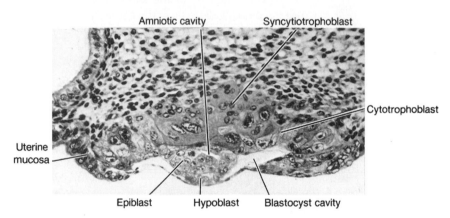

Figure 3.2 Section of a 7.5-day human blastocyst (×100). Note the multinucleated appearance of the syncytiotrophoblast, large cells of the cytotrophoblast, and slit-like amniotic cavity.

amniotic cavity. Epiblast cells adjacent to the cytotrophoblast are called **amnioblasts;** together with the rest of the epiblast, they line the amniotic cavity (Figs. 3.1 and 3.3). The endometrial stroma adjacent to the implantation site is edematous and highly vascular. The large, tortuous glands secrete abundant glycogen and mucus.

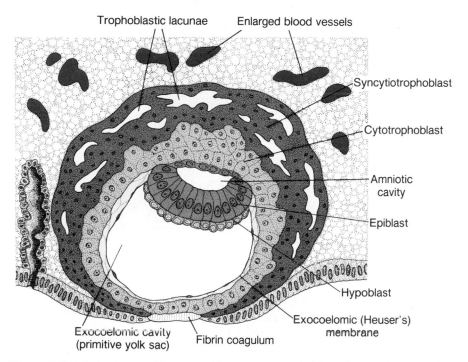

Figure 3.3 A 9-day human blastocyst. The syncytiotrophoblast shows a large number of lacunae. Flat cells form the exocoelomic membrane. The bilaminar disc consists of a layer of columnar epiblast cells and a layer of cuboidal hypoblast cells. The original surface defect is closed by a fibrin coagulum.

Day 9

The blastocyst is more deeply embedded in the endometrium, and the penetration defect in the surface epithelium is closed by a fibrin coagulum (Fig. 3.3). The trophoblast shows considerable progress in development, particularly at the embryonic pole, where vacuoles appear in the syncytium. When these vacuoles fuse, they form large lacunae, and this phase of trophoblast development is thus known as the **lacunar stage** (Fig. 3.3).

At the abembryonic pole, meanwhile, flattened cells probably originating from the hypoblast form a thin membrane, the exocoelomic (Heuser's) membrane, that lines the inner surface of the cytotrophoblast (Fig. 3.3). This membrane, together with the hypoblast, forms the lining of the **exocoelomic cavity, or primitive yolk sac.**

Days 11 and 12

By the 11th to 12th day of development, the blastocyst is completely embedded in the endometrial stroma, and the surface epithelium almost entirely covers

Trophoblastic lacunae   Maternal sinusoids

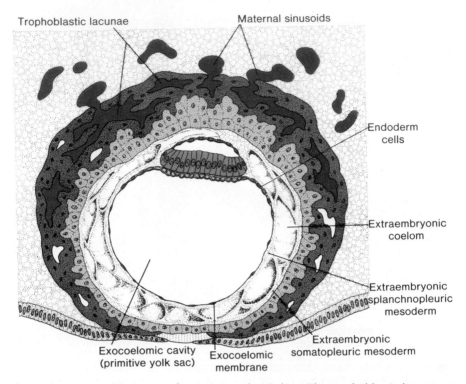

Endoderm
cells

Extraembryonic
coelom

Extraembryonic
splanchnopleuric
mesoderm

Exocoelomic cavity   Exocoelomic   Extraembryonic
(primitive yolk sac)   membrane   somatopleuric mesoderm

Figure 3.4 Human blastocyst of approximately 12 days. The trophoblastic lacunae at the embryonic pole are in open connection with maternal sinusoids in the endometrial stroma. Extraembryonic mesoderm proliferates and fills the space between the exocoelomic membrane and the inner aspect of the trophoblast.

the original defect in the uterine wall (Figs. 3.4 and 3.5). The blastocyst now produces a slight protrusion into the lumen of the uterus. The trophoblast is characterized by lacunar spaces in the syncytium that form an intercommunicating network. This network is particularly evident at the embryonic pole; at the abembryonic pole, the trophoblast still consists mainly of cytotrophoblastic cells (Figs. 3.4 and 3.5).

Concurrently, cells of the syncytiotrophoblast penetrate deeper into the stroma and erode the endothelial lining of the maternal capillaries. These capillaries, which are congested and dilated, are known as **sinusoids.** The syncytial lacunae become continuous with the sinusoids and maternal blood enters the lacunar system (Fig. 3.4). As the trophoblast continues to erode more and more sinusoids, maternal blood begins to flow through the trophoblastic system, establishing the **uteroplacental circulation.**

In the meantime, a new population of cells appears between the inner surface of the cytotrophoblast and the outer surface of the exocoelomic

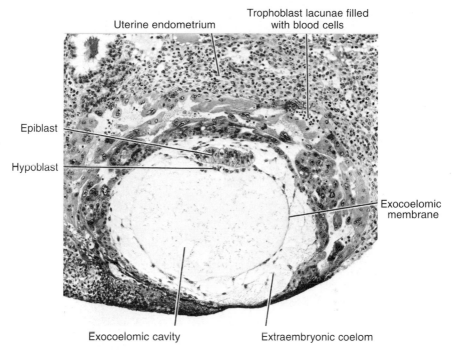

Figure 3.5 Fully implanted 12-day human blastocyst (×100). Note maternal blood cells in the lacunae, the exocoelomic membrane lining the primitive yolk sac, and the hypoblast and epiblast.

cavity. These cells, derived from yolk sac cells, form a fine, loose connective tissue, the **extraembryonic mesoderm,** which eventually fills all of the space between the trophoblast externally and the amnion and exocoelomic membrane internally (Figs. 3.4 and 3.5). Soon, large cavities develop in the extraembryonic mesoderm, and when these become confluent, they form a new space known as the **extraembryonic coelom,** or **chorionic cavity** (Fig. 3.4). This space surrounds the primitive yolk sac and amniotic cavity except where the germ disc is connected to the trophoblast by the connecting stalk (Fig. 3.6). The extraembryonic mesoderm lining the cytotrophoblast and amnion is called the **extraembryonic somatopleuric mesoderm;** the lining covering the yolk sac is known as the **extraembryonic splanchnopleuric mesoderm** (Fig. 3.4).

Growth of the bilaminar disc is relatively slow compared with that of the trophoblast; consequently, the disc remains very small (0.1–0.2 mm). Cells of the endometrium, meanwhile, become polyhedral and loaded with glycogen and lipids; intercellular spaces are filled with extravasate, and the tissue is edematous. These changes, known as the **decidua reaction,** at first are confined to the

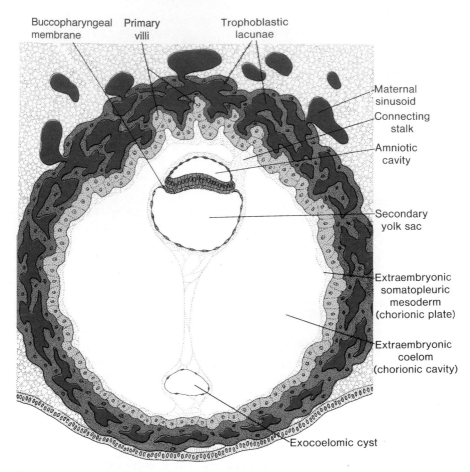

Buccopharyngeal   Primary   Trophoblastic
membrane   villi   lacunae

Maternal
sinusoid

Connecting
stalk

Amniotic
cavity

Secondary
yolk sac

Extraembryonic
somatopleuric
mesoderm
(chorionic plate)

Extraembryonic
coelom
(chorionic cavity)

Exocoelomic cyst

Figure 3.6 A 13-day human blastocyst. Trophoblastic lacunae are present at the embryonic as well as the abembryonic pole, and the uteroplacental circulation has begun. Note the primary villi and the extraembryonic coelom or **chorionic cavity.** The secondary yolk sac is entirely lined with endoderm.

area immediately surrounding the implantation site but soon occur throughout the endometrium.

Day 13

By the 13th day of development, the surface defect in the endometrium has usually healed. Occasionally, however, bleeding occurs at the implantation site as a result of increased blood flow into the lacunar spaces. Because this bleeding occurs near the 28th day of the menstrual cycle, it may be confused with

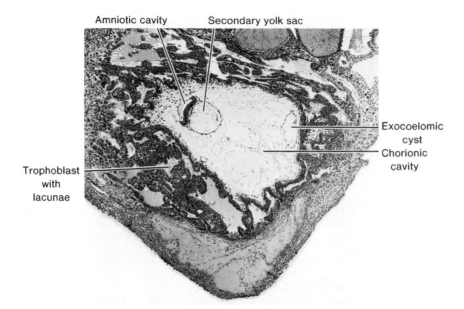

Figure 3.7 Section through the implantation site of a 13-day embryo. Note the amniotic cavity, yolk sac, and exocoelomic cyst in the chorionic cavity. Most of the lacunae are filled with blood.

normal menstrual bleeding and, therefore, cause inaccuracy in determining the expected delivery date.

The trophoblast is characterized by villous structures. Cells of the cytotrophoblast proliferate locally and penetrate into the syncytiotrophoblast, forming cellular columns surrounded by syncytium. Cellular columns with the syncytial covering are known as **primary villi** (Figs. 3.6 and 3.7) (see Chapter 4).

In the meantime, the hypoblast produces additional cells that migrate along the inside of the exocoelomic membrane (Fig. 3.4). These cells proliferate and gradually form a new cavity within the exocoelomic cavity. This new cavity is known as **the secondary yolk sac** or **definitive yolk sac** (Figs. 3.6 and 3.7). This yolk sac is much smaller than the original exocoelomic cavity, or primitive yolk sac. During its formation, large portions of the exocoelomic cavity are pinched off. These portions are represented by **exocoelomic cysts,** which are often found in the extraembryonic coelom or **chorionic cavity** (Figs. 3.6 and 3.7).

Meanwhile, the extraembryonic coelom expands and forms a large cavity, the **chorionic cavity.** The extraembryonic mesoderm lining the inside of the cytotrophoblast is then known as the **chorionic plate.** The only place where extraembryonic mesoderm traverses the chorionic cavity is in the **connecting stalk** (Fig. 3.6). With development of blood vessels, the stalk becomes the **umbilical cord.**

CLINICAL CORRELATES

Abnormal Implantation

The syncytiotrophoblast is responsible for hormone production (see Chapter 6), including **human chorionic gonadotropin** (hCG). By the end of the second week, quantities of this hormone are sufficient to be detected by radioimmunoassays, which serve as the basis for pregnancy testing.

Because 50% of the implanting embryo's genome is derived from the father, it is a foreign body that potentially should be rejected by the maternal system. Recent evidence suggests that a combination of factors protects the conceptus, including production of immunosuppressive cytokines and proteins and the expression of an unusual major histocompatibility complex class IB molecule (HLA-G) that blocks recognition of the conceptus as foreign tissue. If the mother has autoimmune disease, for example systemic lupus erythematosus, antibodies generated by the disease may attack the conceptus and reject it.

Abnormal implantation sites sometimes occur even within the uterus. Normally the human blastocyst implants along the anterior or posterior wall of the body of the uterus. Occasionally the blastocyst implants close to the internal opening os (opening) (Fig. 3.8) of the cervix, so that later in development,

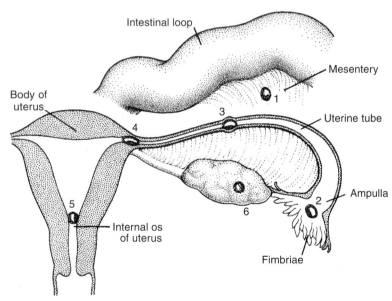

Figure 3.8 Abnormal implantation sites of the blastocyst. *1,* implantation in the abdominal cavity. The ovum most frequently implants in the rectouterine cavity (Douglas' pouch) but may implant at any place covered by peritoneum. *2,* implantation in the ampullary region of the tube. *3,* tubal implantation. *4,* interstitial implantation, that is, in the narrow portion of the uterine tube. *5,* implantation in the region of the internal os, frequently resulting in placenta previa. *6,* ovarian implantation.

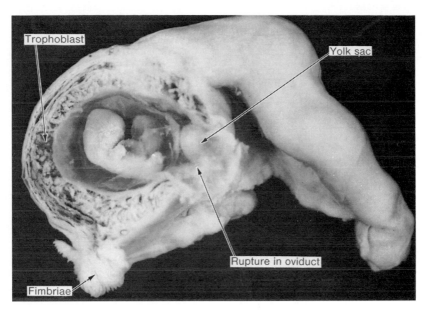

Figure 3.9 Tubal pregnancy. Embryo is approximately 2 months old and is about to escape through a rupture in the tubal wall.

the placenta bridges the opening (**placenta previa**) and causes severe, even life-threatening bleeding in the second part of pregnancy and during delivery.

Occasionally, implantation takes place outside the uterus, resulting in **extrauterine pregnancy,** or **ectopic pregnancy.** Ectopic pregnancies may occur at any place in the abdominal cavity, ovary, or uterine tube (Fig. 3.8). However, 95% of ectopic pregnancies occur in the uterine tube, and most of these are in the ampulla (Fig. 3.9). In the abdominal cavity, the blastocyst most frequently attaches itself to the peritoneal lining of the **rectouterine cavity,** or **Douglas' pouch** (Fig. 3.10). The blastocyst may also attach itself to the peritoneal covering of the intestinal tract or to the omentum. Sometimes the blastocyst develops in the ovary proper, causing a **primary ovarian pregnancy.** In most ectopic pregnancies, the embryo dies about the second month of gestation, causing severe hemorrhaging and abdominal pain in the mother.

Abnormal blastocysts are common. For example, in a series of 26 implanted blastocysts varying in age from 7.5 to 17 days recovered from patients of normal fertility, nine (34.6%) were abnormal. Some consisted of syncytium only; others showed varying degrees of trophoblastic hypoplasia. In two, the embryoblast was absent, and in some, the germ disc showed an abnormal orientation.

It is likely that most abnormal blastocysts would not have produced any sign of pregnancy because their trophoblast was so inferior that the corpus luteum could not have persisted. These embryos probably would have been

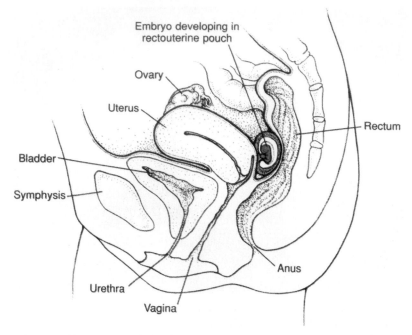

Figure 3.10 Midline section of bladder, uterus, and rectum to show an abdominal pregnancy in the rectouterine (Douglas') pouch.

aborted with the next menstrual flow, and therefore, pregnancy would not have been detected. In some cases, however, the trophoblast develops and forms placental membranes, although little or no embryonic tissue is present. Such a condition is known as a **hydatidiform mole.** Moles secrete high levels of hCG and may produce benign or malignant (**invasive mole, choriocarcinoma**) tumors.

Genetic analysis of hydatidiform moles indicates that although male and female pronuclei may be genetically equivalent, they may be different functionally. This evidence is derived from the fact that while cells of moles are diploid, their entire genome is paternal. Thus, most moles arise from fertilization of an oocyte lacking a nucleus followed by duplication of the male chromosomes to restore the diploid number. These results also suggest that paternal genes regulate most of the development of the trophoblast, since in moles, this tissue differentiates even in the absence of a female pronucleus.

Other examples of functional differences in maternal and paternal genes are provided by the observation that certain genetic diseases depend on whether the defective or missing gene is inherited from the father or the mother. For example, inheritance of a deletion on chromosome 15 from a father produces Prader-Willi syndrome, whereas inheritance of the same defect from the mother results in Angelman syndrome. This phenomenon, in which there is differential modification and/or expression of homologous alleles or chromosome regions, depending on the parent from whom the genetic material is derived, is known as **genomic imprinting.** Imprinting involves autosomes and sex chromosomes (in all female mammals, one X chromosome is inactivated in somatic cells and forms a **chromatin-positive body [Barr body])** and is modulated by deoxyribonucleic acid (DNA) methylation. Certain diseases, such as Huntington's chorea, neurofibromatosis, familial cancer disorders (Wilms' tumors, familial retinoblastoma), and myotonic dystrophy, also involve imprinting. Fragile X syndrome, the leading cause of inherited mental retardation, may be another example of a condition based on imprinting (see Chapter 1).

Preimplantation and postimplantation reproductive failure occurs often. Even in some fertile women under optimal conditions for pregnancy, 15% of oocytes are not fertilized, and 10% to 15% start cleavage but fail to implant. Of the 70% to 75% that implant, only 58% survive until the second week, and 16% of those are abnormal. Hence, when the first expected menstruation is missed, only 42% of the eggs exposed to sperm are surviving. Of this percentage, a number will be aborted during subsequent weeks and a number will be abnormal at the time of birth.

Summary

At the beginning of the second week, the blastocyst is partially embedded in the endometrial stroma. The **trophoblast** differentiates into (a) an inner, actively proliferating layer, the **cytotrophoblast,** and (b) an outer layer, the **syncytiotrophoblast,** which erodes maternal tissues (Fig. 3.1). By day 9, lacunae develop in the syncytiotrophoblast. Subsequently, maternal sinusoids are eroded by the syncytiotrophoblast, maternal blood enters the lacunar network, and by the end of the second week, a primitive **uteroplacental circulation** begins (Fig. 3.6). The cytotrophoblast, meanwhile, forms cellular columns penetrating into and surrounded by the syncytium. These columns are **primary villi.** By the end of the second week, the blastocyst is completely embedded, and the surface defect in the mucosa has healed (Fig. 3.6).

The **inner cell mass** or **embryoblast,** meanwhile, differentiates into (a) the **epiblast** and (b) the **hypoblast,** together forming a **bilaminar disc** (Fig. 3.1). Epiblast cells give rise to **amnioblasts** that line the **amniotic cavity** superior to the epiblast layer. Endoderm cells are continuous with the **exocoelomic membrane,** and together they surround the **primitive yolk sac** (Fig. 3.4). By the end of the second week, extraembryonic mesoderm fills the space between the trophoblast and the amnion and exocoelomic membrane internally. When vacuoles develop in this tissue, the **extraembryonic coelom** or **chorionic cavity** forms (Fig. 3.6). **Extraembryonic mesoderm** lining the cytotrophoblast and amnion is **extraembryonic somatopleuric mesoderm;** the lining surrounding the yolk sac is **extraembryonic splanchnopleuric mesoderm** (Fig. 3.6).

The second week of development is known as the **week of twos:** The trophoblast differentiates into two layers, the cytotrophoblast and syncytiotrophoblast. The embryoblast forms two layers, the epiblast and hypoblast. The extraembryonic mesoderm splits into two layers, the somatopleure and splanchnopleure. And two cavities, the amniotic and yolk sac cavities, form. **Implantation** occurs at the end of the first week. Trophoblast cells invade the epithelium and underlying endometrial stroma with the help of proteolytic enzymes. Implantation may also occur outside the uterus, such as in the rectouterine pouch, on the mesentery, in the uterine tube, or in the ovary (**ectopic pregnancies**).

Problems to Solve

1. *The second week of development is known as the week of twos. Formation of what structures support this statement?*

2. *During implantation, the trophoblast is invading maternal tissues, and because it contains approximately 50% paternal genes, it is a foreign body. Why is the conceptus not rejected by an immunologic response from the mother's system?*

3. *A woman who believes she is pregnant complains of edema and vaginal bleeding. Examination reveals high plasma hCG concentrations and placental tissue, but no evidence of an embryo. How would you explain this condition?*

4. *A young woman who has missed two menstrual periods complains of intense abdominal pain. What might an initial diagnosis be, and how would you confirm it?*

SUGGESTED READING

Aplin JD: Implantation, trophoblast differentiation and hemochorial placentation: mechanistic evidence in vivo and in vitro. *J Cell Sci* 99:681, 1991.

Bianchi DW, Wilkins-Haug LE, Enders AC, Hay ED: Origin of extraembryonic mesoderm in experimental animals: relevance to chorionic mosaicism in humans. *Am J Med Genet* 46:542, 1993.

Cattanack BM, Beechey CV: Autosomal and X-chromosome imprinting. *Dev Suppl* 63, 1990.

Enders AC, King BF: Formation and differentiation of extraembryonic mesoderm in the Rhesus monkey. *Am J Anat* 181:327, 1988.

Enders AC, King BF: Development of the human yolk sac. *In* Nogales FF (ed): *The Human Yolk Sac and Yolk Sac Tumors.* Berlin, Springer Verlag, 1993.

Enders AC, Schlafke S, Hendrickx A: Differentiation of the embryonic disc, amnion, and yolk sac in the rhesus monkey. *Am J Anat* 177:161, 1986.

Hertig AT, Rock J, Adams EC: A description of 34 human ova within the first 17 days of development. *Am J Anat* 98:435, 1956.

Holliday R: Genomic imprinting and allelic exclusion. *Dev Suppl* 125, 1990.

McMaster MT, et al: Human placental HLA-G expression is restricted to differentiated cytotrophoblasts. *J Immunol* 154:3771, 1995.

Monk M, Grant M: Preferential X-chromosome inactivation, DNA methylation and imprinting. *Dev Suppl* 55, 1990.

Roth I, et al: Human placental cytotrophoblasts produce the immunosuppressive cytokine interleukin 10. *J Exp Med* 184:539, 1996.

Rubin GL: Ectopic pregnancy in the United States: 1970 through 1978. *JAMA* 249:1725, 1983.

chapter 4

Third Week of Development: Trilaminar Germ Disc

Gastrulation: Formation of Embryonic Mesoderm and Endoderm

The most characteristic event occurring during the third week of gestation is **gastrulation,** the process that establishes all three **germ layers (ectoderm, mesoderm,** and **endoderm)** in the embryo. Gastrulation begins with formation of the **primitive streak** on the surface of the epiblast (Figs. 4.1–4.3A). Initially, the streak is vaguely defined (Fig. 4.1), but in a 15- to 16-day embryo, it is clearly visible as a narrow groove with slightly bulging regions on either side (Fig. 4.2). The cephalic end of the streak, the **primitive node,** consists of a slightly elevated area surrounding the small **primitive pit** (Fig. 4.3). Cells of the epiblast migrate toward the primitive streak (Fig. 4.3). Upon arrival in the region of the streak, they become flask-shaped, detach from the epiblast, and slip beneath it (Fig. 4.3, B–D). This inward movement is known as **invagination.** Once the cells have invaginated, some displace the hypoblast, creating the embryonic **endoderm,** and others come to lie between the epiblast and newly created endoderm to form **mesoderm.** Cells remaining in the epiblast then form **ectoderm.** Thus, the epiblast, through the process of gastrulation, is the source of all of the germ layers (Fig. 4.3B), and cells in these layers will give rise to all of the tissues and organs in the embryo.

As more and more cells move between the epiblast and hypoblast layers, they begin to spread laterally and cephalad (Fig. 4.3). Gradually,

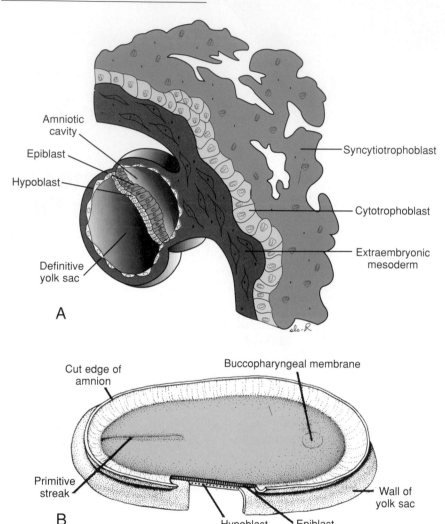

Amniotic cavity

Epiblast

Hypoblast

Definitive yolk sac

Syncytiotrophoblast

Cytotrophoblast

Extraembryonic mesoderm

A

Cut edge of amnion

Buccopharyngeal membrane

Primitive streak

Wall of yolk sac

B

Hypoblast   Epiblast

Figure 4.1 A. Implantation site at the end of the second week. **B.** Representative view of the germ disc at the end of the second week of development. The amniotic cavity has been opened to permit a view on the dorsal side of the epiblast. The hypoblast and epiblast are in contact with each other and the primitive streak forms a shallow groove in the caudal region of the embryo.

they migrate beyond the margin of the disc and establish contact with the extraembryonic mesoderm covering the yolk sac and amnion. In the cephalic direction, they pass on each side of the **prechordal plate.** The prechordal plate itself forms between the tip of the notochord and the **buccopharyngeal membrane** and is derived from some of the first cells that migrate through the node in a cephalic direction. Later, the prechordal plate will be important for

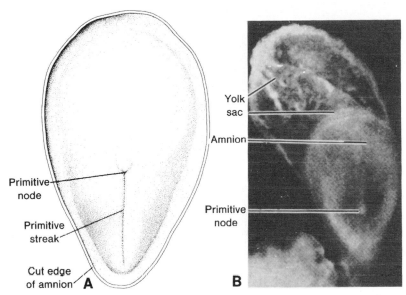

Figure 4.2 A. Dorsal aspect of an 18-day embryo. The embryo has a pear-shaped appearance and shows the primitive streak and node at its caudal end. **B.** Photograph of an 18-day human embryo, dorsal view. Note the primitive node and the notochord extending cranially. The yolk sac has a somewhat mottled appearance. The length of the embryo is 1.25 mm and the greatest width is 0.68 mm.

induction of the forebrain (Figs. 4.3*A* and 4.4*A*). The buccopharyngeal membrane at the cranial end of the disc consists of a small region of tightly adherent ectoderm and endoderm cells that represents the future opening of the oral cavity.

Formation of the Notochord

Prenotochordal cells invaginating in the primitive pit move forward cephalad until they reach the **prechordal plate** (Fig. 4.4). These prenotochordal cells become intercalated in the hypoblast so that, for a short time, the midline of the embryo consists of two cell layers that form the **notochordal plate** (Fig. 4.4, *B* and *C*). As the hypoblast is replaced by endoderm cells moving in at the streak, cells of the notochordal plate proliferate and detach from the endoderm. They then form a solid cord of cells, the **definitive notochord** (Fig. 4.4, *D* and *E*), which underlies the neural tube and serves as the basis for the axial skeleton. Because elongation of the notochord is a dynamic process, the cranial end forms first, and caudal regions are added as the primitive streak assumes a more caudal position. The notochord and prenotochordal cells extend cranially to the prechordal plate (an area just caudal to the buccopharyngeal membrane) and caudally to the primitive pit. At the point where the pit forms an indentation

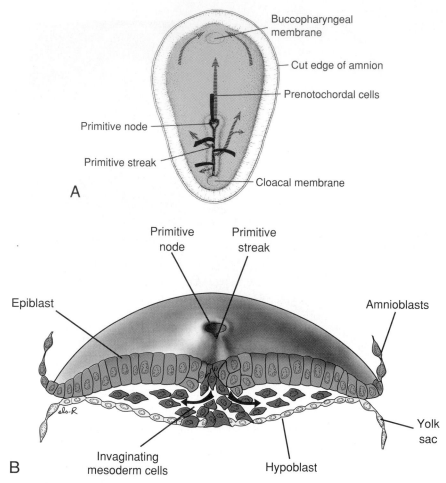

Figure 4.3 A. Dorsal side of the germ disc from a 16-day embryo indicating the movement of surface epiblast cells (*solid black lines*) through the primitive streak and node and the subsequent migration of cells between the hypoblast and epiblast (*broken lines*). **B.** Cross section through the cranial region of the streak at 15 days showing invagination of epiblast cells. The first cells to move inward displace the hypoblast to create the definitive endoderm. Once definitive endoderm is established, inwardly moving epiblast forms mesoderm. **C.** Scanning electron micrograph through the primitive streak of a mouse embryo showing migration of epiblast (*eb*) cells. The node region appears as a shallow pit (*arrow*). **D.** Higher magnification of the section in **C.**

in the epiblast, the **neurenteric canal** temporarily connects the amniotic and yolk sac cavities (Fig. 4.4*A*).

The cloacal membrane is formed at the caudal end of the embryonic disc (Fig. 4.3*A*). This membrane, which is similar in structure to the buccopharyngeal membrane, consists of tightly adherent ectoderm and endoderm cells with no intervening mesoderm. When the cloacal membrane appears, the posterior

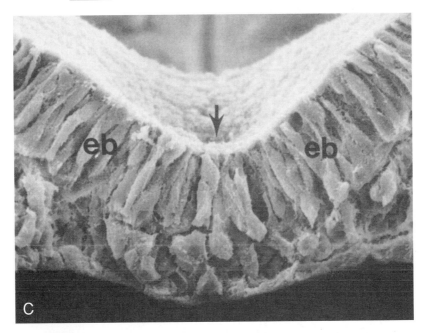

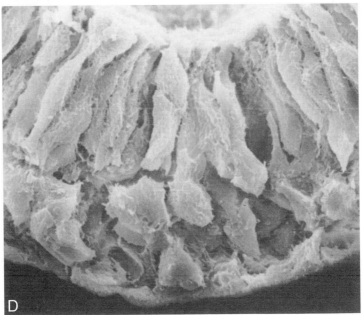

Figure 4.3 *Continued.*

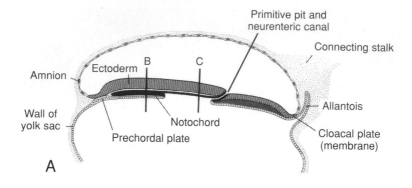

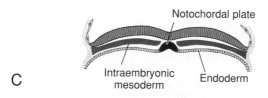

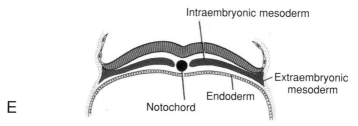

Figure 4.4 Schematic views and scanning electron micrographs illustrating formation of the notochord, whereby prenotochordal cells migrate through the primitive streak, become intercalated in the endoderm to form the notochordal plate, and finally detach from the endoderm to form the definitive notochord. Because these events occur in a cranial-to-caudal sequence, portions of the definitive notochord are established in the head region first. **A.** Drawing of a sagittal section through a 17-day embryo. The most cranial portion of the definitive notochord has formed, while prenotochordal cells caudal to this region are intercalated into the endoderm as the notochordal plate. **B.** Scanning electron micrograph of a mouse embryo showing the region of the buccopharyngeal

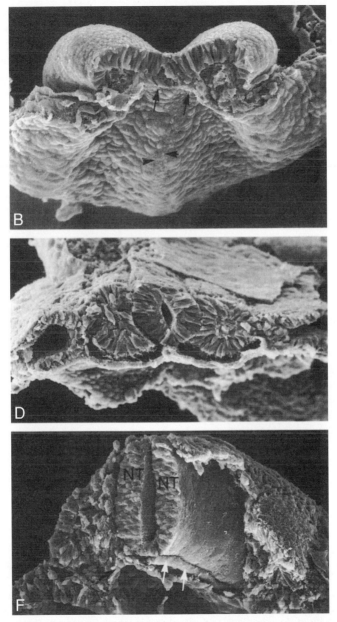

membrane *(arrows)*. Extending posteriorly is the prenotochordal plate *(arrowheads)*. **C.** Schematic cross section through the region of the notochordal plate. Soon the notochordal plate will detach from the endoderm to form the definitive notochord. **D.** Scanning electron micrograph of a mouse embryo showing detachment of the notochordal plate from the endoderm. **E.** Schematic view showing the definitive notochord. **F.** Scanning electron micrograph of a mouse embryo showing the definitive notochord *(arrows)* in close approximation to the neural tube *(NT)*.

wall of the yolk sac forms a small diverticulum that extends into the connecting stalk. This diverticulum, the **allantoenteric diverticulum,** or **allantois,** appears around the 16th day of development (Fig. 4.4*A*). Although in some lower vertebrates the allantois serves as a reservoir for excretion products of the renal system, in humans it remains rudimentary but may be involved in abnormalities of bladder development (see Chapter 14).

Establishment of the Body Axes

Establishment of the body axes, anteroposterior, dorsoventral, and left-right, takes place before and during the period of gastrulation. The anteroposterior axis is signaled by cells at the anterior (cranial) margin of the embryonic disc. This area, the **anterior visceral endoderm (AVE),** expresses genes essential for head formation, including the transcription factors *OTX2, LIM1,* and *HESX1* and the secreted factor **cerberus.** These genes establish the cranial end of the embryo before gastrulation. The primitive streak itself is initiated and maintained by expression of *Nodal,* a member of the *transforming growth factor β (TGF-β)* family (Fig. 4.5). Once the streak is formed, a number of genes regulate formation of dorsal and ventral mesoderm and head and tail structures. Another member of the *TGF-β* family, **bone morphogenetic protein-4 (BMP-4)** is secreted throughout the embryonic disc (Fig. 4.5). In the presence of this protein and **fibroblast growth factor (FGF),** mesoderm will be ventralized to contribute to kidneys (intermediate mesoderm), blood, and body wall mesoderm (lateral plate mesoderm). In fact, all mesoderm would be ventralized if the activity of BMP-4 were not blocked by other genes expressed in the node. For this reason, the node is the **organizer.** It was given that designation by

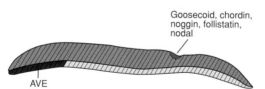

Goosecoid, chordin,
noggin, follistatin,
nodal

AVE

Figure 4.5 Sagittal section through the node and primitive streak showing the expression pattern of genes regulating the craniocaudal and dorsoventral axes. Cells at the prospective cranial end of the embryo in the anterior visceral endoderm (AVE) express the transcription factors *OTX2, LIM1,* and *HESX1* and the secreted factor cerberus that contribute to head development and establish the cephalic region. Once the streak is formed and gastrulation is progressing, bone morphogenetic protein (BMP-4; *hatched areas*), secreted throughout the bilaminar disc, acts with FGF to ventralize mesoderm into intermediate and lateral plate structures. *Goosecoid* regulates *chordin* expression, and this gene product, together with noggin and follistatin, antagonizes the activity of BMP-4, dorsalizing mesoderm into notochord and paraxial mesoderm for the head region. Later, expression of the *Brachyury (T)* gene antagonizes BMP-4 to dorsalize mesoderm in caudal regions of the embryo.

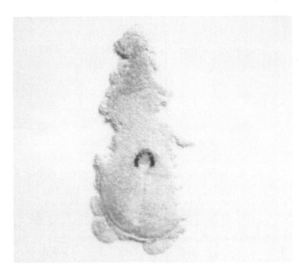

Figure 4.6 Node and primitive streak region removed from a mouse embryo showing expression of nodal using in situ hybridization. *Nodal* is expressed in the node and initiates and maintains the primitive streak.

Hans Spemann, who first described this activity in the dorsal lip of the blastopore, a structure analogous to the node, in *Xenopus* embryos. Thus, ***chordin*** (activated by the transcription factor ***Goosecoid***), ***noggin,*** and *follistatin* antagonize the activity of *BMP-4.* As a result, cranial mesoderm is dorsalized into notochord, somites, and somitomeres (Fig. 4.5). Later, these three genes are expressed in the notochord and are important in neural induction in the cranial region.

As mentioned, ***Nodal*** is involved in initiating and maintaining the primitive streak (Fig. 4.6). Similarly, ***HNF-3β*** maintains the node and later induces regional specificity in the forebrain and midbrain areas. Without *HNF-3β*, embryos fail to gastrulate properly and lack forebrain and midbrain structures. As mentioned previously, *Goosecoid* activates inhibitors of *BMP-4* and contributes to regulation of head development. Overexpression or underexpression of this gene results in severe malformations of the head region, including duplications (Fig. 4.7).

Regulation of dorsal mesoderm formation in mid and caudal regions of the embryo is controlled by the ***Brachyury (T)*** **gene** (Fig. 4.8). Thus, mesoderm formation in these regions depends on this gene product, and its absence results in shortening of the embryonic axis (caudal dysgenesis; see p. 80). The degree of shortening depends upon the time at which the protein becomes deficient.

Left-right sidedness, also established early in development, is orchestrated by a cascade of genes. When the primitive streak appears, **fibroblast growth factor 8 (FGF-8)** is secreted by cells in the node and primitive streak and

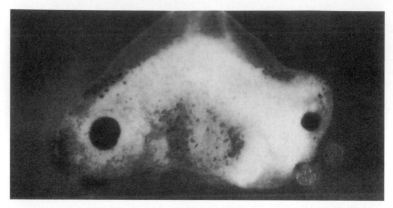

Figure 4.7 Two-headed tadpole produced by injecting additional *Goosecoid* mRNA into frog eggs. Similar results can be obtained by transplanting an additional node region to eggs. *Goosecoid* is normally expressed in the node and is a major regulator of head development.

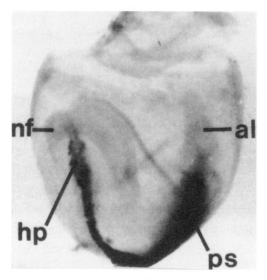

Figure 4.8 Expression pattern of the *Brachyury (T)* gene in the notochord and primitive streak of a mouse embryo. This gene antagonizes the activity of bone morphogenetic protein (BMP-4) in the hindbrain and spinal cord regions and dorsalizes mesoderm to form notochord, somitomeres, and somites (paraxial mesoderm). (Mouse embryos are dorsiflexed into a cup shape during the period of gastrulation and neurulation.) *nf*, neural fold; *hp*, head process; *ps*, primitive streak; *al*, allantois.

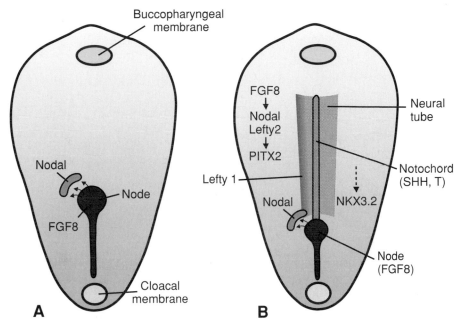

Figure 4.9 Dorsal views of the germ disc showing gene expression patterns responsible for establishing the left right body axis. **A.** Fibroblast growth factor 8 (FGF-8), secreted by the node and primitive streak, establishes expression of *Nodal,* a member of the *transforming growth factor β (TGF-β)* superfamily, on the left side near the node. **B.** Later, as the neural plate is induced, FGF-8 induces expression of *Nodal* and *Lefty-2* in the lateral plate mesoderm, whereas *Lefty-1* is expressed on the left side of the ventral aspect of the neural tube. Products from the *Brachyury (T)* gene, expressed in the notochord, also participate in induction of these three genes. In turn, expression of *Nodal* and *Lefty-2* regulates expression of the transcription factor *PITX 2,* which, through further downstream effectors, establishes left sidedness. *Sonic hedgehog (SHH),* expressed in the notochord, may serve as a midline barrier and also repress expression of left-sided genes on the right. *NKX 3.2* may regulate downstream genes important for establishing right sidedness.

induces expression of **Nodal** but only on the left side of the embryo (Fig. 4.9*A*). Later, as the neural plate is induced, FGF-8 maintains *Nodal* expression in the lateral plate mesoderm (Fig. 4.10), as well as **Lefty-2,** and both of these genes upregulate **PITX2,** a transcription factor responsible for establishing left sidedness (Fig. 4.9*B*). Simultaneously, **Lefty-1** is expressed on the left side of the floor plate of the neural tube and may act as a barrier to prevent left-sided signals from crossing over. **Sonic hedgehog (SHH)** may also function in this role as well as serving as a repressor for left sided gene expression on the right. The **Brachyury(T)** gene, another growth factor secreted by the notochord, is also essential for expression of *Nodal, Lefty-1,* and *Lefty-2* (Fig. 4.9*B*). Genes regulating right-sided development are not as well defined, although expression of the

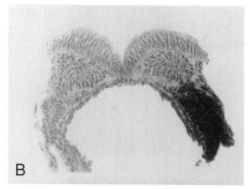

Figure 4.10 Expression pattern of the gene *nodal* in a mouse showing that it is restricted to the left side of the body **(A)** in the lateral plate mesoderm **(B)**. *Nodal*, together with *Lefty*, regulates downstream genes to determine left-right asymmetry.

transcription factor **NKX 3.2** is restricted to the right lateral plate mesoderm and probably regulates effector genes responsible for establishing the right side. Why the cascade is initiated on the left remains a mystery, but the reason may involve cilia on cells in the node that beat to create a gradient of FGF-8 toward the left. Indeed, abnormalities in cilia-related proteins result in laterality defects in mice and some humans with these defects have abnormal ciliary function (see p. 79).

Fate Map Established During Gastrulation

Regions of the epiblast that migrate and ingress through the primitive streak have been mapped and their ultimate fates determined (Fig. 4.11). For example, cells that ingress through the cranial region of the node become notochord; those migrating at the lateral edges of the node and from the cranial end of the streak become **paraxial mesoderm;** cells migrating through the midstreak region become **intermediate mesoderm;** those migrating through the more caudal part of the streak form **lateral plate mesoderm;** and cells migrating through the caudal-most part of the streak contribute to extraembryonic mesoderm (the other source of this tissue is the primitive yolk sac [hypoblast]; see p. 55).

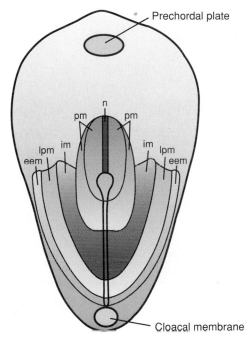

Figure 4.11 Dorsal view of the germ disc showing the primitive streak and a fate map for epiblast cells. Specific regions of the epiblast migrate through different parts of the node and streak to form mesoderm. Thus, cells migrating at the cranial-most part of the node will form the notochord (*n*); those migrating more posteriorly through the node and cranial-most aspect of the streak will form paraxial mesoderm (*pm;* somitomeres and somites); those migrating through the next portion of the streak will form intermediate mesoderm (*im;* urogenital system); those migrating through the more caudal part of the streak will form lateral plate mesoderm (*lpm;* body wall); and those migrating through the most caudal part will contribute to extraembryonic mesoderm (*eem;* chorion).

Growth of the Embryonic Disc

The embryonic disc, initially flat and almost round, gradually becomes elongated, with a broad cephalic and a narrow caudal end (Fig. 4.2). Expansion of the embryonic disc occurs mainly in the cephalic region; the region of the primitive streak remains more or less the same size. Growth and elongation of the cephalic part of the disc are caused by a continuous migration of cells from the primitive streak region in a cephalic direction. Invagination of surface cells in the primitive streak and their subsequent migration forward and laterally continues until the end of the fourth week. At that stage, the primitive streak shows regressive changes, rapidly shrinks, and soon disappears.

That the primitive streak at the caudal end of the disc continues to supply new cells until the end of the fourth week has an important bearing on development of the embryo. In the cephalic part, germ layers begin their specific differentiation by the middle of the third week (Fig. 4.12), whereas in the caudal

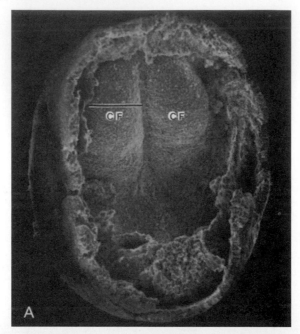

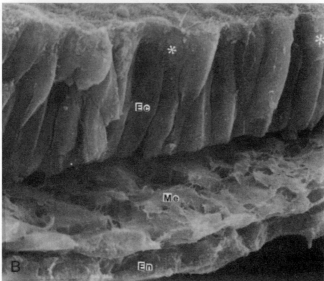

Figure 4.12 A. Scanning electron micrograph (dorsal view) of a mouse embryo (equivalent to human at approximately 18 days), showing initial elevation of the cranial neural folds (*CF*). The primitive streak lies farther caudally and is obscured from view. **B.** Transverse section through the embryo shown in **A** (see line of section). Note three germ layers; pseudostratified columnar cells of the neuroectoderm (*Ec*), flattened endoderm (*En*), and mesoderm (*Me*) sandwiched between these two layers. *Asterisks,* mitotic cells.

part, differentiation begins by the end of the fourth week. Thus gastrulation, or formation of the germ layers, continues in caudal segments while cranial structures are differentiating, causing the embryo to develop cephalocaudally (Fig. 4.12).

CLINICAL CORRELATES

Teratogenesis Associated With Gastrulation

The beginning of the third week of development, when gastrulation is initiated, is a highly sensitive stage for teratogenic insult. At this time, fate maps can be made for various organ systems, such as the eyes and brain anlage, and these cell populations may be damaged by teratogens. For example, high doses of alcohol at this stage kill cells in the anterior midline of the germ disc, producing a deficiency of the midline in craniofacial structures and resulting in **holoprosencephaly.** In such a child, the forebrain is small, the two lateral ventricles often merge into a single ventricle, and the eyes are close together (hypotelorism). Because this stage is reached 2 weeks after fertilization, it is approximately 4 weeks from the last menses. Therefore, the woman may not recognize she is pregnant, having assumed that menstruation is late and will begin shortly. Consequently, she may not take precautions she would normally consider if she knew she was pregnant.

Gastrulation itself may be disrupted by genetic abnormalities and toxic insults. In **caudal dysgenesis (sirenomelia),** insufficient mesoderm is formed in the caudal-most region of the embryo. Because this mesoderm contributes to formation of the lower limbs, urogenital system (intermediate mesoderm), and lumbosacral vertebrae, abnormalities in these structures ensue. Affected individuals exhibit a variable range of defects, including hypoplasia and fusion of the lower limbs, vertebral abnormalities, renal agenesis, imperforate anus, and anomalies of the genital organs (Fig. 4.13). In humans, the condition is associated with maternal diabetes and other causes. In mice, abnormalities of *Brachyury (T), Wnt,* and *engrailed* genes produce a similar phenotype.

Situs inversus is a condition in which transposition of the viscera in the thorax and abdomen occurs. Despite this organ reversal, other structural abnormalities occur only slightly more frequently in these individuals. Approximately 20 % of patients with complete situs inversus also have bronchiectasis and chronic sinusitis because of abnormal cilia **(Kartagener syndrome).** Interestingly, cilia are normally present on the ventral surface of the primitive node and may be involved in left-right patterning during gastrulation. Other conditions of abnormal sidedness are known as **laterality sequences.** Patients with these conditions do not have complete situs inversus but appear to be predominantly bilaterally left sided or right sided. The spleen reflects the differences; those with left-sided bilaterality have polysplenia, and those with right-sided bilaterality have asplenia or hypoplastic spleen. Patients with laterality sequences also are likely to have other malformations, especially heart defects.

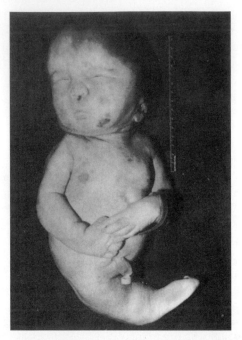

Figure 4.13 Sirenomelia (caudal dysgenesis). Loss of mesoderm in the lumbosacral region has resulted in fusion of the limb buds and other defects.

Tumors Associated With Gastrulation

Sometimes, remnants of the primitive streak persist in the sacrococcygeal region. These clusters of pluripotent cells proliferate and form tumors, known as **sacrococcygeal teratomas,** that commonly contain tissues derived from all three germ layers (Fig. 4.14). This is the most common tumor in newborns, occurring with a frequency of one in 37,000. These tumors may also arise from **primordial germ cells (PGCs)** that fail to migrate to the gonadal ridge (see p. 4).

Further Development of the Trophoblast

By the beginning of the third week, the trophoblast is characterized by **primary villi** that consist of a cytotrophoblastic core covered by a syncytial layer (Figs. 3.6 and 4.15). During further development, mesodermal cells penetrate the core of primary villi and grow toward the decidua. The newly formed structure is known as a **secondary villus** (Fig. 4.15).

By the end of the third week, mesodermal cells in the core of the villus begin to differentiate into blood cells and small blood vessels, forming the villous capillary system (Fig. 4.15). The villus is now known as a **tertiary villus**

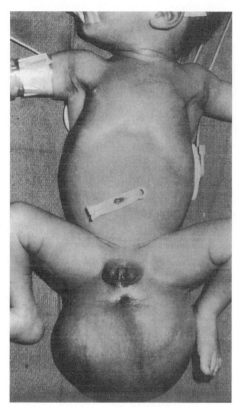

Figure 4.14 Sacrococcygeal teratoma resulting from remnants of the primitive streak. These tumors may become malignant and are most common in females.

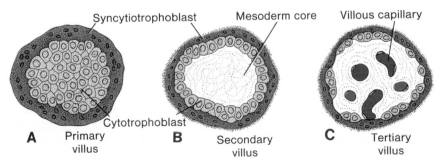

Figure 4.15 Development of a villus. **A.** Transverse section of a primary villus showing a core of cytotrophoblastic cells covered by a layer of syncytium. **B.** Transverse section of a secondary villus with a core of mesoderm covered by a single layer of cytotrophoblastic cells, which in turn is covered by syncytium. **C.** Mesoderm of the villus showing a number of capillaries and venules.

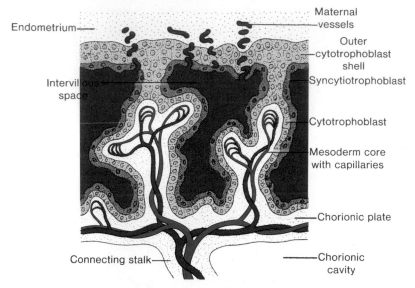

Figure 4.16 Longitudinal section through a villus at the end of the third week of development. Maternal vessels penetrate the cytotrophoblastic shell to enter intervillous spaces, which surround the villi. Capillaries in the villi are in contact with vessels in the chorionic plate and in the connecting stalk, which in turn are connected to intraembryonic vessels.

or **definitive placental villus.** Capillaries in tertiary villi make contact with capillaries developing in mesoderm of the chorionic plate and in the connecting stalk (Figs. 4.16 and 4.17). These vessels, in turn, establish contact with the intraembryonic circulatory system, connecting the placenta and the embryo. Hence, when the heart begins to beat in the fourth week of development, the villous system is ready to supply the embryo proper with essential nutrients and oxygen.

Meanwhile, cytotrophoblastic cells in the villi penetrate progressively into the overlying syncytium until they reach the maternal endometrium. Here they establish contact with similar extensions of neighboring villous stems, forming a thin **outer cytotrophoblast shell** (Figs. 4.16 and 4.17). This shell gradually surrounds the trophoblast entirely and attaches the chorionic sac firmly to the maternal endometrial tissue (Fig. 4.17). Villi that extend from the **chorionic plate** to the **decidua basalis (decidual plate:** the part of the endometrium where the placenta will form; see Chapter 6) are called **stem** or **anchoring villi.** Those that branch from the sides of stem villi are **free (terminal) villi,** through which exchange of nutrients and other factors will occur (Fig. 4.18).

The chorionic cavity, meanwhile, becomes larger, and by the 19th or 20th day, the embryo is attached to its trophoblastic shell by a narrow **connecting stalk** (Fig. 4.17). The connecting stalk later develops into the **umbilical cord,** which forms the connection between placenta and embryo.

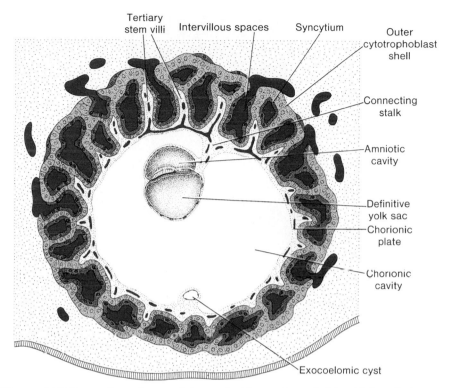

Tertiary stem villi   Intervillous spaces   Syncytium   Outer cytotrophoblast shell

Connecting stalk

Amniotic cavity

Definitive yolk sac

Chorionic plate

Chorionic cavity

Exocoelomic cyst

Figure 4.17 Presomite embryo and the trophoblast at the end of the third week. Tertiary and secondary stem villi give the trophoblast a characteristic radial appearance. Intervillous spaces, which are found throughout the trophoblast, are lined with syncytium. Cytotrophoblastic cells surround the trophoblast entirely and are in direct contact with the endometrium. The embryo is suspended in the chorionic cavity by means of the connecting stalk.

Summary

The most characteristic event occurring during the third week is **gastrulation,** which begins with the appearance of the **primitive streak,** which has at its cephalic end the **primitive node.** In the region of the node and streak, **epiblast** cells move inward **(invaginate)** to form new cell layers, **endoderm** and **mesoderm.** Hence, epiblast gives rise to all three **germ layers** in the embryo. Cells of the **intraembryonic mesodermal germ layer** migrate between the two other germ layers until they establish contact with the extraembryonic mesoderm covering the yolk sac and amnion (Figs. 4.3 and 4.4).

Prenotochordal cells invaginating in the primitive pit move forward until they reach the prechordal plate. They intercalate in the endoderm as the **notochordal plate** (Fig. 4.4). With further development, the plate detaches from the endoderm, and a solid cord, the **notochord,** is formed. It forms a midline axis,

Figure 4.18 Stem villi *(SV)* extend from the chorionic plate *(CP)* to the basal plate *(BP)*. Terminal villi *(arrows)* are represented by fine branches from stem villi.

which will serve as the basis of the axial skeleton (Fig. 4.4). Cephalic and caudal ends of the embryo are established before the primitive streak is formed. Thus, cells in the hypoblast (endoderm) at the cephalic margin of the disc form the **anterior visceral endoderm** that expresses head-forming genes, including *OTX2, LIM1,* and *HESX1* and the secreted factor **cerberus.** *Nodal,* a member of the *TGF-β* family of genes, is then activated and initiates and maintains the integrity of the node and streak. **BMP-4,** in the presence of **FGF,** ventralizes mesoderm during gastrulation so that it forms intermediate and lateral plate mesoderm. **Chordin, noggin,** and **follistatin** antagonize BMP-4 activity and dorsalize mesoderm to form the notochord and somitomeres in the head region. Formation of these structures in more caudal regions is regulated by the *Brachyury (T)* gene. Left-right asymmetry is regulated by a cascade of genes; first, **FGF-8,** secreted by cells in the node and streak, induces *Nodal* and *Lefty-2* expression on the left side. These genes upregulate *PITX2,* a transcription factor responsible for left sidedness.

Epiblast cells moving through the node and streak are predetermined by their position to become specific types of mesoderm and endoderm. Thus, it is possible to construct a fate map of the epiblast showing this pattern (Fig. 4.11).

By the end of the third week, three basic **germ layers,** consisting of **ectoderm, mesoderm,** and **endoderm,** are established in the head region, and the process continues to produce these germ layers for more caudal areas of the embryo until the end of the 4th week. Tissue and organ differentiation has begun, and it occurs in a cephalocaudal direction as gastrulation continues.

In the meantime, the trophoblast progresses rapidly. **Primary villi** obtain a mesenchymal core in which small capillaries arise (Fig. 4.17). When these villous capillaries make contact with capillaries in the chorionic plate and connecting stalk, the villous system is ready to supply the embryo with its nutrients and oxygen (Fig. 4.17).

Problems to Solve

1. *A 22-year-old woman consumes large quantities of alcohol at a party and loses consciousness; 3 weeks later she misses her second consecutive period. A pregnancy test is positive. Should she be concerned about the effects of her binge-drinking episode on her baby?*

2. *An ultrasound scan detects a large mass near the sacrum of a 28-week female fetus. What might the origin of such a mass be, and what type of tissue might it contain?*

3. *On ultrasound examination, it was determined that a fetus had well-developed facial and thoracic regions, but caudal structures were abnormal. Kidneys were absent, lumbar and sacral vertebrae were missing, and the hindlimbs were fused. What process may have been disturbed to cause such defects?*

4. *A child has polysplenia and abnormal positioning of the heart. How might these two abnormalities be linked developmentally, and when would they have originated? Should you be concerned that other defects might be present? What genes might have caused this event, and when during embryogenesis would it have been initiated?*

SUGGESTED READING

Augustine K, Liu ET, Sadler TW: Antisense attenuation of Wnt-1 and Wnt-3a expression in whole embryo culture reveals roles for these genes in craniofacial, spinal cord, and cardiac morphogenesis. *Dev Genet* 14:500, 1993.

Beddington RSP: The origin of the foetal tissues during gastrulation in the rodent. *In* Johnson MH (ed): *Development in Mammals.* New York, Elsevier, 99:1, 1983.

Beddington RSP: Induction of a second neural axis by the mouse node. *Development* 120:613, 1994.

Beddington RSP, Robertson RJ: Anterior patterning in the mouse. *Trends Genet* 14:277, 1998.

Bellairs R: The primitive streak. *Anat Embryol* 174:1, 1986.

De Robertis EM, Sasai Y: A common plan for dorsoventral patterning in *Bilateria. Nature* 380:37, 1996.

Herrmann BG: Expression pattern of the brachyuria gene in whole mount TWis/TWis mutant embryos. *Development* 13:913, 1991.

Holzgreve W, Flake AW, Langer JC: The fetus with sacrococcygeal teratoma. *In* Harrison MR, Gollus MS, Filly RA (eds): *The Unborn Patient. Prenatal Diagnosis and Treatment.* Philadelphia, WB Saunders, 1991.

King BF, Mais JJ: Developmental changes in rhesus monkey placental villi and cell columns. *Anat Embryol* 165:361, 1982.

Meyers EN, Martin GR: Differences in left-right axis pathways in mouse and chick: Functions of FGF-8 and SHH. *Science* 285:403, 1999.

O'Rahilly R: *Developmental Stages in Human Embryos. Part A. Embryos of the First Three Weeks (Stages One to Nine).* Washington, DC, Carnegie Institution of Washington, 1973.

Stott D, Kisbert A, Herrmann BG: Rescue of the tail defect of *Brachyury* mice. *Genes Dev* 7:197, 1993.

Sulik KK, Lauder JM, Dehart DB: Brain malformations in prenatal mice following acute maternal ethanol administration. *Int J Dev Neurosci* 2:203, 1984.

Supp DM, Potter SS, Brueckner M: Molecular motors: The driving force behind mammalian left-right development. *Trends Cell Biol* 10:41, 2000.

Tam PPL, Bedding RSP: The formation of mesodermal tissues in the mouse embryo during gastrulation and early organogenesis. *Development* 99:109, 1987.

chapter 5

Third to Eighth Week: The Embryonic Period

The **embryonic period** or period of **organogenesis,** occurs from the **third to the eighth weeks** of development and is the time when each of the three germ layers, **ectoderm, mesoderm,** and **endoderm,** gives rise to a number of specific tissues and organs. By the end of the embryonic period, the main organ systems have been established, rendering the major features of the external body form recognizable by the end of the second month.

Derivatives of the Ectodermal Germ Layer

At the beginning of the third week of development, the ectodermal germ layer has the shape of a disc that is broader in the cephalic than the caudal region (Fig. 5.1). Appearance of the notochord and prechordal mesoderm induces the overlying ectoderm to thicken and form the **neural plate** (Fig. 5.2). Cells of the plate make up the **neuroectoderm** and their induction represents the initial event in the process of **neurulation.**

MOLECULAR REGULATION OF NEURAL INDUCTION

Blocking the activity of **BMP-4,** a **TGF-β** family member responsible for ventralizing ectoderm and mesoderm, causes induction of the

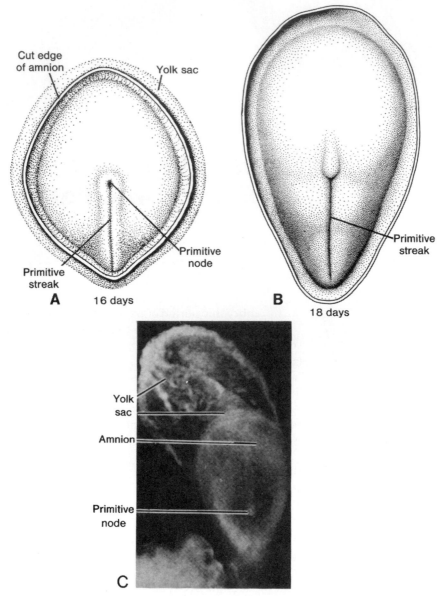

Figure 5.1 A. Dorsal view of a 16-day presomite embryo. The primitive streak and primitive node are visible. **B.** Dorsal view of an 18-day presomite embryo. The embryo is pear-shaped, with its cephalic region somewhat broader than its caudal end. **C.** Dorsal view of an 18-day human embryo. Note the primitive node and, extending forward from it, the notochord. The yolk sac has a somewhat mottled appearance. The length of the embryo is 1.25 mm, and the greatest width is 0.68 mm.

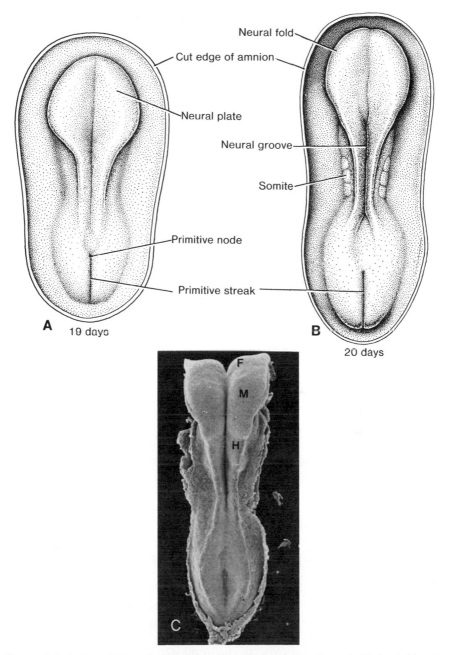

Figure 5.2 A. Dorsal view of a late presomite embryo (approximately 19 days). The amnion has been removed and the neural plate is clearly visible. **B.** Dorsal view of a human embryo at approximately 20 days showing somites and formation of the neural groove and neural folds. **C.** Scanning electron micrograph of a mouse embryo (approximately 20-day human) showing the typical appearance of the neural groove stage. Cranial neural folds have segregated themselves into forebrain (*F*, prosencephalon), midbrain (*M*, mesencephalon), and hindbrain (*H*, rhombencephalon) regions.

neural plate. Thus, in the presence of BMP-4, which permeates the mesoderm and ectoderm of the gastrulating embryo, ectoderm becomes epidermis, and mesoderm forms intermediate and lateral plate mesoderm. If BMP-4 is absent or inactivated, ectoderm becomes neuralized. Secretion of three other molecules, **noggin, chordin,** and **follistatin,** inactivates this protein. These three proteins are present in the organizer (primitive node), notochord, and prechordal mesoderm. They neuralize ectoderm and cause mesoderm to become notochord and paraxial mesoderm (dorsalizes mesoderm). However, these neural inducers induce only forebrain and midbrain types of tissues. Induction of caudal neural plate structures (hindbrain and spinal cord) depends upon two secreted proteins, **WNT-3a** and **FGF (fibroblast growth factor).** In addition, **retinoic acid** appears to play a role in organizing the cranial-to-caudal axis because it can cause respecification of cranial segments into more caudal ones by regulating expression of ***homeobox genes*** (see p. 105).

NEURULATION

Once induction has occurred, the elongated, slipper-shaped neural plate gradually expands toward the primitive streak (Fig. 5.2, *B* and *C*). By the end of the third week, the lateral edges of the neural plate become more elevated to form **neural folds,** and the depressed midregion forms the **neural groove** (Figs. 5.2, 5.3, *A* and *B,* and 5.4). Gradually, the neural folds approach each other in the midline, where they fuse (Fig. 5.3*C*). Fusion begins in the cervical region (fifth somite) and proceeds cranially and caudally (Figs. 5.5 and 5.6). As a result, the **neural tube** is formed. Until fusion is complete, the cephalic and caudal ends of the neural tube communicate with the amniotic cavity by way of the **cranial** and **caudal neuropores,** respectively (Figs. 5.5, 5.6*A,* and 5.7). Closure of the cranial neuropore occurs at approximately day 25 (18- to 20-somite stage), whereas the posterior neuropore closes at day 27 (25-somite stage). Neurulation is then complete, and the central nervous system is represented by a closed tubular structure with a narrow caudal portion, the **spinal cord,** and a much broader cephalic portion characterized by a number of dilations, the **brain vesicles** (see Chapter 19).

As the neural folds elevate and fuse, cells at the lateral border or crest of the neuroectoderm begin to dissociate from their neighbors. This cell population, the **neural crest** (Figs. 5.3 and 5.4), will undergo an epithelial-to-mesenchymal transition as it leaves the neuroectoderm by active migration and displacement to enter the underlying mesoderm. (**Mesoderm** refers to cells derived from the epiblast and extraembryonic tissues. **Mesenchyme** refers to loosely organized embryonic connective tissue regardless of origin.) Crest cells from the trunk region leave the neural folds after closure of the neural tube and migrate along one of two pathways: 1) a dorsal pathway through the dermis, where they will enter the ectoderm through holes in the basal lamina to form **melanocytes** in the skin and hair follicles; and 2) a ventral pathway through the anterior half of each somite to become **sensory ganglia, sympathetic and enteric neurons,**

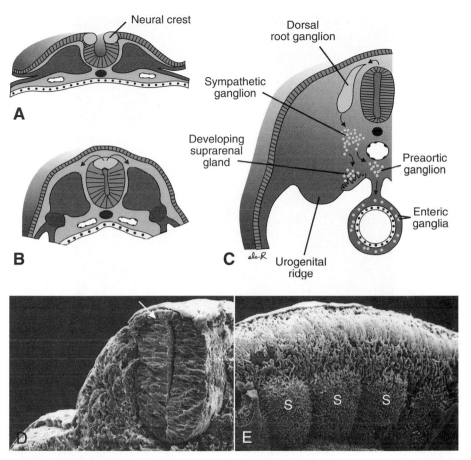

Figure 5.3 Formation and migration of neural crest cells in the spinal cord. **A** and **B.** Crest cells form at the tips of neural folds and do not migrate away from this region until neural tube closure is complete. **C.** After migration, crest cells contribute to a heterogeneous array of structures, including dorsal root ganglia, sympathetic chain ganglia, adrenal medulla, and other tissues (Table 5.1). **D.** In a scanning electron micrograph of a mouse embryo, crest cells at the top of the closed neural tube can be seen migrating away from this area (*arrow*). **E.** In a lateral view with the overlying ectoderm removed, crest cells appear fibroblastic as they move down the sides of the neural tube. (*S,* somites).

Schwann cells, and **cells of the adrenal medulla** (Fig. 5.3). Neural crest cells also form and migrate from cranial neural folds, leaving the neural tube before closure in this region (Fig. 5.4). These cells contribute to the **craniofacial skeleton** as well as **neurons for cranial ganglia, glial cells, melanocytes,** and other cell types (Table 5.1). Induction of neural crest cells requires an interaction between adjacent neural and overlying ectoderm. **Bone morphogenetic proteins (BMPs),** secreted by non-neural ectoderm, appear to initiate the induction

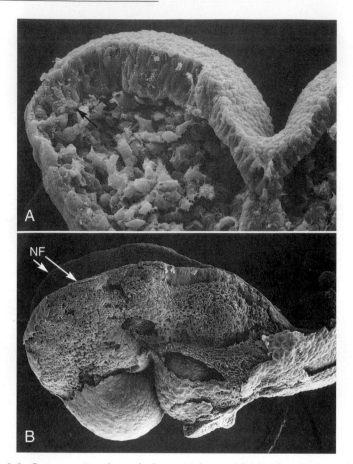

Figure 5.4 A. Cross section through the cranial neural folds of a mouse embryo. Neural crest cells at the tip of the folds (*arrow*) migrate and contribute to craniofacial mesenchyme. **B.** Lateral view of the cranial neural folds of a mouse embryo with the surface ectoderm removed. Numerous neural crest cells can be observed leaving the neural folds (*NF*) and migrating beneath the ectoderm that has been removed. Unlike crest cells of the spinal cord, cranial crest exits the neural folds before they fuse.

process. Crest cells give rise to a heterogeneous array of tissues, as indicated in Table 5.1 (see p. 95).

By the time the neural tube is closed, two bilateral **ectodermal thickenings,** the **otic placodes** and the **lens placodes,** become visible in the cephalic region of the embryo (Fig. 5.8*B*). During further development, the otic placodes invaginate and form the **otic vesicles,** which will develop into structures needed for hearing and maintenance of equilibrium (see Chapter 16). At approximately the same time, the **lens placodes** appear. These placodes also invaginate and, during the fifth week, form the **lenses** of the eyes (see Chapter 17).

In general terms, the ectodermal germ layer gives rise to organs and structures that maintain contact with the outside world: (*a*) the central nervous

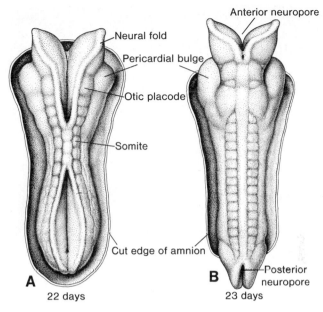

Figure 5.5 A. Dorsal view of a human embryo at approximately day 22. Seven distinct somites are visible on each side of the neural tube. **B.** Dorsal view of a human embryo at approximately day 23. Note the pericardial bulge on each side of the midline in the cephalic part of the embryo.

system; (*b*) the peripheral nervous system; (*c*) the sensory epithelium of the ear, nose, and eye; and (*d*) the epidermis, including the hair and nails. In addition, it gives rise to subcutaneous glands, the mammary glands, the pituitary gland, and enamel of the teeth.

Derivatives of the Mesodermal Germ Layer

Initially, cells of the mesodermal germ layer form a thin sheet of loosely woven tissue on each side of the midline (Fig. 5.9*A*). By approximately the 17th day, however, cells close to the midline proliferate and form a thickened plate of tissue known as **paraxial mesoderm** (Fig. 5.9*B*). More laterally, the mesoderm layer remains thin and is known as the **lateral plate.** With the appearance and coalescence of intercellular cavities in the lateral plate, this tissue is divided into two layers (Fig. 5.9, *B* and *C*): (*a*) a layer continuous with mesoderm covering the amnion, known as the **somatic** or **parietal mesoderm layer;** and (*b*) a layer continuous with mesoderm covering the yolk sac, known as the **splanchnic** or **visceral mesoderm layer** (Figs. 5.9, *C* and *D,* and 5.10). Together, these layers line a newly formed cavity, the **intraembryonic cavity,** which is continuous with the extraembryonic cavity on each side of the embryo. **Intermediate mesoderm** connects paraxial and lateral plate mesoderm (Figs. 5.9, *B* and *D,* and 5.10).

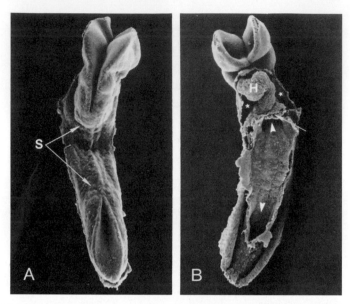

Figure 5.6 Dorsal (**A**) and ventral (**B**) views of a mouse embryo (approximately 22-day human). **A.** The neural groove is closing in cranial and caudal directions and is flanked by pairs of somites (*S*). **B.** The same embryo showing formation of the gut tube with anterior and posterior intestinal portals (*arrowheads*), heart (*H*) in the pericardial cavity (*asterisks*), and the septum transversum (*arrow*) representing the primordium of the diaphragm (see Chapter 11). The neural folds remain open, exposing forebrain and midbrain regions.

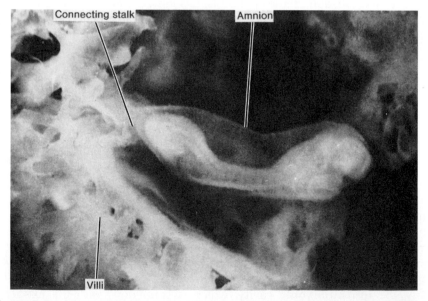

Figure 5.7 A 12- to 13-somite embryo (approximately 23 days). The embryo within its amniotic sac is attached to the chorion by the connecting stalk. Note the well-developed chorionic villi.

TABLE 5.1 **Neural Crest Derivatives**

Connective tissue and bones of the face and skull
Cranial nerve ganglia (see Table 19.2)
C cells of the thyroid gland
Conotruncal septum in the heart
Odontoblasts
Dermis in face and neck
Spinal (dorsal root) ganglia
Sympathetic chain and preaortic ganglia
Parasympathetic ganglia of the gastrointestinal tract
Adrenal medulla
Schwann cells
Glial cells
Arachnoid and pia mater (leptomeninges)
Melanocytes

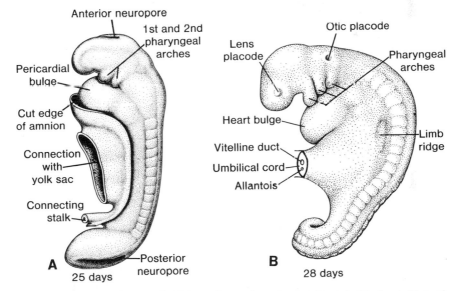

Figure 5.8 A. Lateral view of a 14-somite embryo (approximately 25 days). Note the bulging pericardial area and the first and second pharyngeal arches. **B.** The left side of a 25-somite embryo approximately 28 days old. The first three pharyngeal arches and lens and otic placodes are visible.

PARAXIAL MESODERM

By the beginning of the third week, paraxial mesoderm is organized into segments. These segments, known as **somitomeres,** first appear in the cephalic region of the embryo, and their formation proceeds cephalocaudally. Each somitomere consists of mesodermal cells arranged in concentric whorls around the center of the unit. In the head region, somitomeres form in association with

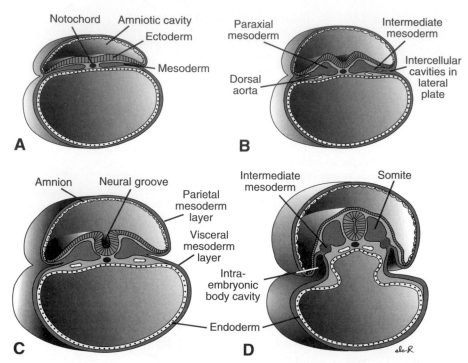

Figure 5.9 Transverse sections showing development of the mesodermal germ layer. **A.** Day 17. **B.** Day 19. **C.** Day 20. **D.** Day 21. The thin mesodermal sheet gives rise to paraxial mesoderm (future somites), intermediate mesoderm (future excretory units), and lateral plate, which is split into parietal and visceral mesoderm layers lining the intraembryonic cavity.

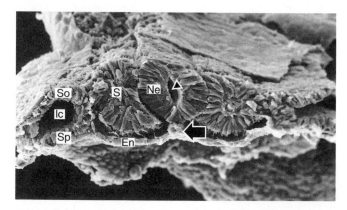

Figure 5.10 Transverse sections through cervical somites of mouse embryos (approximately 21-day human) as visualized by scanning electron microscopy. *Arrow*, notochord; *arrowhead*, neural canal; *En*, endoderm; *Ic*, intraembryonic cavity; *Ne*, neuroectoderm; *S*, Somite; *So*, somatic mesoderm; and *Sp*, splanchnic mesoderm.

segmentation of the neural plate into **neuromeres** and contribute to mesenchyme in the head (see Chapter 15). From the occipital region caudally, somitomeres further organize into somites. The first pair of somites arises in the occipital region of the embryo at approximately the 20th day of development. From here, new somites appear in craniocaudal sequence at a rate of approximately three pairs per day until, at the end of the fifth week, 42 to 44 pairs are present (Figs. 5.3, 5.5, and 5.8). There are four occipital, eight cervical, 12 thoracic, five lumbar, five sacral, and eight to 10 coccygeal pairs. The first occipital and the last five to seven coccygeal somites later disappear, while the remaining somites form the axial skeleton (see Chapter 8). During this period of development, the age of the embryo is expressed in number of somites. Table 5.2 shows the approximate age of the embryo correlated to the number of somites.

By the beginning of the fourth week, cells forming the ventral and medial walls of the somite lose their compact organization, become polymorphous, and shift their position to surround the notochord (Fig. 5.11, *A* and *B*). These cells, collectively known as the **sclerotome,** form a loosely woven tissue, the **mesenchyme.** They will surround the spinal cord and notochord to form the vertebral column (see Chapter 8). Cells at the dorsolateral portion of the somite also migrate as precursors of the limb and body wall musculature (Fig. 5.11*B*). After migration of these muscle cells and cells of the sclerotome, cells at the dorsomedial portion of the somite proliferate and migrate down the ventral side of the remaining dorsal epithelium of the somite to form a new layer, the myotome (Fig. 5.11, *B* and *C*). The remaining dorsal epithelium forms the dermatome, and together these layers constitute the dermomyotome (Fig. 5.11*C*). Each segmentally arranged myotome contributes to muscles of the back (epaxial musculature; see Chapter 9), while dermatomes disperse to form the dermis and subcutaneous tissue of the skin (see Chapter 18). Furthermore, each myotome and dermatome retains its innervation from its segment of origin, no matter where the cells migrate. Hence each somite forms its own

TABLE 5.2 **Number of Somites Correlated to Approximate Age in Days**

Approximate Age (days)	No. of Somites
20	1–4
21	4–7
22	7–10
23	10–13
24	13–17
25	17–20
26	20–23
27	23–26
28	26–29
30	34–35

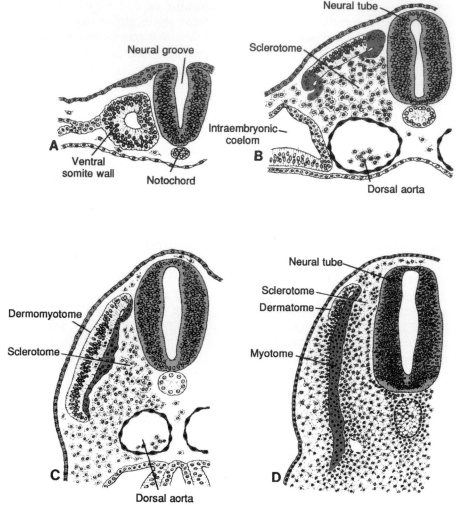

Figure 5.11 Stages in development of a somite. **A.** Mesoderm cells are arranged around a small cavity. **B.** Cells of the ventral and medial walls of the somite lose their epithelial arrangement and migrate in the direction of the notochord. These cells collectively constitute the sclerotome. Cells at the dorsolateral portion of the somite migrate as precursors to limb and body wall musculature. Dorsomedial cells migrate beneath the remaining dorsal epithelium of the somite to form the myotome. **C.** Cells forming the myotome continue to extend beneath the dorsal epithelium. **D.** After ventral extension of the myotome, dermatome cells lose their epithelial configuration and spread out under the overlying ectoderm to form dermis.

sclerotome (the cartilage and bone component), its own **myotome** (providing the segmental muscle component), and its own **dermatome,** the segmental skin component. Each myotome and dermatome also has its own segmental nerve component.

Molecular Regulation of Somite Differentiation

Signals for somite differentiation arise from surrounding structures, including the notochord, neural tube, epidermis, and lateral plate mesoderm (Fig. 5.12). The secreted protein product of the gene *Sonic hedgehog (Shh),* produced by the notochord and floor plate of the neural tube, induces the ventromedial portion of the somite to become sclerotome. Once induced, sclerotome cells express the transcription factor *PAX1,* which initiates the cascade of

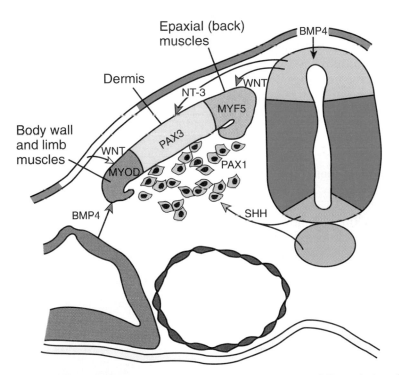

Figure 5.12 Expression patterns of genes that regulate somite differentiation. Sonic hedgehog (SHH), secreted by the notochord and floor plate of the neural tube, causes the ventral part of the somite to form sclerotome and to express *PAX1*, which in turn controls chondrogenesis and vertebrae formation. WNT proteins from the dorsal neural tube activate *PAX3*, which demarcates the dermomyotome. WNT proteins also direct the dorsomedial portion of the somite to form epaxial (back) muscles and to express the muscle-specific gene *MYF5*. The middorsal portion of the somite is directed to become dermis by neurotrophin 3 (NT-3) expressed by the dorsal neural tube. Hypaxial (limb and body wall) musculature is derived from the dorsolateral portion of the somite under the combined influence of activating WNT proteins and inhibitory BMP-4 protein, which together activate *MYOD* expression.

cartilage- and bone-forming genes for vertebral formation. Expression of *PAX3,* regulated by **WNT** proteins from the dorsal neural tube, marks the dermomyotome region of the somite. WNT proteins from the dorsal neural tube also target the dorsomedial portion of the somite, causing it to initiate expression of the muscle-specific gene *MYF5* and to become epaxial musculature. Interplay between the inhibiting protein **BMP-4** (and probably **FGFs**) from the lateral plate mesoderm and activating WNT products from the epidermis directs the dorsolateral portion of the somite to express another muscle-specific gene, *MYOD,* and to form limb and body wall muscles. The midportion of the dorsal epithelium of the somite is directed by **neurotrophin 3 (NT-3),** secreted by the dorsal region of the neural tube, to form dermis.

INTERMEDIATE MESODERM

Intermediate mesoderm, which temporarily connects paraxial mesoderm with the lateral plate (Figs. 5.9*D* and 5.10*A*), differentiates into urogenital structures. In cervical and upper thoracic regions, it forms segmental cell clusters (future **nephrotomes**), whereas more caudally, it forms an unsegmented mass of tissue, the **nephrogenic cord.** Excretory units of the urinary system and the gonads develop from this partly segmented, partly unsegmented intermediate mesoderm (see Chapter 14).

LATERAL PLATE MESODERM

Lateral plate mesoderm splits into parietal and visceral layers, which line the intraembryonic cavity and surround the organs, respectively (Figs. 5.9, *C* and *D*, 5.10, and 5.13*A*). Mesoderm from the parietal layer, together with overlying ectoderm, will form the lateral and ventral body wall. The visceral layer and embryonic endoderm will form the wall of the gut (Fig. 5.13*B*). Mesoderm

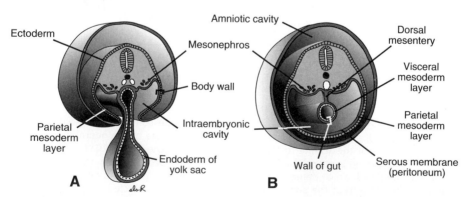

Figure 5.13 A. Transverse section through a 21-day embryo in the region of the mesonephros showing parietal and visceral mesoderm layers. The intraembryonic cavities communicate with the extraembryonic cavity (chorionic cavity). **B.** Section at the end of the fourth week. Parietal mesoderm and overlying ectoderm form the ventral and lateral body wall. Note the peritoneal (serous) membrane.

cells of the parietal layer surrounding the intraembryonic cavity will form thin membranes, the **mesothelial membranes,** or **serous membranes,** which will line the peritoneal, pleural, and pericardial cavities and secrete serous fluid (Fig. 5.13*B*). Mesoderm cells of the visceral layer will form a thin serous membrane around each organ (see Chapter 10).

BLOOD AND BLOOD VESSELS

Blood vessels form in two ways: **vasculogenesis,** whereby vessels arise from blood islands (Fig. 5.14), and **angiogenesis,** which entails sprouting from

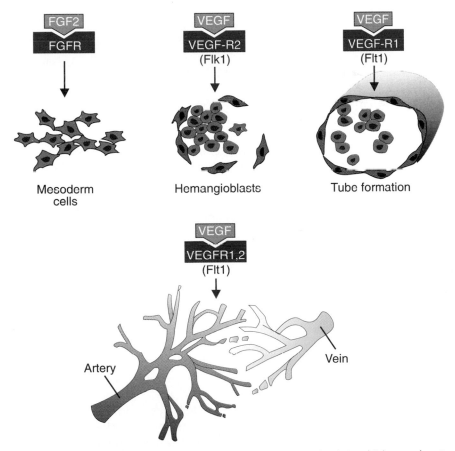

Figure 5.14 Blood vessels form in two ways: vasculogenesis (top), in which vessels arise from blood islands, and angiogenesis (bottom), in which new vessels sprout from existing ones. During vasculogenesis, fibroblast growth factor 2 (FGF-2) binds to its receptor on subpopulations of mesoderm cells and induces them to form hemangioblasts. Then, under the influence of vascular endothelial growth factor (VEGF) acting through two different receptors, these cells become endothelial and coalesce to form vessels. Angiogenesis is also regulated by VEGF, which stimulates proliferation of endothelial cells at points where new vessels will sprout from existing ones. Final modeling and stabilization of the vasculature are accomplished by platelet-derived growth factor (PDGF) and transforming growth factor β (TGF-β).

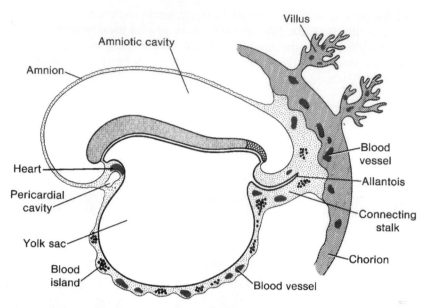

Figure 5.15 Extraembryonic blood vessel formation in the villi, chorion, connecting stalk, and wall of the yolk sac in a presomite embryo of approximately 19 days.

existing vessels. The first blood islands appear in mesoderm surrounding the wall of the yolk sac at 3 weeks of development and slightly later in lateral plate mesoderm and other regions (Fig. 5.15). These islands arise from mesoderm cells that are induced by fibroblast growth factor 2 (FGF-2) to form **hemangioblasts,** a common precursor for vessel and blood cell formation. Hemangioblasts in the center of blood islands form **hematopoietic stem cells,** the precursors of all blood cells, whereas peripheral hemangioblasts differentiate into **angioblasts,** the precursors to blood vessels. These angioblasts proliferate and are eventually induced to form endothelial cells by **vascular endothelial growth factor (VEGF)** secreted by surrounding mesoderm cells (Fig. 5.14). This same factor then regulates coalescence of these endothelial cells into the first primitive blood vessels.

Once the process of vasculogenesis establishes a primary vascular bed, additional vasculature is added by angiogenesis, the sprouting of new vessels (Fig. 5.14). This process is also mediated by VEGF, which stimulates proliferation of endothelial cells at points where new vessels are to be formed. Maturation and modeling of the vasculature are regulated by other growth factors, including platelet-derived growth factor (PDGF) and transforming growth factor β (TGF-β), until the adult pattern is established.

As mentioned, the first blood cells arise in the blood islands of the yolk sac, but this population is transitory. The definitive hematopoietic stem cells arise from mesoderm surrounding the aorta in a site called the

aorta-gonad-mesonephros region (AGM). These cells will colonize the liver, which becomes the major hematopoietic organ of the fetus. Later, stem cells from the liver will colonize the bone marrow, the definitive blood-forming tissue.

Derivatives of the Endodermal Germ Layer

The gastrointestinal tract is the main organ system derived from the endodermal germ layer. This germ layer covers the ventral surface of the embryo and forms the roof of the yolk sac (Fig. 5.16A). With development and growth of the brain vesicles, however, the embryonic disc begins to bulge into the amniotic cavity and to fold **cephalocaudally.** This folding is most pronounced in the regions of the head and tail, where the **head fold** and **tail fold** are formed (Fig. 5.16).

As a result of cephalocaudal folding, a continuously larger portion of the endoderm-lined cavity is incorporated into the body of the embryo proper

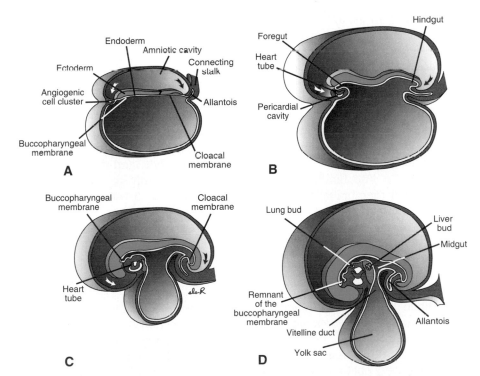

Figure 5.16 Sagittal midline sections of embryos at various stages of development to demonstrate cephalocaudal folding and its effect on position of the endoderm-lined cavity. **A.** Presomite embryo. **B.** Embryo with 7 somites. **C.** Embryo with 14 somites. **D.** End of the first month. Note the angiogenic cell clusters in relation to the buccopharyngeal membrane.

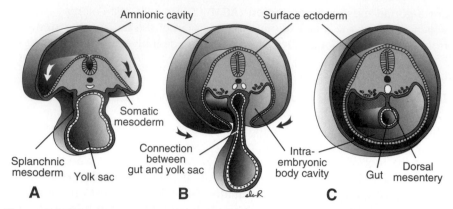

Figure 5.17 Transverse sections through embryos at various stages of development to show the effect of lateral folding on the endoderm-lined cavity. **A.** Folding is initiated. **B.** Transverse section through the midgut to show the connection between the gut and yolk sac. **C.** Section just below the midgut to show the closed ventral abdominal wall and gut suspended from the dorsal abdominal wall by its mesentery.

(Fig. 5.16*C*). In the anterior part, the endoderm forms the **foregut;** in the tail region, it forms the **hindgut.** The part between foregut and hindgut is the **midgut.** The midgut temporarily communicates with the yolk sac by way of a broad stalk, the **vitelline duct** (Fig. 5.16*D*). This duct is wide initially, but with further growth of the embryo, it becomes narrow and much longer (Figs. 5.16*D*, 5.17*B*, and 5.20).

At its cephalic end, the foregut is temporarily bounded by an ectodermal-endodermal membrane called the **buccopharyngeal membrane** (Fig. 5.16, *A* and *C*). In the fourth week, the buccopharyngeal membrane ruptures, establishing an open connection between the amniotic cavity and the primitive gut (Fig. 5.16*D*). The hindgut also terminates temporarily at an ectodermal-endodermal membrane, the **cloacal membrane** (Fig. 5.16*C*), which breaks down in the seventh week to create the opening for the anus.

As a result of rapid growth of the somites, the initial flat embryonic disc also folds laterally, and the embryo obtains a round appearance (Fig. 5.17). Simultaneously, the ventral body wall of the embryo is established except for a small part in the ventral abdominal region where the yolk sac duct and connecting stalk are attached.

While the foregut and hindgut are established, the midgut remains in communication with the yolk sac. Initially, this connection is wide (Fig. 5.17*A*), but as a result of body folding, it gradually becomes long and narrow to form the **vitelline duct** (Figs. 5.17*B* and 5.18). Only much later, when the vitelline duct is obliterated, does the midgut lose its connection with the original endoderm-lined cavity and obtain its free position in the abdominal cavity (Fig. 5.17*C*).

Another important result of cephalocaudal and lateral folding is partial incorporation of the allantois into the body of the embryo, where it forms the

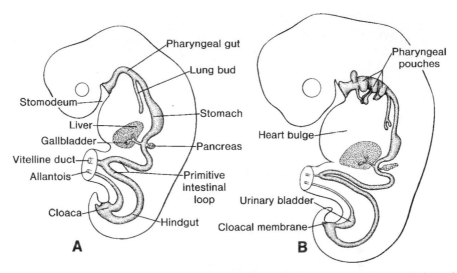

Figure 5.18 Sagittal sections through embryos showing derivatives of the endodermal germ layer. **A.** Pharyngeal pouches, epithelial lining of the lung buds and trachea, liver, gallbladder, and pancreas. **B.** The urinary bladder is derived from the cloaca and, at this stage of development, is in open connection with the allantois.

cloaca (Fig. 5.18*A*). The distal portion of the allantois remains in the connecting stalk. By the fifth week, the yolk sac duct, allantois, and umbilical vessels are restricted to the region of the umbilical ring (Figs. 5.18, 5.19, and 6.15).

In humans, the yolk sac is vestigial and in all probability has a nutritive role only in early stages of development (Fig. 5.20). In the second month of development, it lies in the chorionic cavity (Fig. 5.21).

Hence, the endodermal germ layer initially forms the epithelial lining of the primitive gut and the intraembryonic portions of the allantois and vitelline duct (Fig. 5.18*A*). During further development, it gives rise to (*a*) the epithelial lining of the respiratory tract; (*b*) the **parenchyma** of the thyroid, parathyroids, liver, and pancreas (see Chapters 13 and 15); (*c*) the reticular stroma of the tonsils and thymus; (*d*) the epithelial lining of the urinary bladder and urethra (see Chapter 14); and (*e*) the epithelial lining of the tympanic cavity and auditory tube (see Chapter 16).

Patterning of the Anteroposterior Axis: Regulation by *Homeobox* Genes

Homeobox genes are known for their **homeodomain,** a DNA binding motif, the **homeobox.** They code for transcription factors that activate cascades of genes regulating phenomena such as segmentation and axis formation. Many *homeobox* genes are collected into **homeotic clusters,** although other genes also contain the homeodomain. An important cluster of genes specifying the

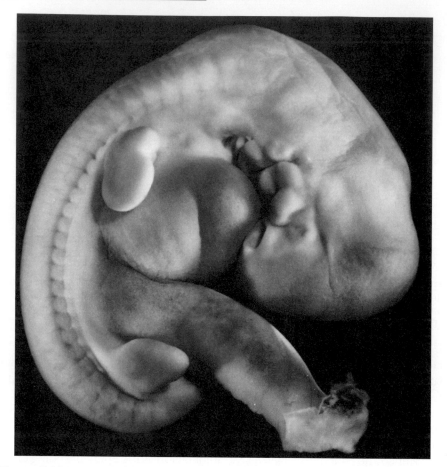

Figure 5.19 Human embryo (CRL 9.8 mm, fifth week) (×29.9). The forelimbs are paddle shaped.

craniocaudal axis is the homeotic gene complex ***Hom-C*** in *Drosophila*. These genes, which contain the ***Antennapedia*** and ***Bithorax*** classes of homeotic genes, are organized on a single chromosome as a functional unit. Thus, genes specifying more cranial structures lie at the 3′ end of the DNA and are expressed first, with genes controlling posterior development expressed sequentially and lying increasingly toward the 5′ end (Fig. 5.22). These genes are **conserved** in humans, existing as four copies, ***HOXA, HOXB, HOXC,*** and ***HOXD,*** which are arranged and expressed like those in *Drosophila*. Thus, each cluster lies on a separate chromosome, and the genes in each group are numbered 1 to 13 (Fig. 5.22). Genes with the same number, but belonging to different clusters form a **paralogous** group, such as *HOXA4, HOXB4, HOXC4,* and *HOXD4*. The pattern of expression of these genes, along with evidence from **knockout**

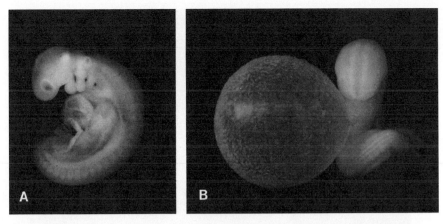

Figure 5.20 A. Lateral view of a 28-somite human embryo. The main external features are the pharyngeal arches and somites. Note the pericardial-liver bulge. **B.** The same embryo taken from a different angle to demonstrate the size of the yolk sac.

experiments in which mice are created that lack one or more of these genes, supports the hypothesis that they play a role in cranial-to-caudal patterning of the derivatives of all three germ layers. For example, an overlapping expression pattern of the *HOX* code exists in the somites and vertebrae, with genes located more 3' in each cluster being expressed in and regulating development of more cranial segments (Fig. 5.22).

External Appearance During the Second Month

At the end of the fourth week, when the embryo has approximately 28 somites, the main external features are the somites and pharyngeal arches (Fig. 5.20). The age of the embryo is therefore usually expressed in somites (Table 5.2). Because counting somites becomes difficult during the second month of development, the age of the embryo is then indicated as the **crown-rump length (CRL)** and expressed in millimeters (Table 5.3). CRL is the measurement from the vertex of the skull to the midpoint between the apices of the buttocks.

During the second month, the external appearance of the embryo is changed by an increase in head size and formation of the limbs, face, ears, nose, and eyes. By the beginning of the fifth week, forelimbs and hindlimbs appear as paddle-shaped buds (Fig. 5.19). The former are located dorsal to the pericardial swelling at the level of the fourth cervical to the first thoracic somites, which explains their innervation by the **brachial plexus.** Hindlimb buds appear slightly later just caudal to attachment of the umbilical stalk at the level of the lumbar and upper sacral somites. With further growth, the terminal portions of

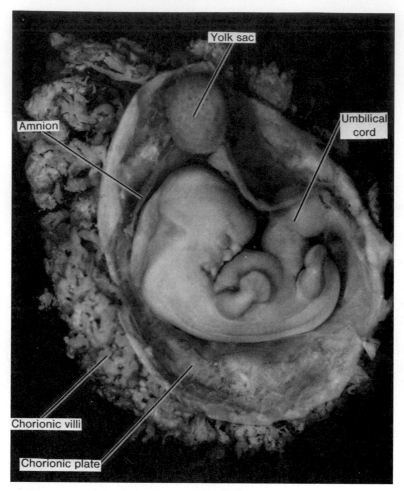

Figure 5.21 Human embryo (CRL 13 mm, sixth week) showing the yolk sac in the chorionic cavity.

the buds flatten and a circular constriction separates them from the proximal, more cylindrical segment (Fig. 5.21). Soon, four radial grooves separating five slightly thicker areas appear on the distal portion of the buds, foreshadowing formation of the digits (Fig. 5.21).

These grooves, known as **rays,** appear in the hand region first and shortly afterward in the foot, as the upper limb is slightly more advanced in development than the lower limb. While fingers and toes are being formed (Fig. 5.23), a second constriction divides the proximal portion of the buds into two segments, and the three parts characteristic of the adult extremities can be recognized (Fig. 5.24).

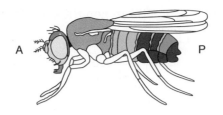

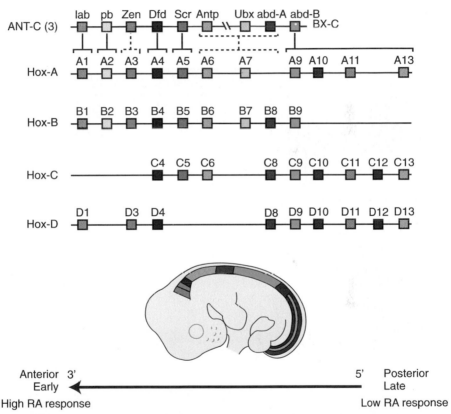

Figure 5.22 Drawing showing the arrangement of homeobox genes of the *Antennapedia* (*ANT-C*) and *Bithorax* (*BX-C*) classes of *Drosophila* and conserved homologous genes of the same classes in humans. During evolution, these genes have been duplicated, such that humans have four copies arranged on four different chromosomes. Homology between *Drosophila* genes and those in each cluster of human genes is indicated by color. Genes with the same number, but positioned on different chromosomes, form a paralogous group. Expression of the genes is in a cranial to caudal direction from the 3′ (expressed early) to the 5′ (expressed later) end as indicated in the fly and mouse embryo diagrams. Retinoic acid modulates expression of these genes with those at the 3′ end being more responsive to the compound.

TABLE 5.3 Crown-Rump Length Correlated to Approximate Age in Weeks

CRL (mm)	Approximate Age (weeks)
5–8	5
10–14	6
17–22	7
28–30	8

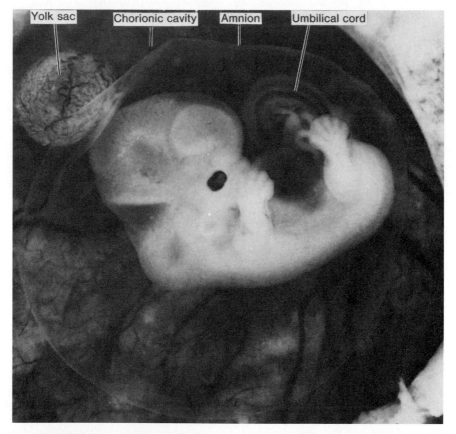

Figure 5.23 Human embryo (CRL 21 mm, seventh week) (×4). The chorionic sac is open to show the embryo in its amniotic sac. The yolk sac, umbilical cord, and vessels in the chorionic plate of the placenta are clearly visible. Note the size of the head in comparison with the rest of the body.

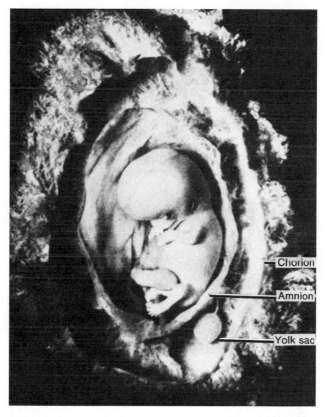

Figure 5.24 Human embryo (CRL 25 mm, seventh to eighth week). The chorion and the amnion have been opened. Note the size of the head, the eye, the auricle of the ear, the well-formed toes, the swelling in the umbilical cord caused by intestinal loops, and the yolk sac in the chorionic cavity.

CLINICAL CORRELATES

Birth Defects

Most major organs and organ systems are formed during the third to eighth week. This period, which is critical for normal development, is therefore called the period of **organogenesis.** Stem cell populations are establishing each of the organ primordia, and these interactions are sensitive to insult from genetic and environmental influences. Thus, this period is when most gross structural birth defects are induced. Unfortunately, the mother may not realize she is pregnant during this critical time, especially during the third and fourth weeks, which are particularly vulnerable. Consequently, she may not avoid harmful influences, such as cigarette smoking and alcohol. Understanding the main events of organogenesis is important for identifying the time that a particular defect was induced and, in turn, determining possible causes for the malformation (see Chapter 7).

Summary

The **embryonic period,** which extends from the third to the eighth weeks of development, is the period during which each of the three germ layers, **ectoderm, mesoderm,** and **endoderm,** gives rise to its own tissues and organ systems. As a result of organ formation, major features of body form are established (Table 5.4).

The **ectodermal germ layer** gives rise to the organs and structures that maintain contact with the outside world: (*a*) **central nervous system;** (*b*) **peripheral nervous system;** (*c*) **sensory epithelium of ear, nose, and eye;** (*d*) **skin, including hair and nails;** and (*e*) **pituitary, mammary,** and **sweat glands** and **enamel** of the teeth. Induction of the neural plate is regulated by inactivation of the growth factor BMP-4. In the cranial region, inactivation is caused by noggin, chordin, and follistatin secreted by the node, notochord, and prechordal mesoderm. Inactivation of BMP-4 in the hindbrain and spinal cord regions is effected by WNT3a and FGF. In the absence of inactivation, BMP-4 causes ectoderm to become epidermis and mesoderm to ventralize to form intermediate and lateral plate mesoderm.

Important components of the **mesodermal germ layer** are **paraxial, intermediate,** and **lateral plate** mesoderm. Paraxial mesoderm forms **somitomeres,** which give rise to mesenchyme of the head and organize into **somites** in occipital and caudal segments. Somites give rise to the **myotome** (muscle tissue), **sclerotome** (cartilage and bone), and **dermatome** (subcutaneous tissue of the skin), which are **all supporting tissues of the body.** Signals for somite differentiation are derived from surrounding structures, including the notochord, neural tube, and epidermis. The notochord and floor plate of the neural tube secrete *Sonic hedgehog,* which induces the sclerotome. **WNT proteins** from the dorsal neural tube cause the dorsomedial portion of the somite to form **epaxial** musculature, while BMP-4, FGFs from the lateral plate mesoderm, and **WNTs** from the epidermis cause the dorsolateral portion to form limb and body wall musculature. The dorsal midportion of the somite becomes dermis under the influence of **neurotrophin 3,** secreted by the dorsal neural tube (Fig. 5.12). Mesoderm also gives rise to the **vascular system,** that is, the heart, arteries, veins, lymph vessels, and all blood and lymph cells. Furthermore, it gives rise to the **urogenital system:** kidneys, gonads, and their ducts (but not the bladder). Finally, the **spleen** and **cortex of the suprarenal glands** are mesodermal derivatives.

The **endodermal germ layer** provides the epithelial lining of the **gastrointestinal tract, respiratory tract,** and **urinary bladder.** It also forms the **parenchyma** of the **thyroid, parathyroids, liver,** and **pancreas.** Finally, the

TABLE 5.4 Summary of Key Events During the Embryonic Period

Days	Somites	Length (mm)	Figure	Characteristic Features
14–15	0	0.2	5.1 A	Appearance of primitive streak
16–18	0	0.4	5.1 B	Notochordal process appears; hemopoietic cells in yolk sac
19–20	0	1.0–2.0	5.2 A	Intraembryonic mesoderm spread under cranial ectoderm; primitive streak continues; umbilical vessels and cranial neural folds beginning to form
20–21	1–4	2.0–3.0	5.2 B, C	Cranial neural folds elevated, and deep neural groove established; embryo beginning to bend
22–23	5–12	3.0–3.5	5.5 A, B; 5.6; 5.7	Fusion of neural folds begins in cervical region; cranial and caudal neuropores open widely; visceral arches 1 and 2 present; heart tube beginning to fold
24–25	13–20	3.0–4.5	5.8 A	Cephalocaudal folding under way; cranial neuropore closing or closed; optic vesicles formed; otic placodes appear
26–27	21–29	3.5–5.0	5.8 B; 5.20 A, B	Caudal neuropore closing or closed; upper limb buds appear; 3 pairs of visceral arches
28–30	30–35	4.0–6.0	5.8 B	Fourth visceral arch formed; hindlimb buds appear; otic vesicle and lens placode
31–35		7.0–10.0	5.19	Forelimbs paddle-shaped; nasal pits formed; embryo tightly C-shaped
36–42		9.0–14.0	5.21	Digital rays in hand and footplates; brain vesicles prominent; external auricle forming from auricular hillocks; umbilical herniation initiated
43–49		13.0–22.0	5.23	Pigmentation of retina visible; digital rays separating; nipples and eyelids formed; maxillary swellings fuse with medial nasal swellings as upper lip forms; prominent umbilical herniation
50–56		21.0–31.0	5.24	Limbs long, bent at elbows, knees; fingers, toes free; face more human-like; tail disappears; umbilical herniation persists to end of third month

epithelial lining of the **tympanic cavity** and **auditory tube** originate in the endodermal germ layer.

Craniocaudal patterning of the embryonic axis is controlled by *homeobox* genes. These genes, conserved from *Drosophila,* are arranged in four clusters, *HOXA, HOXB, HOXC,* and *HOXD,* on four different chromosomes. Genes toward the 3′ end of the chromosome control development of more cranial structures; those more toward the 5′ end regulate differentiation of more posterior structures. Together, they regulate patterning of the hindbrain and axis of the embryo (Fig. 5.22).

As a result of formation of organ systems and rapid growth of the central nervous system, the initial flat embryonic disc begins to fold **cephalocaudally,** establishing the **head** and **tail folds.** The disc also folds **transversely (lateral folds),** establishing the **rounded body form.** Connection with the yolk sac and placenta is maintained through the vitelline duct and umbilical cord, respectively.

Problems to Solve

1. *Why are the third to eighth weeks of embryogenesis so important for normal development and the most sensitive for induction of structural defects?*

SUGGESTED READING

Cossu G, Tajbakhshs S, Buckingham M: How is myogenesis initiated in the embryo? *Trends Genet* 12:218, 1996.

Eichele G: Retinoids and vertebrate limb pattern formation. *Trends Genet* 5:226, 1990.

Hanahan D: Signaling vascular morphogenesis and maintenance. *Science* 277:48, 1997.

Jessell TM, Melton DA: Diffusible factors in vertebrate embryonic induction. *Cell* 68:257, 1992.

Johnson RL, Laufer E, Riddle RD, Tabin C: Ectopic expression of *sonic hedgehog* alters dorsoventral patterning of somites. *Cell* 79:1165, 1994.

Kanzler B, Foreman RK, Lebosky PA, Mallo M: BMP signaling is essential for development of skeletogenic and neurogenic cranial neural crest. *Development* 127:1095, 2000.

Kessel M: Respecification of vertebral identities by retinoic acid. *Development* 115:487, 1992.

Krumlauf R: *Hox* genes and pattern formation in the branchial region of the vertebrate head. *Trends Genet* 9:106, 1993.

Krumlauf R: *Hox* genes in vertebrate development. *Cell* 78:191, 1994.

McGinnis W, Krumlauf R: *Homeobox* genes and axial patterning. Cell 68:283, 1992.

Meier T, Tam PPL: Metameric pattern development in the embryonic axis of the mouse: 1. Differentiation of the cranial segments. *Differentiation* 21:95, 1982.

O'Rahilly R, Muller F: Bidirectional closure of the rostral neuropore. *Am J Anat* 184:259, 1989.

Ordahl CP, Ledouarin N: Two myogenic lineages within the developing somite. *Development* 114:339, 1992.

Risau W: Mechanisms of angiogenesis. *Nature* 386:671, 1997.

Sadler TW: Mechanisms of neural tube closure and defects. *Ment Retard Dev Disabilities Res Rev* 4:247, 1998.

Sasai Y, DeRobertis EM: Ectodermal patterning in vertebrate embryos. *Dev Biol* 182:5, 1997.

Schoenwolf G, Bortier H, Vakaet L: Fate mapping the avian neural plate with quail-chick chimeras: origin of prospective median wedge cells. *J Exp Zool* 249:271, 1989.

Slack JM: Embryonic induction. *Mech Dev* 41:91, 1993.

Smith JL, Schoenwolf GC: Neurulation: coming to closure. *Trends Neurosci* 20:510, 1997.

Stern HM, Brown AMC, Hauschka SD: Myogenesis in paraxial mesoderm: preferential induction by dorsal neural tube and by cells expressing Wnt-1. *Development* 121:3675, 1995.

Streeter GL: Developmental horizons in human embryos: age group XI, 13–20 somites, and age group XII, 21–29 somites. *Contrib Embryol* 30:211, 1942.

Streeter GL: Developmental horizons in human embryos: age group XIII, embryos 4 or 5 mm long, and age group XIV, indentation of lens vesicle. *Contrib Embryol* 31:26, 1945.

Tam PPL, Beddington RSP: The formation of mesodermal tissues in the mouse embryo during gastrulation and early organogenesis. *Development* 99:109, 1987.

Tam PPL, Meier S, Jacobson AG: Differentiation of the metameric pattern in the embryonic axis of the mouse: 2. Somitomeric organization of the pre-somitic mesoderm. *Differentiation* 21:109, 1982.

Zon LI: Developmental biology of hematopoiesis. *Blood* 8:2876, 1995.

c h a p t e r   6

Third Month to Birth: The Fetus and Placenta

Development of the Fetus

The period from the beginning of the ninth week to birth is known as the **fetal period.** It is characterized by maturation of tissues and organs and rapid growth of the body. The length of the fetus is usually indicated as the **crown-rump length** (CRL) (sitting height) or as the **crown-heel length** (CHL), the measurement from the vertex of the skull to the heel (standing height). These measurements, expressed in centimeters, are correlated with the age of the fetus in weeks or months (Table 6.1; p. 118). Growth in length is particularly striking during the third, fourth, and fifth months, while an increase in weight is most striking during the last 2 months of gestation. In general, **the length of pregnancy is considered to be 280 days, or 40 weeks after the onset of the last normal menstrual period (LNMP) or more accurately, 266 days or 38 weeks after fertilization.** For the purposes of the following discussion, age is calculated from the time of fertilization and is expressed in weeks or calendar months.

MONTHLY CHANGES

One of the most striking changes taking place during fetal life is the relative slowdown in growth of the head compared with the rest of the body. At the beginning of the third month the head constitutes

TABLE 6.1 Growth in Length and Weight During the Fetal Period

Age (Weeks)	CRL (cm)	Weight (g)
9–12	5–8	10–45
13–16	9–14	60–200
17–20	15–19	250–450
21–24	20–23	500–820
25–28	24–27	900–1300
29–32	28–30	1400–2100
33–36	31–34	2200–2900
37–38	35–36	3000–3400

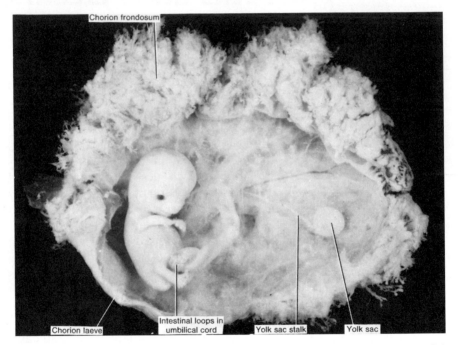

Figure 6.1 A 9-week fetus. Note the large size of the head compared with that of the rest of the body. The yolk sac and long vitelline duct are visible in the chorionic cavity. Note the umbilical cord and herniation of intestinal loops. One side of the chorion has many villi (chorion frondosum), while the other side is almost smooth (chorion laeve).

approximately half of the CRL (Figs. 6.1 and 6.2). By the beginning of the fifth month, the size of the head is about one-third of the CHL, and at birth it is approximately one-fourth of the CHL (Fig. 6.2). Hence, over time, growth of the body accelerates but that of the head slows down.

During the **third month** the face becomes more human looking (Figs. 6.3 and 6.4). The eyes, initially directed laterally, move to the ventral aspect of the

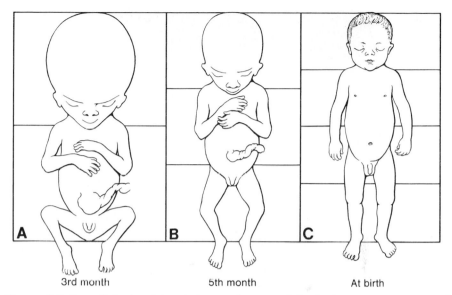

3rd month   5th month   At birth

Figure 6.2 Size of the head in relation to the rest of the body at various stages of development.

face, and the ears come to lie close to their definitive position at the side of the head (Fig. 6.3). The limbs reach their relative length in comparison with the rest of the body, although the lower limbs are still a little shorter and less well developed than the upper extremities. **Primary ossification centers** are present in the long bones and skull by the 12th week. Also by the 12th week, external genitalia develop to such a degree that the sex of the fetus can be determined by external examination (ultrasound). During the 6th week **intestinal loops cause a large swelling (herniation) in the umbilical cord,** but by the 12th week the loops withdraw into the abdominal cavity. At the end of the third month, reflex activity can be evoked in aborted fetuses, indicating muscular activity.

During the **fourth** and **fifth months** the fetus lengthens rapidly (Fig. 6.5 and Table 6.1), and at the end of the first half of intrauterine life its CRL is approximately 15 cm, that is, about half the total length of the newborn. The weight of the fetus increases little during this period and by the end of the fifth month is still less than 500 g. The fetus is covered with fine hair, called **lanugo hair;** eyebrows and head hair are also visible. **During the fifth month movements of the fetus can be felt by the mother.**

During the **second half of intrauterine life,** weight increases considerably, particularly during the last 2.5 months, when 50 % of the full-term weight (approximately 3200 g) is added. During the **sixth month,** the skin of the fetus is reddish and has a wrinkled appearance because of the lack of underlying connective tissue. A fetus born early in the sixth month has great difficulty

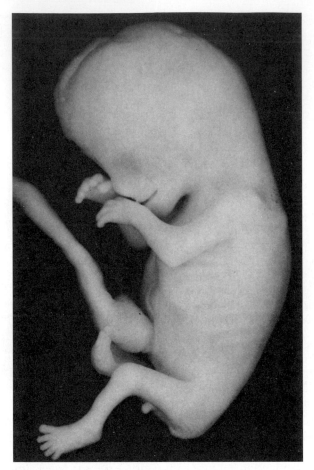

Figure 6.3 An 11-week fetus. The umbilical cord still shows a swelling at its base, caused by herniated intestinal loops. Toes are developed, and the sex of the fetus can be seen. The skull of this fetus lacks the normal smooth contours.

surviving. Although several organ systems are able to function, the respiratory system and the central nervous system have not differentiated sufficiently, and coordination between the two systems is not yet well established. By 6.5 to 7 months, the fetus has a length of about 25 cm and weighs approximately 1100 g. If born at this time, the infant has a 90% chance of surviving. Some developmental events occurring during the first 7 months are indicated in Table 6.2.

During the last 2 months, the fetus obtains well-rounded contours as the result of deposition of subcutaneous fat (Fig. 6.6). By the end of intrauterine life, the skin is covered by a whitish, fatty substance (**vernix caseosa**) composed of secretory products from sebaceous glands.

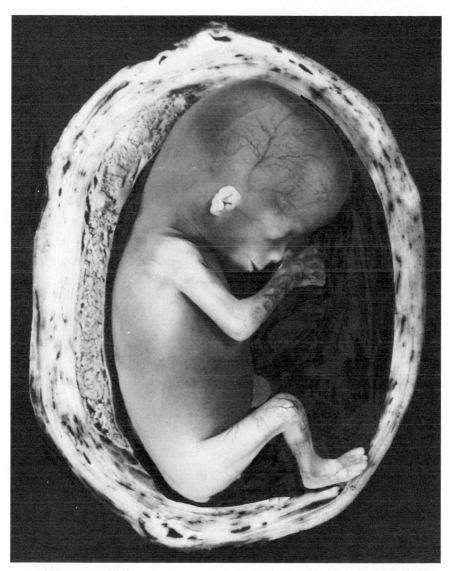

Figure 6.4 A 12-week fetus in utero. Note the extremely thin skin and underlying blood vessels. The face has all of the human characteristics, but the ears are still primitive. Movements begin at this time but are usually not felt by the mother.

At the end of the **ninth month** the skull has the largest circumference of all parts of the body, an important fact with regard to its passage through the birth canal. At the time of birth the weight of a normal fetus is 3000 to 3400 g; its CRL is about 36 cm; and its CHL is about 50 cm. Sexual characteristics are pronounced, and the testes should be in the scrotum.

TABLE 6.2 Developmental Horizons during Fetal Life

Event	Age (Weeks)
Tastebuds appear	7
Swallowing	10
Respiratory movements	14–16
Sucking movements	24
Some sounds can be heard	24–26
Eyes sensitive to light*	28

*Recognition of form and color occurs postnatally.

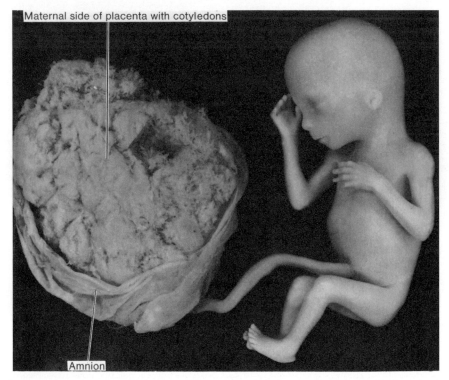

Figure 6.5 An 18-week-old fetus connected to the placenta by its umbilical cord. The skin of the fetus is thin because of lack of subcutaneous fat. Note the placenta with its cotyledons and the amnion.

TIME OF BIRTH

The date of birth is most accurately indicated as 266 days, or 38 weeks, after fertilization. The oocyte is usually fertilized within 12 hours of ovulation. However, sperm deposited in the reproductive tract up to 6 days prior to ovulation can survive to fertilize oocytes. Thus, most pregnancies occur when sexual intercourse occurs within a 6-day period that ends on the day of ovulation. A pregnant woman usually will see her obstetrician when she has missed two successive menstrual bleeds. By that time, her recollection about coitus is usually vague, and it is readily understandable that the day of fertilization is difficult to determine.

The obstetrician calculates the date of birth as 280 days or 40 weeks from the first day of the LNMP. In women with regular 28-day menstrual periods the method is fairly accurate, but when cycles are irregular, substantial miscalculations may be made. An additional complication occurs when the woman has some bleeding about 14 days after fertilization as a result of erosive activity by the implanting blastocyst (see Chapter 3). Hence the day of delivery is not always easy to determine. Most fetuses are born within 10 to 14 days of the calculated delivery date. If they are born much earlier, they are categorized as **premature;** if born later, they are considered **postmature.**

Occasionally the age of an embryo or small fetus must be determined. By combining data on the onset of the last menstrual period with fetal length, weight, and other morphological characteristics typical for a given month of development, a reasonable estimate of the age of the fetus can be formulated. A valuable tool for assisting in this determination is **ultrasound,** which can provide an accurate (1 to 2 days) measurement of CRL during the 7th to 14th weeks. Measurements commonly used in the 16th to 30th weeks are **biparietal diameter (BPD),** head and abdominal circumference, and femur length. An accurate determination of fetal size and age is important for managing pregnancy, especially if the mother has a small pelvis or the baby has a birth defect.

CLINICAL CORRELATES

Low Birth Weight

There is considerable variation in fetal length and weight, and sometimes these values do not correspond with the calculated age of the fetus in months or weeks. Most factors influencing length and weight are genetically determined, but environmental factors also play an important role.

Intrauterine growth restriction (IUGR) is a term applied to infants who are at or below the 10th percentile for their expected birth weight at a given gestational age. Sometimes these infants are described as small for dates, **small for gestational age (SGA),** fetally malnourished, or dysmature. Approximately 1 in 10 babies have IUGR and therefore an increased risk of neurological deficiencies, congenital malformations, meconium aspiration, hypoglycemia, hypocalcemia, and respiratory distress syndrome (RDS). The incidence is higher in blacks than in whites. Causative factors include chromosomal abnormalities (10%); teratogens; congenital infections (rubella, cytomegalovirus, toxoplasmosis, and syphilis); poor maternal health (hypertension and renal and cardiac disease); the mother's nutritional status and socioeconomic level; her use of cigarettes, alcohol, and other drugs; placental insufficiency; and multiple births (e.g., twins, triplets). Fetuses that weigh less

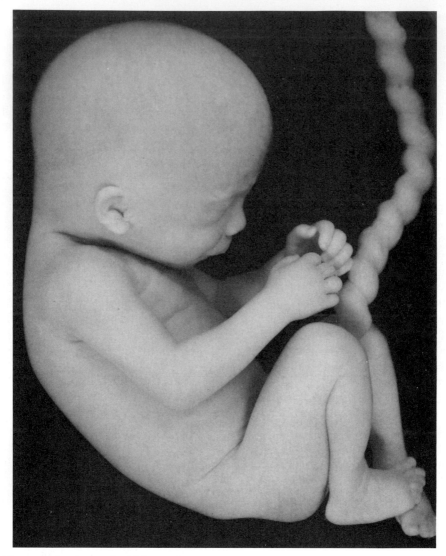

Figure 6.6 A 7-month-old fetus. This fetus would be able to survive. It has well-rounded contours as a result of deposition of subcutaneous fat. Note the twisting of the umbilical cord.

than 500 g seldom survive, while those that weigh 500 to 1000 g may live if provided with expert care. However, approximately 50% of babies born weighing less than 1000 g who survive will have severe neurological deficits. Infants may be full term, but small because of IUGR or small because they are born prematurely.

The major growth-promoting factor during development before and after birth is **insulinlike growth factor-I (IGF-I), which has mitogenic and anabolic effects.** Fetal tissues express IGF-I and serum levels are correlated with fetal growth. Mutations in the *IGF-I* gene result in IUGR and this growth retardation is continued after birth. In contrast to the prenatal period, postnatal growth depends upon **growth hormone (GH).** This hormone binds to its receptor (GHR), activating a signal transduction pathway and resulting in synthesis and secretion of IGF-I. Mutations in the GHR result in **Laron dwarfism,** which is characterized by growth retardation, midfacial hypoplasia, blue sclera, and limited elbow extension. These individuals show little or no IUGR, since IGF-I production does not depend upon GH during fetal development.

Fetal Membranes and Placenta

As the fetus grows, its demands for nutritional and other factors increase causing major changes in the placenta. Foremost among these is an increase in surface area between maternal and fetal components to facilitate exchange. The disposition of fetal membranes is also altered as production of amniotic fluid increases.

CHANGES IN THE TROPHOBLAST

By the beginning of the second month, the **trophoblast** is characterized by a great number of secondary and tertiary villi that give it a radial appearance (Fig. 6.7). The villi are anchored in the mesoderm of the **chorionic plate** and are attached peripherally to the maternal decidua by way of the outer **cytotrophoblast shell.** The surface of the villi is formed by the syncytium, resting on a layer of cytotrophoblastic cells that in turn cover a core of vascular mesoderm (Fig. 6.8, *A* and *C*). The capillary system developing in the core of the villous stems soon comes in contact with capillaries of the chorionic plate and connecting stalk, thus giving rise to the extraembryonic vascular system (see Fig. 5.15).

During the following months, numerous small extensions sprout from existing villous stems into the surrounding **lacunar** or **intervillous spaces.** Initially these newly formed villi are primitive (Fig. 6.8C), but by the beginning of the fourth month, cytotrophoblastic cells and some connective tissue cells disappear. The syncytium and endothelial wall of the blood vessels are then the only layers that separate the maternal and fetal circulations (Fig. 6.8, *B* and *D*). Frequently the syncytium becomes very thin, and large pieces containing several nuclei may break off and drop into the intervillous blood lakes. These pieces, known as **syncytial knots,** enter the maternal circulation and usually degenerate without causing any symptoms. Disappearance of cytotrophoblastic cells progresses from the smaller to larger villi, and although some always persist in large villi, they do not participate in the exchange between the two circulations.

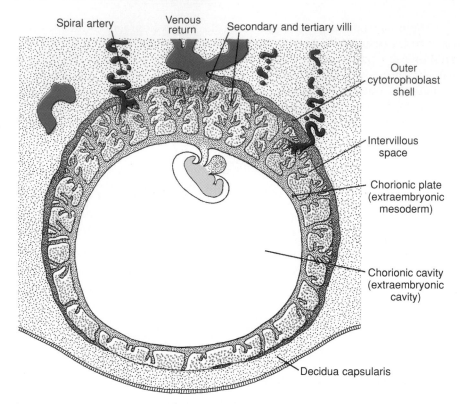

Spiral artery   Venous return   Secondary and tertiary villi

Outer cytotrophoblast shell

Intervillous space

Chorionic plate (extraembryonic mesoderm)

Chorionic cavity (extraembryonic cavity)

Decidua capsularis

Figure 6.7 Human embryo at the beginning of the second month of development. At the embryonic pole, villi are numerous and well formed; at the abembryonic pole they are few in number and poorly developed.

CHORION FRONDOSUM AND DECIDUA BASALIS

In the early weeks of development, villi cover the entire surface of the chorion (Fig. 6.7). As pregnancy advances, villi on the embryonic pole continue to grow and expand, giving rise to the **chorion frondosum** (bushy chorion). Villi on the abembryonic pole degenerate and by the third month this side of the chorion, now known as the **chorion laeve,** is smooth (Figs. 6.9 and 6.10*A*).

The difference between the embryonic and abembryonic poles of the chorion is also reflected in the structure of the **decidua,** the functional layer of the endometrium, which is shed during parturition. The decidua over the chorion frondosum, the **decidua basalis,** consists of a compact layer of large cells, **decidual cells,** with abundant amounts of lipids and glycogen. This layer, the **decidual plate,** is tightly connected to the chorion. The decidual layer over the abembryonic pole is the **decidua capsularis** (Fig. 6.10*A*). With growth of the chorionic vesicle, this layer becomes stretched and degenerates. Subsequently, the chorion laeve comes into contact with the uterine wall (**decidua parietalis**)

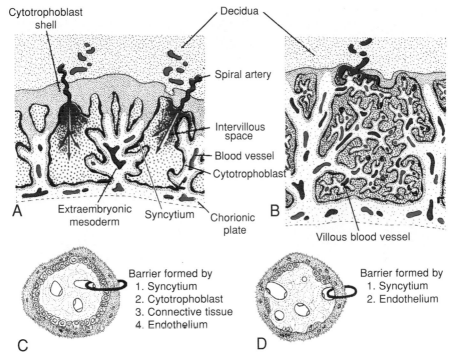

Figure 6.8 Structure of villi at various stages of development. **A.** During the fourth week. The extraembryonic mesoderm penetrates the stem villi in the direction of the decidual plate. **B.** During the fourth month. In many small villi the wall of the capillaries is in direct contact with the syncytium. **C** and **D.** Enlargement of the villus as shown in **A** and **B**, respectively.

on the opposite side of the uterus and the two fuse (Figs. 6.10 to 6.12), obliterating the uterine lumen. Hence the only portion of the chorion participating in the exchange process is the chorion frondosum, which, together with the decidua basalis, makes up the **placenta.** Similarly, fusion of the amnion and chorion to form the **amniochorionic membrane** obliterates the chorionic cavity (Fig. 6.10, *A* and *B*). It is this membrane that ruptures during labor (breaking of the water).

Structure of the Placenta

By the beginning of the fourth month, the placenta has two components: (*a*) a **fetal portion,** formed by the chorion frondosum; and (*b*) a **maternal portion,** formed by the decidua basalis (Fig. 6.10*B*). On the fetal side, the placenta is bordered by the **chorionic plate** (Fig. 6.13); on its maternal side, it is bordered by the decidua basalis, of which the **decidual plate** is most intimately incorporated

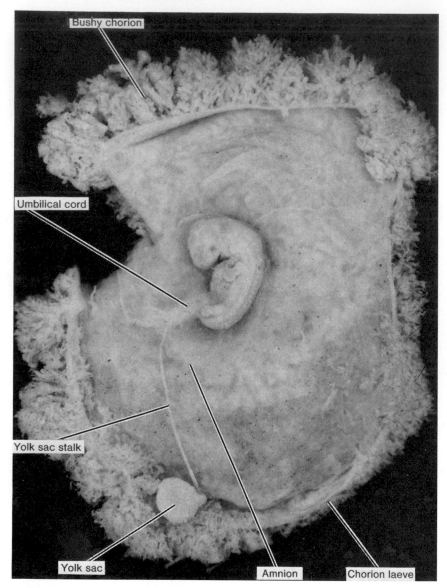

Figure 6.9 A 6-week embryo. The amniotic sac and chorionic cavity have been opened to expose the embryo showing the bushy appearance of the trophoblast at the embryonic pole in contrast to small villi at the abembryonic pole and the connecting stalk and yolk sac with its extremely long duct.

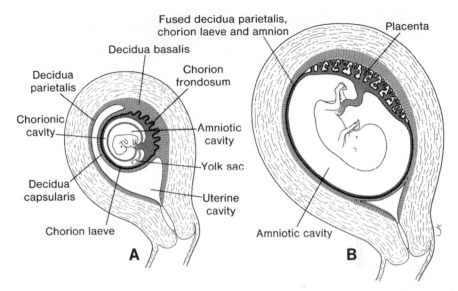

Figure 6.10 Relation of fetal membranes to wall of the uterus. **A.** End of the second month. Note the yolk sac in the chorionic cavity between the amnion and chorion. At the abembryonic pole, villi have disappeared (chorion laeve). **B.** End of the third month. The amnion and chorion have fused, and the uterine cavity is obliterated by fusion of the chorion laeve and the decidua parietalis.

into the placenta. In the **junctional zone,** trophoblast and decidua cells intermingle. This zone, characterized by decidual and syncytial giant cells, is rich in amorphous extracellular material. By this time most cytotrophoblast cells have degenerated. Between the chorionic and decidual plates are the intervillous spaces, which are filled with maternal blood. They are derived from lacunae in the syncytiotrophoblast and are lined with syncytium of fetal origin. The villous trees grow into the intervillous blood lakes (Figs. 6.17 and 6.13).

During the fourth and fifth months the decidua forms a number of **decidual septa,** which project into intervillous spaces but do not reach the chorionic plate (Fig. 6.13). These septa have a core of maternal tissue, but their surface is covered by a layer of syncytial cells, so that at all times a syncytial layer separates maternal blood in intervillous lakes from fetal tissue of the villi. As a result of this septum formation, the placenta is divided into a number of compartments, or **cotyledons** (Fig. 6.14). Since the decidual septa do not reach the chorionic plate, contact between intervillous spaces in the various cotyledons is maintained.

As a result of the continuous growth of the fetus and expansion of the uterus, the placenta also enlarges. Its increase in surface area roughly parallels that of the expanding uterus and throughout pregnancy it covers approximately 15 to 30% of the internal surface of the uterus. The increase in thickness of the placenta results from arborization of existing villi and is not caused by further penetration into maternal tissues.

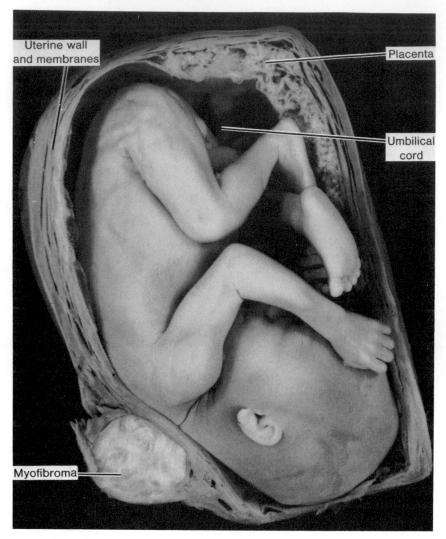

Figure 6.11 A 19-week-old fetus in its natural position in the uterus showing the umbilical cord and placenta. The lumen of the uterus is obliterated. In the wall of the uterus is a large growth, a myofibroma.

FULL-TERM PLACENTA

At full term, the placenta is discoid with a diameter of 15 to 25 cm, is approximately 3 cm thick, and weighs about 500 to 600 g. At birth, it is torn from the uterine wall and, approximately 30 minutes after birth of the child, is expelled from the uterine cavity. After birth, when the placenta is viewed from the **maternal side,** 15 to 20 slightly bulging areas, the **cotyledons,** covered by a thin layer of decidua basalis, are clearly recognizable (Fig. 6.14*B*). Grooves between the cotyledons are formed by decidual septa.

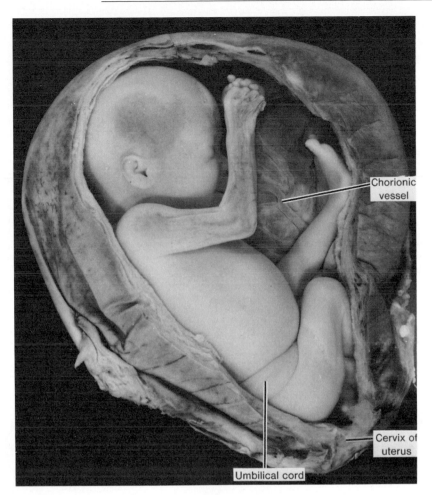

Chorionic vessel

Cervix of uterus

Umbilical cord

Figure 6.12 A 23-week-old fetus in the uterus. Portions of the wall of the uterus and the amnion have been removed to show the fetus. In the background are placental vessels converging toward the umbilical cord. The umbilical cord is tightly wound around the abdomen, possibly causing abnormal fetal position in the uterus (breech position).

The **fetal surface** of the placenta is covered entirely by the chorionic plate. A number of large arteries and veins, the **chorionic vessels,** converge toward the umbilical cord (Fig. 6.14*A*). The chorion, in turn, is covered by the amnion. Attachment of the umbilical cord is usually eccentric and occasionally even marginal. Rarely, however, does it insert into the chorionic membranes outside the placenta (**velamentous insertion**).

CIRCULATION OF THE PLACENTA

Cotyledons receive their blood through 80 to 100 spiral arteries that pierce the decidual plate and enter the intervillous spaces at more or less regular

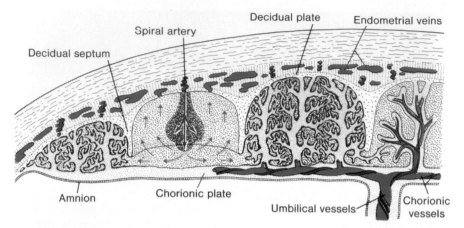

Figure 6.13 The placenta in the second half of pregnancy. The cotyledons are partially separated by the decidual (maternal) septa. Most of the intervillous blood returns to the maternal circulation by way of the endometrial veins. A small portion enters neighboring cotyledons. The intervillous spaces are lined by syncytium.

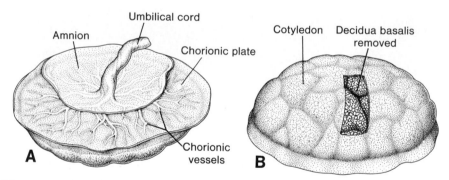

Figure 6.14 A full-term placenta. **A.** Fetal side. The chorionic plate and umbilical cord are covered by amnion. **B.** Maternal side showing the cotyledons. In one area the decidua has been removed. The maternal side of the placenta is always carefully inspected at birth, and frequently one or more cotyledons with a whitish appearance are present because of excessive fibrinoid formation and infarction of a group of intervillous lakes.

intervals (Fig. 6.13). The lumen of the spiral artery is narrow, so blood pressure in the intervillous space is high. This pressure forces the blood deep into the intervillous spaces and bathes the numerous small villi of the villous tree in oxygenated blood. As the pressure decreases, blood flows back from the chorionic plate toward the decidua, where it enters the endometrial veins (Fig. 6.13). Hence, blood from the intervillous lakes drains back into the maternal circulation through the endometrial veins.

Collectively, the intervillous spaces of a mature placenta contain approximately 150 ml of blood, which is replenished about 3 or 4 times per minute. This blood moves along the chorionic villi, which have a surface area of 4 to 14 m^2. However, placental exchange does not take place in all villi, only in those whose fetal vessels are in intimate contact with the covering syncytial membrane. In these villi, the syncytium often has a brush border consisting of numerous microvilli, which greatly increases the surface area and consequently the exchange rate between maternal and fetal circulations (Fig. 6.8*D*). The **placental membrane,** which separates maternal and fetal blood, is initially composed of four layers: (*a*) the endothelial lining of fetal vessels; (*b*) the connective tissue in the villus core; (*c*) the cytotrophoblastic layer; and (*d*) the syncytium (Fig. 6.8*C*). From the fourth month on, however, the placental membrane thins, since the endothelial lining of the vessels comes in intimate contact with the syncytial membrane, greatly increasing the rate of exchange (Fig. 6.8*D*). Sometimes called the **placental barrier,** the placental membrane is not a true barrier, since many substances pass through it freely. Because the maternal blood in the intervillous spaces is separated from the fetal blood by a chorionic derivative, the human placenta is considered to be of the **hemochorial** type.

FUNCTION OF THE PLACENTA

Main functions of the placenta are (*a*) **exchange of metabolic and gaseous products** between maternal and fetal bloodstreams and (*b*) **production of hormones.**

Exchange of Gases

Exchange of gases, such as oxygen, carbon dioxide, and carbon monoxide, is accomplished by simple diffusion. At term, the fetus extracts 20 to 30 ml of oxygen per minute from the maternal circulation and even a short-term interruption of the oxygen supply is fatal to the fetus. Placental blood flow is critical to oxygen supply, since the amount of oxygen reaching the fetus primarily depends on delivery, not diffusion.

Exchange of Nutrients and Electrolytes

Exchange of nutrients and electrolytes, such as amino acids, free fatty acids, carbohydrates, and vitamins, is rapid and increases as pregnancy advances.

Transmission of Maternal Antibodies

Immunological competence begins to develop late in the first trimester, by which time the fetus makes all of the components of **complement.** Immunoglobulins consist almost entirely of **maternal immunoglobulin G (IgG)** that begins to be transported from mother to fetus at approximately 14 weeks. In this manner, the fetus gains passive immunity against various infectious diseases. Newborns begin to produce their own IgG, but adult levels are not attained until the age of 3 years.

CLINICAL CORRELATES

Erythroblastosis Fetalis and Fetal Hydrops

Over 400 red blood cell antigens have been identified, and although most do not cause problems during pregnancy, some can stimulate a maternal antibody response against fetal blood cells. This process is an example of **isoimmunozation,** and if the maternal response is sufficient, the antibodies will attack and hemolyze fetal red blood cells resulting in **hemolytic disease of the newborn (erythroblastosis fetalis)**. The anemia may become so severe that **fetal hydrops** (edema and effusions into the body cavities) occurs, leading to fetal death. Most severe cases are caused by antigens from the **CDE (Rhesus)** blood group system. The D or **Rh antigen** is the most dangerous, since immunization can result from a single exposure and occurs earlier and with greater severity with each succeeding pregnancy. The antibody response occurs in cases where the fetus is D(Rh) positive and the mother is D(Rh) negative and is elicited when fetal red blood cells enter the maternal system due to small areas of bleeding at the surface of placental villi or at birth. Analysis of amniotic fluid for bilirubin, a breakdown product of hemoglobin, serves as a measure of the degree of red cell hemolysis. Treatment for the affected fetus involves intrauterine or postnatal transfusions. However, the disease is prevented by identifying women at risk using an antibody screen and treating them with anti-D-immunoglobulin.

Antigens from the **ABO blood group** can also elicit an antibody response, but the effects are much milder than those produced by the CDE group. About 20% of all infants have an ABO maternal incompatibility, but only 5% will be clinically affected. These can be effectively treated postnatally.

Hormone Production

By the end of the fourth month the placenta produces **progesterone** in sufficient amounts to maintain pregnancy if the corpus luteum is removed or fails to function properly. In all probability, all hormones are synthesized in the syncytial trophoblast. In addition to progesterone, the placenta produces increasing amounts of **estrogenic hormones,** predominantly **estriol,** until just before the end of pregnancy, when a maximum level is reached. These high levels of estrogens stimulate uterine growth and development of the mammary glands.

During the first two months of pregnancy, the syncytiotrophoblast also produces **human chorionic gonadotropin (hCG),** which maintains the corpus luteum. This hormone is excreted by the mother in the urine, and in the early stages of gestation, its presence is used as an indicator of pregnancy. Another hormone produced by the placenta is **somatomammotropin** (formerly **placental lactogen**). It is a growth hormone-like substance that gives the fetus priority on maternal blood glucose and makes the mother somewhat diabetogenic. It also promotes breast development for milk production.

CLINICAL CORRELATES

The Placental Barrier

Most maternal hormones do not cross the placenta. The hormones that do cross, such as thyroxine, do so only at a slow rate. Some synthetic progestins rapidly cross the placenta and may masculinize female fetuses. Even more dangerous was the use of the synthetic estrogen **diethylstilbestrol,** which easily crosses the placenta. This compound produced carcinoma of the vagina and abnormalities of the testes in individuals who were exposed to it during their intrauterine life (see Chapter 7).

Although the placental barrier is frequently considered to act as a protective mechanism against damaging factors, many viruses, such as rubella, cytomegalovirus, Coxsackie, variola, varicella, measles, and poliomyelitis virus, traverse the placenta without difficulty. Once in the fetus, some viruses cause infections, which may result in cell death and birth defects (see Chapter 7).

Unfortunately, most drugs and drug metabolites traverse the placenta without difficulty, and many cause serious damage to the embryo (see Chapter 7). In addition, maternal use of heroin and cocaine can cause habituation in the fetus.

Amnion and Umbilical Cord

The oval line of reflection between the amnion and embryonic ectoderm (**amnio-ectodermal junction**) is the **primitive umbilical ring.** At the fifth week of development, the following structures pass through the ring (Fig. 6.15, *A* and *C*): (*a*) the **connecting stalk,** containing the allantois and the umbilical vessels, consisting of two arteries and one vein; (*b*) the **yolk stalk (vitelline duct),** accompanied by the vitelline vessels; and (*c*) the **canal connecting the intraembryonic and extraembryonic cavities** (Fig. 6.15*C*). The yolk sac proper occupies a space in the **chorionic cavity,** that is, the space between the amnion and chorionic plate (Fig. 6.15*B*).

During further development, the amniotic cavity enlarges rapidly at the expense of the chorionic cavity, and the amnion begins to envelop the connecting and yolk sac stalks, crowding them together and giving rise to the **primitive umbilical cord** (Fig. 6.15*B*). Distally the cord contains the yolk sac stalk and umbilical vessels. More proximally it contains some intestinal loops and the remnant of the allantois (Fig. 6.15, *B* and *D*). The yolk sac, found in the chorionic cavity, is connected to the umbilical cord by its stalk. At the end of the third month, the amnion has expanded so that it comes in contact with the chorion, obliterating the chorionic cavity (Fig. 6.10*B*). The yolk sac then usually shrinks and is gradually obliterated.

The abdominal cavity is temporarily too small for the rapidly developing intestinal loops and some of them are pushed into the extraembryonic space in the umbilical cord. These extruding intestinal loops form a **physiological**

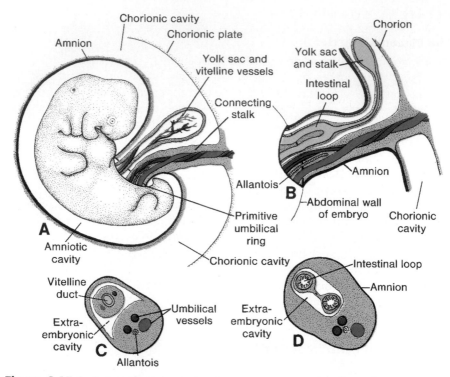

Figure 6.15 A. A 5-week embryo showing structures passing through the primitive umbilical ring. **B.** The primitive umbilical cord of a 10-week embryo. **C.** Transverse section through the structures at the level of the umbilical ring. **D.** Transverse section through the primitive umbilical cord showing intestinal loops protruding in the cord.

umbilical hernia (see Chapter 13). At approximately the end of the third month, the loops are withdrawn into the body of the embryo and the cavity in the cord is obliterated. When the allantois and the vitelline duct and its vessels are also obliterated, all that remains in the cord are the umbilical vessels surrounded by the **jelly of Wharton.** This tissue, which is rich in proteoglycans, functions as a protective layer for the blood vessels. The walls of the arteries are muscular and contain many elastic fibers, which contribute to a rapid constriction and contraction of the umbilical vessels after the cord is tied off.

CLINICAL CORRELATES

Umbilical Cord Abnormalities

At birth, the umbilical cord is approximately 2 cm in diameter and 50 to 60 cm long. It is tortuous, causing **false knots.** An extremely long cord may encircle

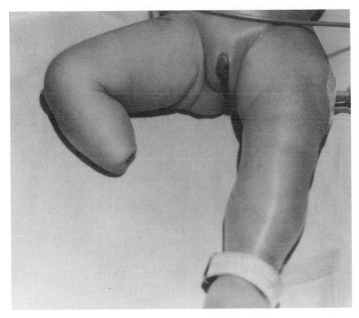

Figure 6.16 Infant showing limb amputation resulting from amniotic bands.

the neck of the fetus, usually without increased risk, whereas a short one may cause difficulties during delivery by pulling the placenta from its attachment in the uterus.

Normally there are two arteries and one vein in the umbilical cord. In 1 in 200 newborns, however, only one artery is present, and these babies have approximately a 20% chance of having cardiac and other vascular defects. The missing artery either fails to form (agenesis) or degenerates early in development.

Amniotic Bands

Occasionally, tears in the amnion result in **amniotic bands** that may encircle part of the fetus, particularly the limbs and digits. Amputations, **ring constrictions,** and other abnormalities, including craniofacial deformations, may result (Fig. 6.16). Origin of the bands is probably from infection or toxic insults that involve either the fetus, fetal membranes, or both. Bands then form from the amnion, like scar tissue, constricting fetal structures.

Placental Changes at the End of Pregnancy

At the end of pregnancy, a number of changes that occur in the placenta may indicate reduced exchange between the two circulations. These changes include (*a*) an increase in fibrous tissue in the core of the villus, (*b*) thickening of basement membranes in fetal capillaries, (*c*) obliterative changes in small capillaries of the villi, and (*d*) deposition of fibrinoid on the surface of the villi in the junctional zone and in the chorionic plate. Excessive fibrinoid formation frequently causes infarction of an intervillous lake or sometimes of an entire cotyledon. The cotyledon then assumes a whitish appearance.

Amniotic Fluid

The amniotic cavity is filled with a clear, watery fluid that is produced in part by amniotic cells but is derived primarily from maternal blood. The amount of fluid increases from approximately 30 ml at 10 weeks of gestation to 450 ml at 20 weeks to 800 to 1000 ml at 37 weeks. During the early months of pregnancy, the embryo is suspended by its umbilical cord in this fluid, which serves as a protective cushion. The fluid (*a*) absorbs jolts, (*b*) prevents adherence of the embryo to the amnion, and (*c*) allows for fetal movements. The volume of amniotic fluid is replaced every 3 hours. From the beginning of the fifth month, the fetus swallows its own amniotic fluid and it is estimated that it drinks about 400 ml a day, about half of the total amount. Fetal urine is added daily to the amniotic fluid in the fifth month, but this urine is mostly water, since the placenta is functioning as an exchange for metabolic wastes. During childbirth, the amnio-chorionic membrane forms a hydrostatic wedge that helps to dilate the cervical canal.

CLINICAL CORRELATES

Amniotic Fluid

Hydramnios or **polyhydramnios** is the term used to describe an excess of amniotic fluid (1500–2000 ml), whereas **oligohydramnios** refers to a decreased amount (less than 400 ml). Both conditions are associated with an increase in the incidence of birth defects. Primary causes of hydramnios include idiopathic causes (35%), maternal diabetes (25%), and congenital malformations, including central nervous system disorders (e.g., anencephaly) and gastrointestinal defects (atresias, e.g., esophageal) that prevent the infant from swallowing the fluid. Oligohydramnios is a rare occurrence that may result from renal agenesis.

Premature rupture of the amnion, the most common cause of preterm labor, occurs in 10% of pregnancies. Furthermore, clubfoot and lung hypoplasia may be caused by **oligohydramnios** following amnion rupture. Causes of rupture are largely unknown, but in some cases trauma plays a role.

Fetal Membranes in Twins

Arrangement of fetal membranes in twins varies considerably, depending on the type of twins and on the time of separation of **monozygotic twins.**

DIZYGOTIC TWINS

Approximately two-thirds of twins are **dizygotic,** or **fraternal,** and their incidence of 7 to 11 per 1000 births increases with maternal age. They result from simultaneous shedding of two oocytes and fertilization by different spermatozoa. Since the two zygotes have totally different genetic constitutions, the twins have no more resemblance than any other brothers or sisters. They may or may not be of different sex. The zygotes implant individually in the uterus, and usually each develops its own placenta, amnion, and chorionic sac (Fig. 6.17A). Sometimes, however, the two placentas are so close together that they fuse. Similarly, the walls of the chorionic sacs may also come into close apposition and fuse (Fig. 6.17B). Occasionally, each dizygotic twin possesses red blood cells of two different types (**erythrocyte mosaicism**), indicating that fusion of the two placentas was so intimate that red cells were exchanged.

MONOZYGOTIC TWINS

The second type of twins, which develops from a single fertilized ovum, is **monozygotic,** or **identical, twins.** The rate for monozygotic twins is 3 to 4 per 1000. They result from splitting of the zygote at various stages of development. The earliest separation is believed to occur at the two-cell stage, in which case two separate zygotes develop. The blastocysts implant separately, and each embryo has its own placenta and chorionic sac (Fig. 6.18A). Although the arrangement of the membranes of these twins resembles that of dizygotic twins, the two can be recognized as partners of a monozygotic pair by their strong resemblance in blood groups, fingerprints, sex, and external appearance, such as eye and hair color.

Splitting of the zygote usually occurs at the early blastocyst stage. The inner cell mass splits into two separate groups of cells within the same blastocyst cavity (Fig. 6.18B). The two embryos have a common placenta and a common chorionic cavity, but separate amniotic cavities (Fig. 6.18B). In rare cases the separation occurs at the bilaminar germ disc stage, just before the appearance of the primitive streak (Fig. 6.18C). This method of splitting results in formation of two partners with a single placenta and a common chorionic and amniotic sac. Although the twins have a common placenta, blood supply is usually well balanced.

Although triplets are rare (about 1/7600 pregnancies), birth of quadruplets, quintuplets, and so forth is rarer. In recent years multiple births have occurred more frequently in mothers given gonadotropins (fertility drugs) for ovulatory failure.

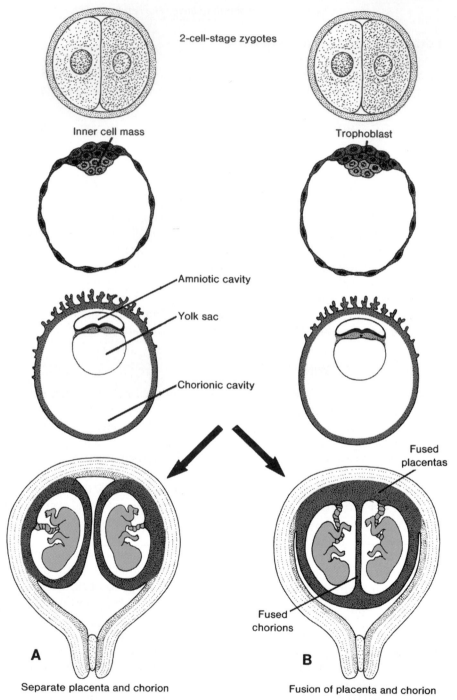

2-cell-stage zygotes

Inner cell mass

Trophoblast

Amniotic cavity

Yolk sac

Chorionic cavity

Fused placentas

Fused chorions

A

Separate placenta and chorion

B

Fusion of placenta and chorion

Figure 6.17 Development of dizygotic twins. Normally each embryo has its own amnion, chorion, and placenta, **A**, but sometimes the placentas are fused, **B**. Each embryo usually receives the appropriate amount of blood, but on occasion large anastomoses shunt more blood to one of the partners than to the other.

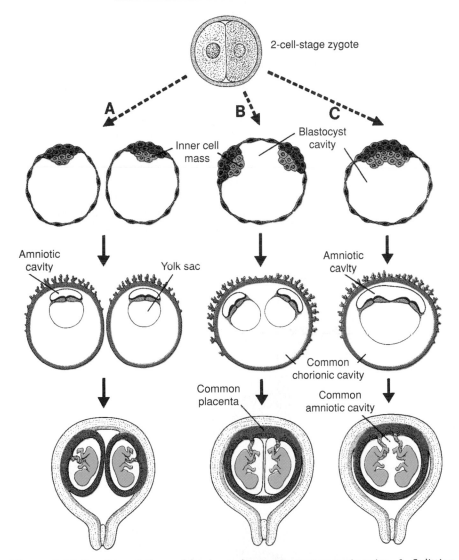

Figure 6.18 Possible relations of fetal membranes in monozygotic twins. **A.** Splitting occurs at the two-cell stage, and each embryo has its own placenta, amniotic cavity, and chorionic cavity. **B.** Splitting of the inner cell mass into two completely separated groups. The two embryos have a common placenta and a common chorionic sac but separate amniotic cavities. **C.** Splitting of the inner cell mass at a late stage of development. The embryos have a common placenta, a common amniotic cavity, and a common chorionic cavity.

CLINICAL CORRELATES

Twin Defects

Twin pregnancies have a high incidence of perinatal mortality and morbidity and a tendency toward preterm delivery. Approximately 12% of premature infants are twins and twins are usually small at birth. Low birth weight and prematurity place infants of twin pregnancies at great risk, and approximately 10 to 20% of them die, compared with only 2% of infants from single pregnancies.

The incidence of twinning may be much higher, since twins are conceived more often than they are born. Many twins die before birth and some studies indicate that only 29% of women pregnant with twins actually give birth to two infants. The term **vanishing twin** refers to the death of one fetus. This disappearance, which occurs in the first trimester or early second trimester, may result from resorption or formation of a **fetus papyraceus** (Fig. 6.19).

Another problem leading to increased mortality among twins is the **twin transfusion syndrome,** which occurs in 5 to 15% of monochorionic

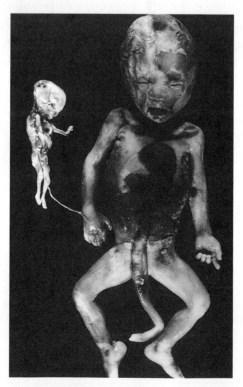

Figure 6.19 Fetus papyraceus. One twin is larger, and the other has been compressed and mummified, hence the term papyraceus.

monozygotic pregnancies. In this condition, placental vascular anastomoses, which occur in a balanced arrangement in most monochorionic placentas, are formed so that one twin receives most of the blood flow and flow to the other is compromised. As a result, one twin is larger than the other (Fig. 6.20). The outcome is poor, with the death of both twins occurring in 60 to 100% of cases.

At later stages of development, partial splitting of the primitive node and streak may result in formation of **conjoined (Siamese) twins.** These twins are classified according to the nature and degree of union as **thoracopagus** (*pagos*, fastened); **pygopagus;** and **craniopagus** (Figs. 6.21 and 6.22). Occasionally, monozygotic twins are connected only by a common skin bridge or by a common liver bridge. The type of twins formed depends upon when and to what extent abnormalities of the node and streak occurred. Mis-expression of genes, such as *Goosecoid,* may also result in conjoined twins. Many conjoined twins have survived, including the most famous pair, Chang and Eng, who were joined at the abdomen and who traveled to England and the United States on exhibitions in the mid-1800s. Finally settling in North Carolina, they farmed and fathered 21 children with their two wives.

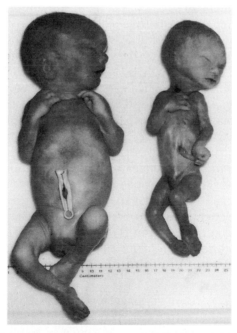

Figure 6.20 Monozygotic twins with twin transfusion syndrome. Placental vascular anastomoses produced unbalanced blood flow to the two fetuses.

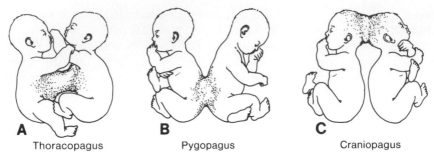

A Thoracopagus   **B** Pygopagus   **C** Craniopagus

Figure 6.21 Thoracopagus, pygopagus, and craniopagus twins. Conjoined twins can be separated only if they have no vital parts in common.

Parturition (Birth)

For the first 34 to 38 weeks of gestation, the uterine myometrium does not respond to signals for **parturition (birth)**. However, during the last 2 to 4 weeks of pregnancy, this tissue undergoes a transitional phase in preparation for the onset of **labor.** Ultimately, this phase ends with a thickening of the myometrium in the upper region of the uterus and a softening and thinning of the lower region and cervix.

Labor itself is divided into three stages: 1) **effacement** (thinning and short-ening) and dilatation of the cervix; this stage ends when the cervix is fully dilated; 2) **delivery of the fetus;** and 3) **delivery of the placenta and fetal membranes.** Stage 1 is produced by uterine contractions that force the am-niotic sac against the cervical canal like a wedge or, if the membranes have ruptured, then pressure will be exerted by the presenting part of the fetus, usually the head. Stage 2 is also assisted by uterine contractions, but the most important force is provided by increased intra-abdominal pressure from con-traction of abdominal muscles. Stage 3 requires uterine contractions and is aided by increasing intra-abdominal pressure.

As the uterus contracts, the upper part retracts creating a smaller and smaller lumen, while the lower part expands, thereby producing direction to the force. Contractions usually begin about 10 minutes apart; then, during the sec-ond stage of labor, they may occur less than 1 minute apart and last from 30 to 90 seconds. Their occurrence in pulses is essential to fetal survival, since they are of sufficient force to compromise uteroplacental blood flow to the fetus.

CLINICAL CORRELATES

Preterm Birth

Factors initiating labor are not known and may involve: **"retreat from mainte-nance of pregnancy"** in which pregnancy supporting factors (e.g., hormones,

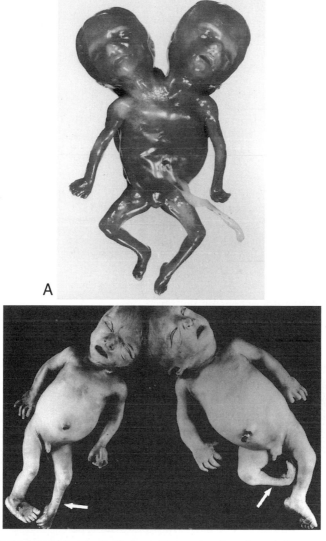

Figure 6.22 Conjoined twins. **A.** Twins with two heads, a broad thorax, two spines, two partially fused hearts, four lungs, and a duplicated gut down to the ileum. **B.** Twins joined at the head (craniopagus) with multiple deformations of the limbs.

etc.) are withdrawn; or **active induction** caused by stimulatory factors targeting the uterus. Probably, components of both phenomena are involved. Unfortunately, a lack of knowledge about these factors has restricted progress in preventing **preterm birth.** Preterm birth (delivery before 34 weeks) of premature infants is the second leading cause of infant mortality in the United States and also contributes significantly to morbidity. It is due to premature rupture of the membranes, premature onset of labor, or pregnancy complications requiring premature delivery. Maternal hypertension and diabetes as well as abruptio placenta are risk factors. Maternal infections, including bacterial vaginosis, are also associated with an increased risk.

Summary

The **fetal period extends from the ninth week of gestation until birth** and is characterized by rapid growth of the body and maturation of organ systems. Growth in length is particularly striking during the third, fourth, and fifth months (approximately 5 cm per month), while increase in weight is most striking during the last 2 months of gestation (approximately 700 g per month) (Table 6.1; p. 118).

A striking change is the relative slowdown in the growth of the head. In the third month, it is about half of CRL. By the fifth month the size of the head is about one-third of CHL, and at birth it is one-fourth of CHL (Fig. 6.2).

During the fifth month, fetal movements are clearly recognized by the mother, and the fetus is covered with fine, small hair.

A fetus born during the sixth or the beginning of the seventh month has difficulty surviving, mainly because the respiratory and central nervous systems have not differentiated sufficiently.

In general, the **length of pregnancy** for a full-term fetus is considered to be **280 days, or 40 weeks after onset of the last menstruation** or, **more accurately, 266 days or 38 weeks after fertilization.**

The **placenta** consists of two components: (*a*) a fetal portion, derived from the **chorion frondosum** or **villous chorion,** and (*b*) a maternal portion, derived from the **decidua basalis.** The space between the chorionic and decidual plates is filled with **intervillous lakes** of maternal blood. **Villous trees** (fetal tissue) grow into the maternal blood lakes and are bathed in them. The fetal circulation is at all times separated from the maternal circulation by (*a*) a syncytial membrane (a chorion derivative) and (*b*) endothelial cells from fetal capillaries. Hence the human placenta is of the **hemochorial** type.

Intervillous lakes of the fully grown placenta contain approximately 150 ml of maternal blood, which is renewed 3 or 4 times per minute. The villous area varies from 4 to 14 m^2, facilitating exchange between mother and child.

Main functions of the placenta are (*a*) exchange of gases; (*b*) exchange of nutrients and electrolytes; (*c*) transmission of maternal antibodies, providing the fetus with passive immunity; (*d*) production of hormones, such as progesterone, estradiol, and estrogen (in addition, it produces hCG and somatomammotropin); and (*e*) detoxification of some drugs.

The **amnion** is a large sac containing amniotic fluid in which the fetus is suspended by its umbilical cord. The fluid (*a*) absorbs jolts, (*b*) allows for fetal movements, and (*c*) prevents adherence of the embryo to surrounding tissues. The fetus swallows amniotic fluid, which is absorbed through its gut and cleared by the placenta. The fetus adds urine to the amniotic fluid, but this is mostly water. An excessive amount of amniotic fluid (**hydramnios**) is associated with anencephaly and esophageal atresia, whereas an insufficient amount (**oligohydramnios**) is related to renal agenesis.

The **umbilical cord,** surrounded by the amnion, contains (*a*) two umbilical arteries, (*b*) one umbilical vein, and (*c*) Wharton's jelly, which serves as a protective cushion for the vessels.

Fetal membranes in twins vary according to their origin and time of formation. Two-thirds of twins are **dizygotic,** or **fraternal;** they have two amnions, two chorions, and two placentas, which sometimes are fused. **Monozygotic twins** usually have two amnions, one chorion, and one placenta. In cases of **conjoined twins,** in which the fetuses are not entirely split from each other, there is one amnion, one chorion, and one placenta.

Signals initiating **parturition** (birth) are not clear, but preparation for labor usually begins between 34 and 38 weeks. Labor itself consists of three stages: 1) effacement and dilatation of the cervix; 2) delivery of the fetus; and 3) delivery of the placenta and fetal membranes.

Problems to Solve

1. *An ultrasound at 7 months of gestation shows too much space (fluid accumulation) in the amniotic cavity. What is this condition called, and what are its causes?*

2. *Later in her pregnancy a woman realizes that she was probably exposed to toluene in the workplace during the third week of gestation but tells a fellow worker that she is not concerned about her baby because the placenta protects her infant from toxic factors by acting as a barrier. Is she correct?*

SUGGESTED READING

Amselem S, et al.: Laron dwarfism and mutations in the growth hormone receptor gene. *N Engl J Med* 321:989, 1989.

Barnea ER, Hustin J, Jauniaux E (eds): *The First Twelve Weeks of Gestation.* Berlin, Springer-Verlag, 1992.

Bassett JM: Current perspectives on placental development and its integration with fetal growth. *Proc Nutr Soc* 50:311, 1991.

Benirschke K, Kaufman P: *The Pathology of the Human Placenta.* Berlin, Springer-Verlag, 1990.

Cunningham FG, Gant NF, Leveno KJ, Gilstrap LC, Hauth JC, Wenstrom KD: Fetal growth and development. *In Williams Obstetrics.* New York, McGraw Hill, 2001.

Cunningham FG, Gant NF, Leveno KJ, Gilstrap LC, Hauth JC, Wenstrom KD: Parturition. *In Williams Obstetrics.* New York, McGraw Hill, 2001.

Levi S: Ultrasonic assessment of the high rate of human multiple pregnancy in the first trimester. *J Clin Ultrasound* 4:3, 1976.

Naeye RL: *Disorders of the Placenta, Fetus, and Neonate.* St Louis, Mosby–Year Book, 1992.

Nyberg DA, Callan PW: Ultrasound evaluation of the placenta. *In* Callen PW (ed): *Ultrasonography in Obstetrics and Gynecology.* 2nd ed. Philadelphia, WB Saunders, 1988.

Peipert JF, Donnenfeld AE: Oligohydramnios: a review. *Obstet Gynecol* 46:325, 1991.

Petraglia F, et al.: Neuroendocrine mechanisms regulating placental hormone production. *Contrib Gynecol Obstet* 18:147, 1991.

Schnaufer L: Conjoined twins. *In* Raffensperger JG (ed): *Swenson's Pediatric Surgery.* 5th ed. Norwalk, CT, Appleton & Lange, 1990.

Spencer R: Conjoined twins: theoretical embryologic basis. *Teratology* 45:591, 1992.

Wallace I, Wallace A: *The Two.* New York, Simon and Schuster, 1978.

Wilcox AJ, Weinberg CR, Baird DD: Timing of sexual intercourse in relation to ovulation. *N Engl J Med* 333:1517, 1995.

chapter 7

Birth Defects and Prenatal Diagnosis

Birth Defects

Birth defect, congenital malformation, and **congenital anomaly** are synonymous terms used to describe structural, behavioral, functional, and metabolic disorders present at birth. The science that studies these disorders is **teratology** (Gr. *teratos,* monster). Major structural anomalies occur in 2 to 3% of liveborn infants, and an additional 2 to 3% are recognized in children by age 5 years, for a total of 4 to 6%. Birth defects are the leading cause of infant mortality, accounting for approximately 21% of infant deaths. They are the fifth leading cause of years of potential life lost prior to age 65 and a major contributor to disabilities. They are also nondiscriminatory; mortality rates produced by birth defects are the same for Asians, African Americans, Latin Americans, whites, and Native Americans.

In 40 to 60% of persons with birth defects, the cause is unknown. Genetic factors, such as chromosome abnormalities and mutant genes, account for approximately 15%; environmental factors produce approximately 10%; a combination of genetic and environmental influences (multifactorial inheritance) produces 20 to 25%; and twinning causes 0.5 to 1%.

Minor anomalies occur in approximately 15% of newborns. These structural abnormalities, such as microtia (small ears), pigmented spots, and short palpebral fissures, are not themselves detrimental to health

149

but, in some cases, are associated with major defects. For example, infants with one minor anomaly have a 3% chance of having a major malformation; those with two minor anomalies have a 10% chance; and those with three or more minor anomalies have a 20% chance. Therefore, minor anomalies serve as clues for diagnosing more serious underlying defects. In particular, ear anomalies are easily recognizable indicators of other defects and are observed in virtually all children with syndromic malformations.

TYPES OF ABNORMALITIES

Malformations occur during formation of structures, for example during organogenesis. They may result in complete or partial absence of a structure or in alterations of its normal configuration. Malformations are caused by environmental and/or genetic factors acting independently or in concert. Most malformations have their origin during the **third to eighth weeks of gestation.**

Disruptions result in morphological alterations of already formed structures and are due to destructive processes. Vascular accidents leading to bowel atresias (see Chapter 13; p. 296) and defects produced by amniotic bands are examples of destructive factors that produce disruptions.

Deformations are due to mechanical forces that mold a part of the fetus over a prolonged period. Clubfeet, for example, are due to compression in the amniotic cavity. Deformations often involve the musculoskeletal system and may be reversible postnatally.

A **syndrome** is a group of anomalies occurring together that have a specific common cause. This term indicates that a diagnosis has been made and that the risk of recurrence is known. In contrast, **association** is the nonrandom appearance of two or more anomalies that occur together more frequently than by chance alone, but whose cause has not been determined. Examples include **CHARGE** (**C**olobomas, **H**eart defects, **A**tresia of the choanae, **R**etarded growth, **G**enital anomalies, and **E**ar abnormalities) and **VACTERL** (**V**ertebral, **A**nal, **C**ardiac, **T**racheo **E**sophageal, **R**enal, and **L**imb anomalies). Although they do not constitute a diagnosis, associations are important because recognition of one or more of the components promotes the search for others in the group.

ENVIRONMENTAL FACTORS

Until the early 1940s it was assumed that congenital defects were caused primarily by hereditary factors. With the discovery by Gregg that German measles affecting a mother during early pregnancy caused abnormalities in the embryo, it suddenly became evident that congenital malformations in humans could also be caused by environmental factors. In 1961 observations by Lenz linked limb defects to the sedative **thalidomide** and made it clear that drugs could also cross the placenta and produce birth defects (Fig. 7.1). Since that

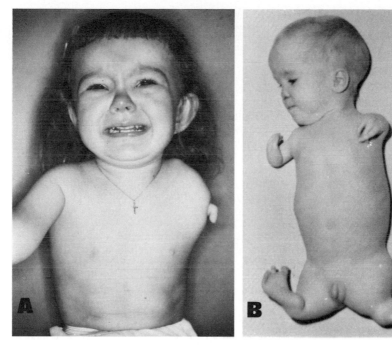

Figure 7.1 A. Child with unilateral amelia. **B.** Child with meromelia. The hand is attached to the trunk by an irregularly shaped bone. Both infants were born to mothers who took thalidomide.

time many agents have been identified as **teratogens** (factors that cause birth defects) (Table 7.1).

Principles of Teratology

Factors determining the capacity of an agent to produce birth defects have been defined and set forth as the **principles of teratology.** They include the following:

1. Susceptibility to teratogenesis depends on the **genotype of the conceptus** and the manner in which this genetic composition interacts with the environment. The **maternal genome** is also important with respect to drug metabolism, resistance to infection, and other biochemical and molecular processes that affect the conceptus.

2. Susceptibility to teratogens varies with the **developmental stage at the time of exposure.** The most sensitive period for inducing birth defects is the **third to eighth weeks** of gestation, the period of **embryogenesis.** Each organ system may have one or more stages of susceptibility. For example, cleft palate can be induced at the blastocyst stage (day 6),

TABLE 7.1 Teratogens Associated With Human Malformations

Teratogen	Congenital Malformations
Infectious agents	
Rubella virus	Cataracts, glaucoma, heart defects, deafness, teeth
Cytomegalovirus	Microcephaly, blindness, mental retardation, fetal death
Herpes simplex virus	Microphthalmia, microcephaly, retinal dysplasia
Varicella virus	Limb hypoplasia, mental retardation, muscle atrophy
HIV	Microcephaly, growth retardation
Toxoplasmosis	Hydrocephalus, cerebral calcifications, microphthalmia
Syphilis	Mental retardation, deafness
Physical agents	
X-rays	Microcephaly, spina bifida, cleft palate, limb defects
Hyperthermia	Anencephaly, spina bifida, mental retardation, facial defects
Chemical agents	
Thalidomide	Limb defects, heart malformations
Aminopterin	Anencephaly, hydrocephaly, cleft lip and palate
Diphenylhydantoin (phenytoin)	Fetal hydantoin syndrome: facial defects, mental retardation
Valproic acid	Neural tube defects, heart, craniofacial, and limb anomalies
Trimethadione	Cleft palate, heart defects, urogenital and skeletal abnormalities
Lithium	Heart malformations
Amphetamines	Cleft lip and palate, heart defects
Warfarin	Chondrodysplasia, microcephaly
ACE inhibitors	Growth retardation, fetal death
Cocaine	Growth retardation, microcephaly, behavioral abnormalities, gastroschisis
Alcohol	Fetal alcohol syndrome, short palpebral fissures, maxillary hypoplasia, heart defects, mental retardation
Isotretinoin (vitamin A)	Vitamin A embryopathy: small, abnormally shaped ears, mandibular hypoplasia, cleft palate, heart defects
Industrial solvents	Low birth weight, craniofacial and neural tube defects
Organic mercury	Neurological symptoms similar to those of cerebral palsy
Lead	Growth retardation, neurological disorders
Hormones	
Androgenic agents (ethisterone, norethisterone)	Masculinization of female genitalia: fused labia, clitoral hypertrophy
Diethylstilbestrol (DES)	Malformation of the uterus, uterine tubes, and upper vagina; vaginal cancer; malformed testes
Maternal diabetes	Variety of malformations; heart and neural tube defects most common

during gastrulation (day 14), at the early limb bud stage (fifth week), or when the palatal shelves are forming (seventh week). Furthermore, while most abnormalities are produced during embryogenesis, defects may also be induced before or after this period; no stage of development is completely safe.

3. Manifestations of abnormal development depend **on dose and duration of exposure** to a teratogen.

4. Teratogens act in specific ways (**mechanisms**) on developing cells and tissues to initiate abnormal embryogenesis (**pathogenesis**). Mechanisms may involve inhibition of a specific biochemical or molecular process; pathogenesis may involve cell death, decreased cell proliferation, or other cellular phenomena.

5. Manifestations of abnormal development are **death, malformation, growth retardation,** and **functional disorders.**

Infectious Agents

Infectious agents that cause birth defects (Table 7.1) include a number of viruses. **Rubella** used to be a major problem, but the ability to detect serum antibody titers and development of a vaccine have significantly lowered the incidence of birth defects from this cause. Today approximately 85% of women are immune.

Cytomegalovirus is a serious threat. Often, the mother has no symptoms, but the effects on the fetus can be devastating. The infection is often fatal, and if it is not, meningoencephalitis caused by the virus produces mental retardation.

Herpes simplex, varicella, and **human immunodeficiency (HIV) viruses** can cause birth defects. Herpes-induced abnormalities are rare, and usually infection is transmitted as a venereal disease to the child during delivery. Similarly, HIV (the cause of acquired immunodeficiency syndrome, or AIDS) appears to have a low teratogenic potential. Infection with varicella causes a 20% incidence of birth defects.

Other Viral Infections and Hyperthermia

Malformations following maternal infection with measles, mumps, hepatitis, poliomyelitis, ECHO virus, Coxsackie virus, and influenza virus have been described. Prospective studies indicate that the malformation rate following exposure to these agents is low if not nonexistent.

A complicating factor introduced by these and other infectious agents is that most are **pyrogenic,** and elevated body temperature (**hyperthermia**) is teratogenic. Defects produced by exposure to elevated temperatures include anencephaly, spina bifida, mental retardation, micropthalmia, and facial abnormalities. In addition to febrile illnesses, use of hot tubs and saunas can produce sufficient temperature elevations to cause birth defects.

Toxoplasmosis and **syphilis** cause birth defects. Poorly cooked meat; domestic animals, especially cats; and feces in contaminated soil can carry the protozoan parasite *Toxoplasmosis gondii*. A characteristic feature of fetal toxoplasmosis infection is cerebral calcifications.

Radiation

Ionizing radiation kills rapidly proliferating cells, so it is a potent teratogen, producing virtually any type of birth defect depending upon the dose and stage of development of the conceptus at the time of exposure. Radiation from nuclear explosions is also teratogenic. Among women survivors pregnant at the time of the atomic bomb explosions over Hiroshima and Nagasaki, 28% aborted, 25% gave birth to children who died in their first year of life, and 25% had severe birth defects involving the central nervous system. Radiation is also a mutagenic agent and can lead to genetic alterations of germ cells and subsequent malformations.

Chemical Agents

The role of chemical agents and pharmaceutical drugs in the production of abnormalities in humans is difficult to assess for two reasons: (*a*) most studies are retrospective, relying on the mother's memory for a history of exposure; and (*b*) pregnant women take a large number of pharmaceutical drugs. A National Institutes of Health study discovered that pregnant women took 900 different drugs, for an average of 4 per woman. Only 20% of pregnant women used no drugs during their pregnancy. Even with this widespread use of chemical agents, relatively few of the many drugs used during pregnancy have been positively identified as being teratogenic. One example is **thalidomide,** an antinauseant and sleeping pill. In 1961 it was noted in West Germany that the frequency of **amelia** and **meromelia** (total or partial absence of the extremities), a rare hereditary abnormality, had suddenly increased (Fig. 7.1). This observation led to examination of the prenatal histories of affected children and to the discovery that many mothers had taken thalidomide early in pregnancy. The causal relation between thalidomide and meromelia was discovered only because the drug produced such an unusual abnormality. If the defect had been a more common type, such as cleft lip or heart malformation, the association with the drug might easily have been overlooked.

Other drugs with teratogenic potential include the anticonvulsants **diphenylhydantoin (phenytoin), valproic acid,** and **trimethadione,** which are used by **epileptic** women. Specifically, trimethadione and diphenylhydantoin produce a broad spectrum of abnormalities that constitute distinct patterns of dysmorphogenesis known as the **trimethadione** and **fetal hydantoin syndromes.** Facial clefts are particularly common to these syndromes. Valproic acid also causes craniofacial abnormalities but has a particular propensity for producing neural tube defects.

Antipsychotic and **antianxiety agents** (major and minor tranquilizers, respectively) are suspected producers of congenital malformations. The antipsychotics **phenothiazine** and **lithium** have been implicated as teratogens. Although evidence for the teratogenicity of phenothiazines is conflicting, that concerning lithium is better documented. In any case, it has been strongly suggested that use of these agents during pregnancy carries a high risk.

Similar observations have been made for the antianxiety agents **meprobamate, chlordiazepoxide**, and **diazepam** (**Valium**). A prospective study showed that severe anomalies occurred in 12% of fetuses exposed to meprobamate and 11% in those exposed to chlordiazepoxide, compared with 2.6% of controls. Likewise, retrospective studies demonstrate up to a fourfold increase in cleft lip with or without cleft palate in offspring whose mothers took diazepam during pregnancy.

The **anticoagulant warfarin** is teratogenic, whereas **heparin** does not appear to be. **Antihypertensive agents** that inhibit **angiotensin-converting enzyme** (ACE) **inhibitors** produce growth retardation, renal dysfunction, fetal death, and oligohydramnios.

Caution has also been expressed regarding a number of other compounds that may damage the embryo or fetus. The most prominent among these are propylthiouracil and potassium iodide (goiter and mental retardation), streptomycin (deafness), sulfonamides (kernicterus), the antidepressant imipramine (limb deformities), tetracyclines (bone and tooth anomalies), amphetamines (oral clefts and cardiovascular abnormalities), and quinine (deafness). Finally, there is increasing evidence that **aspirin** (salicylates), the most commonly ingested drug during pregnancy, may harm the developing offspring when used in large doses.

One of the increasing problems in today's society is the effect of social drugs, such as LSD (lysergic acid diethylamide), PCP (phencyclidine, or "angel dust"), marijuana, alcohol, and cocaine. In the case of LSD, limb abnormalities and malformations of the central nervous system have been reported. A comprehensive review of more than 100 publications, however, led to the conclusion that pure LSD used in moderate doses is not teratogenic and does not cause genetic damage. A similar lack of conclusive evidence for teratogenicity has been described for marijuana and PCP. **Cocaine** has been reported to cause a number of birth defects, possibly due to its action as a vasoconstrictor that causes hypoxia.

There is a well-documented association between maternal **alcohol** ingestion and congenital abnormalities, and these defects, together with mental retardation and growth deficiency, make up the **fetal alcohol syndrome** (FAS) (Fig. 7.2). Even moderate alcohol consumption during pregnancy may be detrimental to embryonic development. The central nervous system is particularly sensitive to alcohol, and **alcohol-related neurodevelopmental disorder** (ARND) may result from exposure. The incidence of FAS and ARND together is 1 in 100 live births. Furthermore, **alcohol is the leading cause of mental retardation.**

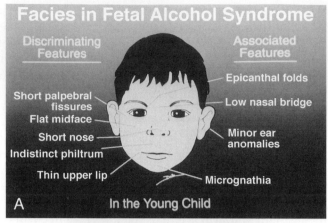

Facies in Fetal Alcohol Syndrome

Discriminating Features

Associated Features

Epicanthal folds

Short palpebral fissures

Low nasal bridge

Flat midface

Short nose

Minor ear anomalies

Indistinct philtrum

Thin upper lip

Micrognathia

A   In the Young Child

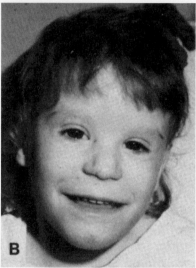

B

Figure 7.2 A. Characteristic features of a child with fetal alcohol syndrome. **B.** Child with fetal alcohol syndrome illustrating many of the features in the drawing. These children may also have cardiovascular and limb defects.

Cigarette smoking has not been linked to major birth defects, but it does contribute to intrauterine growth retardation and premature delivery. There is also evidence that it causes behavioral disturbances.

Isotretinoin (13-*cis*-retinoic acid), an analogue of **vitamin A,** has been shown to cause a characteristic pattern of malformations known as the **isotretinoin embryopathy** or **vitamin A embryopathy.** The drug is prescribed for the treatment of cystic acne and other chronic dermatoses, but it is highly teratogenic and can produce virtually any type of malformation. Even topical retinoids, such as etretinate, may have the potential to cause abnormalities. With the recent support for the use of multivitamins containing folic acid, there is concern that overuse of vitamin supplements could be harmful, since most contain approximately 8,000 IU of vitamin A. How much is potentially harmful is controversial, but most scientists agree that 25,000 IU is a threshold level for teratogenicity.

Hormones

Androgenic Agents. In the past synthetic progestins were frequently used during pregnancy to prevent abortion. The progestins ethisterone and norethisterone have considerable androgenic activity, and many cases of masculinization of the genitalia in female embryos have been reported. The abnormalities consist of an enlarged clitoris associated with varying degrees of fusion of the labioscrotal folds.

Endocrine Disrupters. Endocrine disrupters are exogenous agents that interfere with the normal regulatory actions of hormones controlling developmental processes. Most commonly these agents interfere with the action of estrogen through its receptor to cause developmental abnormalities of the central nervous system and reproductive tract. For some time it has been known that the synthetic estrogen **diethylstilbestrol,** which was used to prevent abortion, raised the incidence of carcinomas of the vagina and cervix in women exposed to the drug in utero. Furthermore, a high percentage of these women had reproductive dysfunction due in part to congenital malformations of the uterus, uterine tubes, and upper vagina. Male embryos exposed in utero can also be affected, as evidenced by an increase in malformations of the testes and abnormal sperm analysis among these individuals. In contrast to females, however, males do not demonstrate an increased risk of developing carcinomas of the genital system.

Today environmental estrogens are a concern and numerous studies to determine their effects on the unborn are under way. Decreasing sperm counts and increasing incidences of testicular cancer, hypospadias, and other abnormalities of the reproductive tract in humans, together with documented central nervous system abnormalities (masculinization of female brains and feminization of male brains) in other species with high environmental exposures, have raised awareness of the possible harmful effects of these agents. Many are formed from chemicals used for industrial purposes and from pesticides.

Oral Contraceptives. Birth control pills, containing estrogens and progestogens, appear to have a low teratogenic potential. Since other hormones, such as diethylstilbestrol, produce abnormalities, however, use of oral contraceptives should be discontinued if pregnancy is suspected.

Cortisone. Experimental work has repeatedly shown that cortisone injected into mice and rabbits at certain stages of pregnancy causes a high percentage of cleft palates in the offspring. However, it has been impossible to implicate cortisone as an environmental factor causing cleft palate in humans.

Maternal Disease

Diabetes. Disturbances in carbohydrate metabolism during pregnancy in diabetic mothers cause a high incidence of stillbirths, neonatal deaths, abnormally large infants, and congenital malformations. The risk of congenital anomalies in children of diabetic mothers is 3 to 4 times that for the offspring of nondiabetic mothers and has been reported to be as high as 80 % in the offspring of diabetics with long-standing disease. The variety of observed malformations includes caudal dysgenesis (sirenomelia).

Factors responsible for these abnormalities have not been delineated, although evidence suggests that altered glucose levels play a role and that **insulin is not teratogenic.** In this respect, a significant correlation exists between the severity and duration of the mother's disease and the incidence of malformations. Also, strict control of maternal metabolism with aggressive **insulin therapy** prior to conception reduces the occurrence of malformations. Such therapy, however, increases the frequency and severity of **hypoglycemic episodes.** Numerous animal studies have shown that during gastrulation and neurulation, mammalian embryos depend on glucose as an energy source, so that even brief episodes of low blood glucose are teratogenic. Therefore, caution must be exercised in managing the pregnant diabetic woman. In the case of non-insulin-dependent diabetes, **oral hypoglycemic agents** may be employed. These agents include the sulfonylureas and biguanides. Both classes of agents have been implicated as teratogens.

Phenylketonuria. Mothers with **phenylketonuria** (PKU), in which the enzyme phenylalanine hydroxylase is deficient, resulting in increased serum concentrations of phenylalanine, are at risk for having infants with mental retardation, microcephaly, and cardiac defects. Women with PKU who maintain their low phenylalanine diet prior to conception reduce the risk to their infants to that observed in the general population.

Nutritional Deficiencies

Although many nutritional deficiencies, particularly vitamin deficiencies, have been proven to be teratogenic in laboratory animals, the evidence in humans is sparse. Thus, with the exception of **endemic cretinism,** which is related to **iodine** deficiency, no analogies to animal experiments have been discovered. However, the evidence suggests that poor maternal nutrition prior to and during pregnancy contributes to low birth weight and birth defects.

Obesity

Prepregnancy obesity, defined as having a **body mass index (BMI)** $>29\text{kg/m}^2$, is associated with a two- to three-fold increased risk for having a child with a neural tube defect. Causation has not been determind but may relate to maternal metabolic disturbances affecting glucose, insulin, or other factors.

Hypoxia

Hypoxia induces congenital malformations in a great variety of experimental animals. Whether the same is valid for humans remains to be seen. Although children born at relatively high altitudes are usually lighter in weight and smaller than those born near or at sea level, no increase in the incidence of congenital malformations has been noted. In addition, women with cyanotic cardiovascular disease often give birth to small infants, but usually without gross congenital malformations.

Heavy Metals

Several years ago, researchers in Japan noted that a number of mothers with diets consisting mainly of fish had given birth to children with multiple neurological symptoms resembling cerebral palsy. Further examination revealed that the fish contained an abnormally high level of **organic mercury,** which was spewed into Minamata Bay and other coastal waters of Japan by large industries. Many of the mothers did not show any symptoms themselves, indicating that the fetus was more sensitive to mercury than the mother. In the United States, similar observations were made when seed corn sprayed with a mercury-containing fungicide was fed to hogs and the meat was subsequently eaten by a pregnant woman. Similarly in Iraq, several thousand babies were affected after mothers ate grain treated with mercury-containing fungicides.

   Lead has been associated with increased abortions, growth retardation, and neurological disorders.

MALE-MEDIATED TERATOGENESIS

A number of studies have indicated that exposures to chemicals and other agents, such as ethylnitrosourea and radiation, can cause mutations in male germ cells. Epidemiological investigations have linked paternal occupational and environmental exposures to mercury, lead, solvents, alcohol, cigarette smoking, and other compounds to spontaneous abortion, low birth weight, and birth defects. Advanced paternal age is a factor for an increased risk of limb and neural tube defects, Down syndrome, and new autosomal dominant mutations. Interestingly, men younger than 20 also have a relatively high risk of fathering a child with a birth defect. Even transmission of paternally mediated toxicity is possible through seminal fluid and from household contamination from chemicals brought home on workclothes by the father. Studies also show that males with birth defects themselves have a greater than twofold risk of having an affected child.

CLINICAL CORRELATES

Prevention of Birth Defects

Many birth defects can be prevented. For example, supplementation of salt or water supplies with iodine eliminates mental retardation and bone deformities resulting from **cretinism.** Placing women with diabetes or PKU under strict metabolic control prior to conception reduces the incidence of birth defects in their offspring. **Folate supplementation** lowers the incidence of neural tube defects, such as spina bifida and anencephaly. Avoidance of alcohol and other drugs during **all** stages of pregnancy reduces the incidence of birth defects. A common denominator for all prevention strategies is to initiate interventions **prior to conception.** Such an approach also helps prevent low-birth-weight babies.

It is important for physicians prescribing drugs to women of childbearing age to consider the possibility of pregnancy and the potential teratogenicity of the compounds. Recently hundreds of children have been born with severe craniofacial, cardiac, and neural tube defects produced by **retinoids (vitamin A embryopathy).** These compounds are used for the treatment of cystic acne (isotretinoin, 13-*cis*-retinoic acid) but are also effective topically (**Retin-A**) for common acne and reducing wrinkles. Oral preparations are highly teratogenic, and recent evidence suggests that topical applications may also cause abnormalities. Since patients with acne are usually young and may be sexually active, these agents must be used cautiously.

Prenatal Diagnosis

The perinatologist has several approaches for assessing growth and development of the fetus in utero, including **ultrasound, amniocentesis, chorionic villus sampling,** and **maternal serum screening.** In combination, these techniques are designed to detect malformations, genetic abnormalities, overall fetal growth, and complications of pregnancy, such as placental or uterine abnormalities. Their use and development of in utero therapies have heralded a new concept in which the fetus is now a patient.

ULTRASONOGRAPHY

Ultrasonography is a relatively noninvasive technique that uses high-frequency sound waves reflected from tissues to create images. The approach may be transabdominal or transvaginal, with the latter producing images with higher resolution (Fig. 7.3). In fact, the technique, which was first developed in the 1950s, has advanced to a degree where detection of blood flow in major vessels,

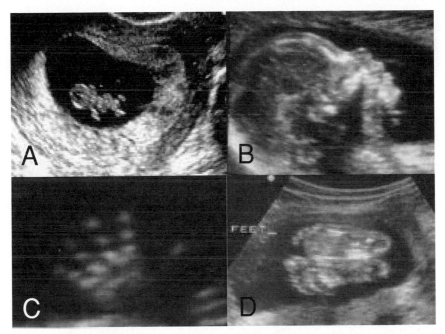

Figure 7.3 Examples of the effectiveness of ultrasound in imaging the embryo and fetus. **A.** 6-week embryo. **B.** Lateral view of the fetal face. **C.** Hand. **D.** Feet.

movement of heart valves, and flow of fluid in the trachea and bronchi is possible. The technique is safe and commonly used, with approximately 80% of pregnant women in the United States receiving at least one scan.

Important parameters revealed by ultrasound include: characteristics of fetal age and growth; presence or absence of congenital anomalies; status of the uterine environment, including the amount of amniotic fluid (Fig. 7.4A); placental position and umbilical blood flow; and whether multiple gestations are present (Fig.7.4B). All of these factors are then used to determine proper approaches for management of the pregnancy.

Determination of the fetal age and growth is crucial in planning pregnancy management, especially for low-birth-weight infants. In fact, studies show that ultrasound screened and managed pregnancies with low-birth-weight babies reduced the mortality rate by 60% compared with an unscreened group. Fetal age and growth are assessed by **crown-rump length** during the fifth to tenth weeks of gestation. After that, a combination of measurements, including the **biparietal diameter (BPD)** of the skull, **femur length,** and **abdominal circumference** are used (Fig. 7.5). Multiple measures of these parameters over time improve the ability to determine the extent of fetal growth.

Congenital malformations that can be determined by ultrasound include: the neural tube defects anencephaly and spina bifida (see Chapter 19);

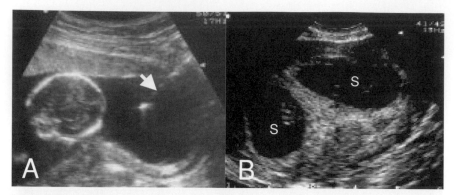

Figure 7.4 A. Ultrasound image showing position of the fetal skull and placement of the needle into the amniotic cavity (arrow) during amniocentesis. **B.** Twins. Ultrasound showing the presence of two gestational sacs (S).

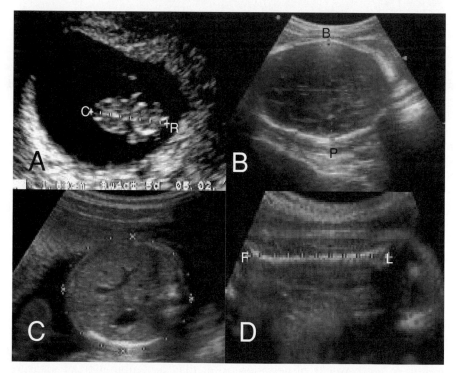

Figure 7.5 Ultrasounds showing measures used to assess embryonic and fetal growth. **A.** Crown-rump (C-R) length in a 7-week embryo. **B.** Biparietal (B-P) diameter of the skull. **C.** Abdominal circumference. **D.** Femur length (F-L).

abdominal wall defects, such as omphalocele and gastroschisis (see Chapter 13); and heart, (see Chapter 11) and facial defects, including cleft lip and palate (see Chapter 15).

MATERNAL SERUM SCREENING

A search for biochemical markers of fetal status led to development of **maternal serum screening tests.** One of the first of these tests assessed serum **alphafetoprotein (AFP)** concentrations. AFP is produced normally by the fetal liver, peeks at approximately 14 weeks, and "leaks" into the maternal circulation via the placenta. Thus, AFP concentrations increase in maternal serum during the second trimester and then begin a steady decline after 30 weeks of gestation. In cases of neural tube defects and several other abnormalities, including omphalocele, gastroschisis, bladder exstrophy, amniotic band syndrome, sacrococcygeal teratoma, and intestinal atresia, AFP levels increase in amniotic fluid and maternal serum. In other instances, AFP concentrations decrease as, for example, in Down syndrome, trisomy 18, sex chromosome abnormalities, and triploidy. These conditions are also associated with lower serum concentrations of **human chorionic gonadotropin (hCG)** and **unconjugated estriol.** Therefore, maternal serum screening provides a relatively noninvasive technique for an initial assessment of fetal well being.

AMNIOCENTESIS

During amniocentesis, a needle is inserted transabdominally into the amniotic cavity (identified by ultrasound; Fig. 7.4A) and approximately 20 to 30 ml of fluid are withdrawn. Because of the amount of fluid required, the procedure is not usually performed before 14 weeks gestation, when sufficient quantities are available without endangering the fetus. The risk of fetal loss as a result of the procedure is 1%, but it is less in centers skilled in the technique.

The fluid itself is analyzed for biochemical factors, such as AFP and acetylcholinesterase. In addition, fetal cells, sloughed into the amniotic fluid, can be recovered and used for metaphase karyotyping and other genetic analyses (see Chapter 1). Unfortunately, the harvested cells are not rapidly dividing, and therefore, cell cultures containing mitogens must be established to provide sufficient metaphase cells for analysis. This culture period requires 8 to 14 days, and consequently, making a diagnosis is delayed. Once chromosomes are obtained, major chromosomal alterations, such as translocations, breaks, trisomies, and monosomies, can be identified. With special stains (Giemsa) and high-resolution techniques, chromosome banding patterns can be determined. Furthermore, now that the human genome has been sequenced, more sophisticated molecular analyses using polymerase chain reaction (PCR) and genotyping assays will increase the level of detection for genetic abnormalities.

CHORIONIC VILLUS SAMPLING (CVS)

Chorionic villus sampling (CVS) involves inserting a needle transabdominally or transvaginally into the placental mass and aspirating approximately 5 to 30 mg of villus tissue. Cells may be analyzed immediately, but accuracy of results is problematic because of the high frequency of chromosomal errors in the normal placenta. Therefore, cells from the mesenchymal core are isolated by trypsinization of the external trophoblast and cultured. Because of the large number of cells obtained, only 2 to 3 days in culture are necessary to permit genetic analysis. Thus, the time for genetic characterization of the fetus is reduced compared with amniocentesis. However, the risk of fetal loss from CVS is approximately twofold greater than with amniocentesis, and there have been indications that the procedure carries an increased risk for limb reduction defects.

Generally, these prenatal diagnostic tests are not used on a routine basis (although ultrasonography is approaching routine use), being reserved instead for high-risk pregnancies. Indications for using the tests include: 1) advanced maternal age (35 years and older); 2) history of neural tube defects in the family; 3) previous gestation with a chromosome abnormality, such as Down syndrome; 4) chromosome abnormalities in either parent; and 5) a mother who is a carrier for an X-linked disorder.

Fetal Therapy

FETAL TRANSFUSION

In cases of fetal anemia produced by maternal antibodies or other causes, blood transfusions for the fetus can be performed. Ultrasound is used to guide insertion of a needle into the umbilical cord vein, and blood is transfused directly into the fetus.

FETAL MEDICAL TREATMENT

Treatment for infections, fetal cardiac arrhythmias, compromised thyroid function, and other medical problems is usually provided to the mother and reaches the fetal compartment after crossing the placenta. In some cases, however, agents may be administered to the fetus directly by intramuscular injection into the gluteal region or via the umbilical vein.

FETAL SURGERY

Because of advances in ultrasound and surgical procedures, operating on fetuses has become possible. However, because of risks to the mother, infant, and subsequent pregnancies, procedures are only performed in centers with well-trained teams and only when there are no reasonable alternatives. Several types of surgeries may be performed, including placing shunts to remove fluid from organs and cavities. For example, in obstructive urinary disease of the urethra, a pigtail shunt may be inserted into the fetal bladder. One problem is diagnosing the condition early enough to prevent renal damage. Ex utero surgery, where the uterus is opened and the fetus operated upon directly, has been used for repairing congenital diaphragmatic hernias, removing cystic (adenomatoid) lesions in the lung, and repairing spina bifida defects. Repairs of hernias and lung lesions have good outcomes if proper selection criteria for cases are employed, and one of these is the fact that without surgery, fetal demise is almost certain. Surgery for neural tube defects is more controversial because the abnormalities are not life threatening. Also, the evidence is not conclusive that repair of the lesion improves neurological function, although it does alleviate the accompanying hydrocephalus by freeing the tethered spinal cord and preventing herniation of the cerebellum into the foramen magnum (see Chapter 19; p. 445).

STEM CELL TRANSPLANTATION AND GENE THERAPY

Because the fetus does not develop any immunocompetence before 18 weeks gestation, it may be possible to transplant tissues or cells before this time without rejection. Research in this field is focusing on hematopoietic stem cells for treatment of immunodeficiency and hematologic disorders. Gene therapy for inherited metabolic diseases, such as Tay-Sachs and cystic fibrosis, is also being investigated.

Summary

A variety of agents (Table 7.1; p. 152) are known to produce congenital malformations in approximately 2 to 3% of all live-born infants. These agents include viruses, such as rubella and cytomegalovirus; radiation; drugs, such as thalidomide, aminopterin, anticonvulsants, antipsychotics, and antianxiety compounds; social drugs, such as PCP, cigarettes, and alcohol; hormones, such as diethylstilbestrol; and maternal diabetes. Effects of teratogens depend on the **maternal and fetal genotype,** the **stage of development** when exposure occurs, and the **dose and duration of exposure** of the agent. Most major malformations are produced during the **period of embryogenesis (teratogenic period; third to eighth weeks),** but in stages before and after this time, the fetus is also susceptible, so that no period of gestation is completely free of risk. **Prevention** of many birth defects is possible, but it depends on beginning preventative measures before conception and increasing physicians' and women's awareness of the risks.

A variety of techniques are available to assess the growth and developmental status of the fetus. **Ultrasound** can accurately determine fetal age and growth parameters and detect many malformations. **Maternal serum screening** for alpha-fetoprotein can indicate the presence of a neural tube defect or other abnormalities. **Amniocentesis** is a procedure in which a needle is placed into the amniotic cavity and a fluid sample is withdrawn. This fluid can be analyzed biochemically and also provides cells for culture and genetic analysis. **Chorionic villus sampling (CVS)** involves aspirating a tissue sample directly from the placenta to obtain cells for genetic analysis. Because many of these procedures involve a potential risk to the fetus and mother, they are generally only used for higher risk pregnancies (the exception is ultrasound). These risk factors include advanced maternal age (35 years and older); a history of neural tube defects in the family; previous gestation with a chromosome abnormality; chromosome abnormalities in either parent; and a mother who is a carrier for an X-linked disorder.

Modern medicine has also made the fetus a patient who can receive treatment, such as transfusions, medications for disease, fetal surgery, and gene therapy.

Problems to Solve

1. *Amniocentesis reveals an elevated AFP level. What should be included in a differential diagnosis, and how would a definitive one be made?*

2. *A 40-year-old woman is approximately 8 weeks pregnant. What tests are available to determine whether her unborn child has Down syndrome? What are the risks and advantages of each technique?*

3. *Why is it important to determine the status of an infant prenatally? What maternal or family factors might raise your concern about the well-being of an unborn infant?*

4. *What factors influence the action of a teratogen?*

5. *A young woman in only the third week of her pregnancy develops a fever of 104° but refuses to take any medication because she is afraid that drugs will harm her baby. Is she correct?*

6. *A young woman who is planning a family seeks advice about folic acid and other vitamins. Should she take such a supplement, and if so, when and how much?*

7. *A young insulin-dependent diabetic woman who is planning a family is concerned about the possible harmful effects of her disease on her unborn child. Are her concerns valid, and what would you recommend?*

SUGGESTED READING

Barlow S, Kavlock RJ, Moore JA, Shantz S, Sheehan DL, Shuey DL, Lary JM: Teratology Society Public Affairs Committee Position Paper: developmental toxicity of endocrine disruptors to humans. *Teratology* 60:365, 1999.

Barnea ER, Hustin J, Jauniaux E (eds): *The First Twelve Weeks of Gestation*. Berlin, Springer-Verlag, 1992.

Bendich A, et al.: Influence of maternal nutrition on pregnancy outcome: public policy issues. Introduction to Part V. *Ann N Y Acad Sci* 678:284, 1993.

Boehm CE, Kazazian HH Jr: Prenatal diagnosis by DNA analysis. *In* Harrison MR, Golbus MS, Filly RA (eds): *The Unborn Patient: Prenatal Diagnosis and Treatment.* 2nd ed. Philadelphia, WB Saunders, 1991.

Brent RL, Beckman DA: Angiotensin-converting enzyme inhibitors, an embryopathic class of drugs with unique properties: information for clinical teratology counselors. *Teratology* 43:543, 1991.

Brent RL, Holmes LB: Clinical and basic science from the thalidomide tragedy: what have we learned about the causes of limb defects? *Teratology* 38:241, 1988.

Buehler BA, Rao V, Finnell RH: Biochemical and molecular teratology of fetal hydantoin syndrome. *Ped Neuro Genet* 12:741, 1994.

Centers for Disease Control. Contribution of birth defects to infant mortality—United States, 1986. *MMWR Morb Mortal Wkly Rep* 38(37):633, 1989.

Colborn T, Dumanoski D, Myers JP: *Our Stolen Future.* New York, Dutton, 1996.

Cooper RL, Kavlock RJ: Endocrine disrupters and reproductive development: a weight of evidence overview [review]. *J Endocrinol* 152(2):159, 1997.

Cunningham FG, Gant NE, Leveno KJ, Gilstrap LC, Hauth JC, Wenstrom KD: Fetal abnormalities and acquired disorders. *In Williams Obstetrics.* 21st ed. New York, McGraw Hill, 2001.

Dansky LV, Finnell RH: Parental epilepsy, anticonvulsant drugs, and reproductive outcome: epidemiologic and experimental findings spanning three decades; 2: human studies. *Reprod Toxicol* 5:301, 1991.

Generoso WM, et al.: Mutagen induced fetal anomalies and death following treatment of females within hours after mating. *Mutat Res* 199:175, 1988.

Gorlin RJ, Cohen MM, Levin LS (eds): *Syndromes of the Head and Neck.* 3rd ed. New York, Oxford University Press, 1990.

Graham JM, Edwards MJ: Teratogen update: gestational effects of maternal hyperthermia due to febrile illnesses and resultant pattern of defects in humans. *Teratology* 58:209, 1998.

Gray LE, Ostby J: Effects of pesticides and toxic substances on behavioral and morphological reproductive development: endocrine versus nonendocrine mechanisms. *Toxicol Ind Health* 14:159, 1998.

Gregg NM: Congenital cataract following German measles in mothers. *Trans Ophthalmol Soc Aust* 3:35, 1941.

Hales BF, Robaire B: Paternally mediated effects on development. *In* Hood RD (ed): *Handbook of Developmental Toxicology.* New York, CRC Press, 1997.

Jones KL (ed): *Smith's Recognizable Patterns of Human Malformation.* 4th ed. Philadelphia, WB Saunders, 1988.

Jones KL, Smith DW, Ulleland CN, et al: Pattern of malformation in offspring of chronic alcoholic mothers. *Lancet* 1:1267, 1973.

Kaufman RH, Binder GS, Gray PM, Adam E: Upper genital tract changes associated with exposure in utero to diethylstilbestrol. *Am J Obstet Gynecol* 128:51, 1977.

Khatta KS, Moghtader GK, McMartin K, Berrera M, Kennedy D, Koren G: Pregnancy outcome following gestational exposure to organic solvents. *JAMA* 281:1106, 1999.

Lammer EJ, et al: Retinoic acid embryopathy. *N Engl J Med* 313:837, 1985.

Lenke RR, Levy HL: Maternal phenylketonuria and hyperphenylalaninemia: an international survey of untreated and treated pregnancies. *N Engl J Med* 303:1202, 1980.

Lenz W: A short history of thalidomide embryopathy. *Teratology* 38:203, 1988.

Lie RT, Wilcox AJ, Skjaerven R: Survival and reproduction among males with birth defects and risk of recurrence in their children. *JAMA* 285:755, 2001.

Manning FA: General principles and applications of ultrasonography. *In* Creasy RK, Resnik R (eds.): *Maternal Fetal Medicine.* 4th ed. philadelphia, WB Saunders, 1999.

McIntosh GC, Olshan AF, Baird PA: Paternal age and the risk of birth defects in offspring. *Epidemiology* 6:282, 1995.

Nash JE, Persaud TVN: Embryopathic risks of cigarette smoking. *Exp Pathol* 33:65, 1988.

Sadler TW, Denno KM, Hunter ES III: Effects of altered maternal metabolism during gastrulation and neurulation stages of development. *Ann N Y Acad Sci* 678:48, 1993.

Sampson PD, et al.: Incidence of fetal alcohol syndrome and prevalence of alcohol-related neurodevelopmental disorder. *Teratology* 56:317, 1997.

Schmidt RR, Johnson EM: Principles of teratology. *In* Hood RD (ed): *Handbook of Developmental Toxicology.* New York, CRC Press, 1997.

Scioscia AL: Prenatal genetic diagnosis. In Creasy RK, Resnik R (eds.): *Maternal Fetal Medicine.* 4th ed. Philadelphia, WB Saunders, 1999.

Shaw GM, Todoroff K, Finnell RH, Lammer EJ: Spina bifida phenotypes in infants of fetuses of obese mothers. *Teratology* 61:376,2000.

Shenefelt RE: Morphogenesis of malformations in hamsters caused by retinoic acid: relation to dose and stage of development. *Teratology* 5:103, 1972.

Shepard TH: *Catalog of Teratogenic Agents.* 7th ed. Baltimore, Johns Hopkins University Press, 1992.

Spirt BA, Fordon LP, Oliphant M: *Prenatal Ultrasound: A Color Atlas With Anatomic and Pathologic Correlation.* New York, Churchill Livingstone, 1987.

Stevenson RE, Hall JG, Goodman RM (eds): *Human Malformations and Related Anomalies,* vols 1 and 2. New York, Oxford University Press, 1993.

Wald N: Folic acid and prevention of neural tube defects. *Ann N Y Acad Sci* 678:112, 1993.

Weaver DD: Inborn errors of metabolism. *In* Weaver DD (ed): *Catalogue of Prenatally Diagnosed Conditions.* Baltimore, Johns Hopkins University Press, 1989.

Werler MM, Prober BR, Holmes LB: Smoking and pregnancy. *In* Sever JL, Brent RL (eds): *Teratogen Update: Environmentally Induced Birth Defect Risks.* New York, Alan R Liss, 1986.

Wilcox AJ, Weinberg CR, Baird DD: Timing of sexual intercourse in relation to ovulation. *N Engl J Med* 333:1517, 1995.

Wilson JG, Fraser FC: *Handbook of Teratology,* vols 1–3. New York, Plenum, 1977.

Woods KA, Camach-Hubner C, Savage MO, Clark AJL: Intrauterine growth retardation and postnatal growth failure associated with deletion of the insulin-like growth factor I gene. *N Engl J Med* 335:1363, 1996.

part two

Special
Embryology

Skeletal System

The skeletal system develops from **paraxial** and **lateral plate (somatic layer) mesoderm** and from **neural crest.** Paraxial mesoderm forms a segmented series of tissue blocks on each side of the neural tube, known as **somitomeres** in the head region and **somites** from the occipital region caudally. Somites differentiate into a ventromedial part, the **sclerotome,** and a dorsolateral part, the **dermomyotome.** At the end of the fourth week sclerotome cells become polymorphous and form a loosely woven tissue, the **mesenchyme,** or embryonic connective tissue (Fig. 8.1). It is characteristic for mesenchymal cells to migrate and to differentiate in many ways. They may become fibroblasts, chondroblasts, or **osteoblasts (bone-forming cells).**

The bone-forming capacity of mesenchyme is not restricted to cells of the sclerotome, but occurs also in the somatic mesoderm layer of the body wall, which contributes mesoderm cells for formation of the pelvic and shoulder girdles and the long bones of the limbs. Neural crest cells in the head region also differentiate into mesenchyme and participate in formation of bones of the face and skull. Occipital somites and somitomeres also contribute to formation of the cranial vault and base of the skull. In some bones, such as the flat bones of the skull, mesenchyme in the dermis differentiates directly into bone, a process known as **intramembranous ossification** (Fig. 8.2). In most bones, however, mesenchymal cells first give rise to **hyaline cartilage models,** which in turn become ossified by **endochondral ossification** (see Figs. 8.5 and

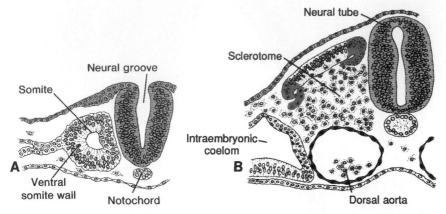

Figure 8.1 Development of the somite. **A.** Paraxial mesoderm cells are arranged around a small cavity. **B.** As a result of further differentiation, cells in the ventromedial wall lose their epithelial arrangement and become mesenchymal. Collectively they are called the sclerotome. Cells in the dorsolateral wall of the somite form limb and body wall musculature, while cells at the dorsomedial portion migrate beneath the remaining dorsal epithelium (the dermatome) to form the myotome.

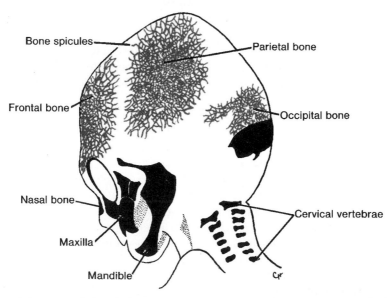

Figure 8.2 Bones of the skull of a 3-month-old fetus showing the spread of bone spicules from primary ossification centers in the flat bones of the skull.

8.13). The following paragraphs discuss development of the most important bony structures and some of their abnormalities.

Skull

The skull can be divided into two parts: the **neurocranium,** which forms a protective case around the brain, and the **viscerocranium,** which forms the skeleton of the face.

NEUROCRANIUM

The neurocranium is most conveniently divided into two portions: (*a*) the membranous part, consisting of **flat bones,** which surround the brain as a vault; and (*b*) the **cartilaginous part,** or **chondrocranium,** which forms bones of the base of the skull.

Membranous Neurocranium

The membranous portion of the skull is derived from neural crest cells and paraxial mesoderm as indicated in Figure 8.3. Mesenchyme from these two sources invests the brain and undergoes **membranous ossification.** The result is formation of a number of flat, membranous bones that are characterized by the presence of needle-like **bone spicules.** These spicules progressively radiate

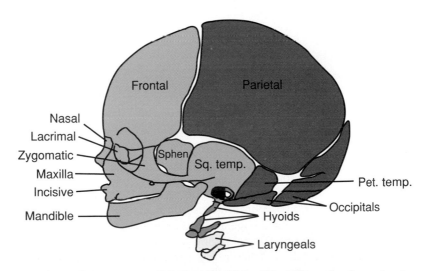

Figure 8.3 Skeletal structures of the head and face. Mesenchyme for these structures is derived from neural crest (*blue*), lateral plate mesoderm (*yellow*), and paraxial mesoderm (somites and somitomeres) (*red*).

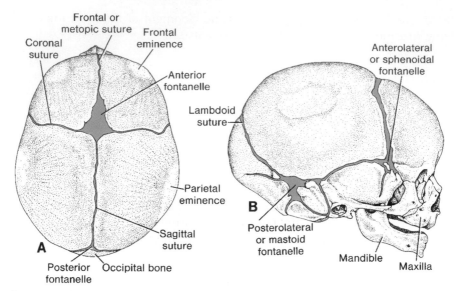

Figure 8.4 Skull of a newborn, seen from above (**A**) and the right side (**B**). Note the anterior and posterior fontanelles and sutures. The posterior fontanelle closes about 3 months after birth; the anterior fontanelle, about the middle of the second year. Many of the sutures disappear during adult life.

from primary ossification centers toward the periphery (Fig. 8.2). With further growth during fetal and postnatal life, membranous bones enlarge by apposition of new layers on the outer surface and by simultaneous osteoclastic resorption from the inside.

Newborn Skull

At birth the flat bones of the skull are separated from each other by narrow seams of connective tissue, the **sutures,** which are also derived from two sources: neural crest cells (sagittal suture) and paraxial mesoderm (coronal suture). At points where more than two bones meet, sutures are wide and are called **fontanelles** (Fig. 8.4). The most prominent of these is the **anterior fontanelle,** which is found where the two parietal and two frontal bones meet. Sutures and fontanelles allow the bones of the skull to overlap (**molding**) during birth. Soon after birth membranous bones move back to their original positions, and the skull appears large and round. In fact, the size of the vault is large compared with the small facial region (Fig. 8.4*B*).

Several sutures and fontanelles remain membranous for a considerable time after birth. The bones of the vault continue to grow after birth, mainly because the brain grows. Although a 5- to 7-year-old child has nearly all of its cranial capacity, some sutures remain open until adulthood. In the first few years after birth palpation of the anterior fontanelle may give valuable information

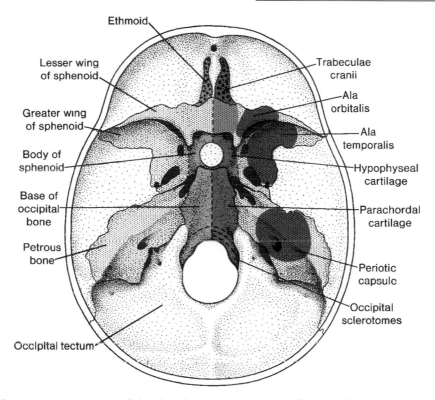

Figure 8.5 Dorsal view of the chondrocranium, or base of the skull, in the adult. On the right side are the various embryonic components participating in formation of the median part of the chondrocranium (*blue*) and components of the lateral part (*red*). On the left are the names of the adult structures. Bones that form rostral to the rostral half of the sella turcica arise from neural crest and constitute the prechordal (in front of the notochord) chondrocranium. Those forming posterior to this landmark arise from paraxial mesoderm (chordal chondrocranium).

as to whether ossification of the skull is proceeding normally and whether intracranial pressure is normal.

Cartilaginous Neurocranium or Chondrocranium

The cartilaginous neurocranium or chondrocranium of the skull initially consists of a number of separate cartilages (Fig. 8.5). Those that lie in front of the rostral limit of the notochord, which ends at the level of the pituitary gland in the center of the sella turcica, are derived from neural crest cells. They form the **prechordal chondrocranium.** Those that lie posterior to this limit arise from paraxial mesoderm and form the **chordal chondrocranium.** The base of the skull is formed when these cartilages fuse and ossify by endochondral ossification.

The base of the occipital bone is formed by the **parachordal cartilage** and the bodies of three **occipital sclerotomes** (Fig. 8.5). Rostral to the occipital base plate are the **hypophyseal cartilages** and **trabeculae cranii.** These cartilages soon fuse to form the body of the **sphenoid** and **ethmoid,** respectively. In this manner an elongated median plate of cartilage extending from the nasal region to the anterior border of the **foramen magnum** forms.

A number of other mesenchymal condensations arise on either side of the median plate. The most rostral, the **ala orbitalis,** forms the lesser wing of the sphenoid bone. Caudally it is followed by the **ala temporalis,** which gives rise to the greater wing of the sphenoid. A third component, the **periotic capsule,** gives rise to the petrous and mastoid parts of the temporal bone. These components later fuse with the median plate and with each other, except for openings through which cranial nerves leave the skull (Fig. 8.5).

VISCEROCRANIUM

The viscerocranium, which consists of the bones of the face, is formed mainly from the first two pharyngeal arches (see Chapter 15). The first arch gives rise to a dorsal portion, the **maxillary process,** which extends forward beneath the region of the eye and gives rise to the **maxilla, the zygomatic bone,** and **part of the temporal bone** (Fig. 8.6). The ventral portion, the **mandibular process,**

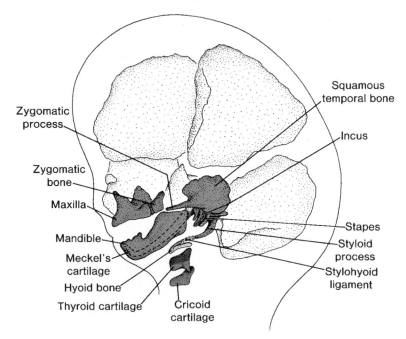

Figure 8.6 Lateral view of the head and neck region of an older fetus, showing derivatives of the arch cartilages participating in formation of bones of the face.

contains **Meckel's cartilage.** Mesenchyme around Meckel's cartilage condenses and ossifies by membranous ossification to give rise to the **mandible.** Meckel's cartilage disappears except in the **sphenomandibular** ligament. The dorsal tip of the mandibular process, along with that of the second pharyngeal arch, later gives rise to the **incus,** the **malleus,** and the **stapes** (Fig. 8.6). Ossification of the three ossicles begins in the fourth month, making these the first bones to become fully ossified. Mesenchyme for formation of the bones of the face is derived from neural crest cells, including the nasal and lacrimal bones (Fig. 8.3).

At first the face is small in comparison with the neurocranium. This appearance is caused by (*a*) virtual absence of the paranasal air sinuses and (*b*) the small size of the bones, particularly the jaws. With the appearance of teeth and development of the air sinuses, the face loses its babyish characteristics.

CLINICAL CORRELATES

Craniofacial Defects and Skeletal Dysplasias

Neural Crest Cells

Neural crest cells originating in the neuroectoderm form the facial skeleton and most of the skull. These cells also constitute a vulnerable population as they leave the neuroectoderm; they are often a target for teratogens. Therefore, it is not surprising that craniofacial abnormalities are common birth defects (see Chapter 15).

Cranioschisis

In some cases the cranial vault fails to form (**cranioschisis**), and brain tissue exposed to amniotic fluid degenerates, resulting in **anencephaly.** Cranioschisis is due to failure of the cranial neuropore to close. (Fig. 8.7*A*). Children with such severe skull and brain defects cannot survive. Children with relatively small defects in the skull through which meninges and/or brain tissue herniate (**cranial meningocele** and **meningoencephalocele,** respectively) (Fig. 8.7*B*) may be treated successfully. In such cases the extent of neurological deficits depends on the amount of damage to brain tissue.

Craniosynostosis and Dwarfism

Another important category of cranial abnormalities is caused by premature closure of one or more sutures. These abnormalities are collectively known as **craniosynostosis,** which occurs in 1 in 2500 births and is a feature of over 100 genetic syndromes. The shape of the skull depends on which of the sutures closed prematurely. Early closure of the sagittal suture (57% of cases) results in frontal and occipital expansion, and the skull becomes long and narrow (**scaphocephaly**) (Fig. 8.8*A*). Premature closure of the coronal suture results in a short, high skull, known as **acrocephaly,** or **tower skull** (Fig. 8.8*B*). If the coronal and lambdoid sutures close prematurely on one side only, asymmetric

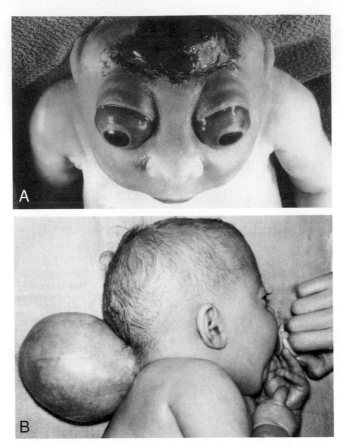

Figure 8.7 A. Child with anencephaly. Cranial neural folds fail to elevate and fuse, leaving the cranial neuropore open. The skull never forms, and brain tissue degenerates. **B.** Patient with meningocele. This rather common abnormality frequently can be successfully repaired.

craniosynostosis, known as **plagiocephaly,** results (Fig. 8.8*C*). Regulation of suture closure involves secretion of various isoforms of transforming growth factor β (TGFβ).

One of the exciting breakthroughs in molecular biology and genetics is the discovery of the role of the **fibroblast growth factors** (FGFs) and **fibroblast growth factor receptors** (FGFRs) in skeletal dysplasias. There are nine members of the FGF family and four receptors. Together they regulate cellular events, including proliferation, differentiation, and migration. Signaling is mediated by the receptors, which are **transmembrane tyrosine kinase** receptors, each of which has three extracellular immunoglobulin domains, a transmembrane segment, and a cytoplasmic tyrosine kinase domain. **FGFR-1** and **FGFR-2** are coexpressed in prebone and precartilage regions, including craniofacial

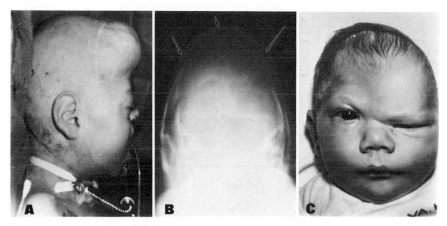

Figure 8.8 A. Child with scaphocephaly caused by early closure of the sagittal suture. Note the frontal and occipital bossing. **B.** Radiograph of a child with acrocephaly caused by early closure of the coronal suture. **C.** Child with plagiocephaly resulting from early closure of coronal and lambdoid sutures on one side of the skull (see sutures in Fig. 8.4).

structures; **FGFR-3** is expressed in the cartilage growth plates of long bones. In general, FGFR-2 increases proliferation; FGFR-1 promotes osteogenic differentiation; while the role of FGFR-3 is unclear, but expression is increased in the occipital region. Mutations in these receptors, which often involve only a single amino acid substitution, have been linked to specific types of **craniosynostosis (FGFR-1 and FGFR-2)** and several forms of **dwarfism (FGFR-3)** (Fig. 8.9; and Table 8.1, p. 181). In addition to these genes, mutations in the transcription factor *MSX2,* a regulator of parietal bone growth, causes Boston type craniosynostosis, which can affect a number of bones and sutures. The *TWIST* gene codes for a DNA binding protein and plays a role in regulating proliferation. Mutations in this gene result in proliferation and premature differentiation in the coronal suture causing craniosynostosis.

    Achondroplasia (ACH), the most common form of dwarfism (1/26,000 live births), primarily affects the long bones. Other skeletal defects include a large skull with a small midface, short fingers, and accentuated spinal curvature (Fig. 8.10). ACH is inherited as an autosomal dominant, and 80% of cases appear sporadically. **Thanatophoric dysplasia** is the most common neonatal lethal form of dwarfism (1/20,000 live births). There are two types; both are autosomal dominant. Type I is characterized by short, curved femurs with or without cloverleaf skull; type II individuals have straight, relatively long femurs and severe cloverleaf skull caused by craniosynostosis (Fig. 8.11). **Hypochondroplasia,** another autosomal dominant form of dwarfism, appears to be a milder type of ACH. In common to all of these forms of skeletal dysplasias are

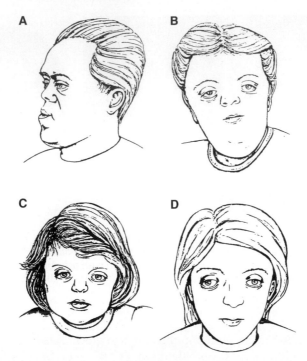

Figure 8.9 Faces of children with achondroplasia and different types of craniosynostosis. **A.** Achondroplasia. **B.** Apert syndrome. **C.** Pfeiffer syndrome. **D.** Crouzon syndrome. Underdevelopment of the midfacial area (**A**) is common to all individuals affected with these syndromes.

mutations in FGFR-3 causing abnormal endochondral bone formation so that growth of the long bones and base of the skull is adversely affected.

Acromegaly is caused by congenital hyperpituitarism and excessive production of growth hormone. It is characterized by disproportional enlargement of the face, hands, and feet. Sometimes, it causes more symmetrical excessive growth and gigantism.

Microcephaly

Microcephaly is usually an abnormality in which the brain fails to grow and the skull fails to expand. Many children with microcephaly are severely retarded.

Limbs

LIMB GROWTH AND DEVELOPMENT

At the end of the fourth week of development, limb buds become visible as outpocketings from the ventrolateral body wall (Fig. 8.12*A*). Initially they

TABLE 8.1 Genes Associated With Skeletal Defects

Gene	Chromosome	Abnormality	Phenotype
FGFR1	8p12	Pfeiffer syndrome	Craniosynostosis, broad great toes and thumbs, cloverleaf skull, underdeveloped face
FGFR2	10q26	Pfeiffer syndrome	Same
		Apert syndrome	Craniosynostosis, underdeveloped face, symmetric syndactyly of hands and feet
		Jackson-Weiss syndrome	Craniosynostosis, underdeveloped face, foot anomalies, hands usually spared
		Crouzon syndrome	Craniosynostosis, underdeveloped face, no foot or hand defects
FGFR3	4p16	Achondroplasia	Short-limb dwarfism, underdeveloped face
		Thanatophoric dysplasia (type I)	Curved short femurs, with or without cloverleaf skull
		Thanatophoric dysplasia (type II)	Relatively long femurs, severe cloverleaf skull
		Hypochondroplasia	Milder form of achondroplasia with normal craniofacial features
MSX2	5q35	Boston-type craniosynostosis	Craniosynostosis
TWIST	7p21	Saethre-Chotzen syndrome	Craniosynostosis, midfacial hypoplasia, cleft palate, vertebral anomalies, hand and foot abnormalities
HOXA13		Hand-foot-genital syndrome	Small, short digits, divided uterus, hypospadius
HOXD13	2q31	Synpolydactyly	Fused, multiple digits

consist of a mesenchymal core derived from the somatic layer of lateral plate mesoderm that will form the bones and connective tissues of the limb, covered by a layer of cuboidal ectoderm. Ectoderm at the distal border of the limb thickens and forms the **apical ectodermal ridge** (AER) (Fig. 8.13*A*). This ridge exerts an inductive influence on adjacent mesenchyme, causing it to remain as a population of undifferentiated, rapidly proliferating cells, the **progress zone**. As the limb grows, cells farther from the influence of the AER begin to differentiate into cartilage and muscle. In this manner development of the limb proceeds proximodistally.

In 6-week-old embryos the terminal portion of the limb buds becomes flattened to form the **handplates** and **footplates** and is separated from the

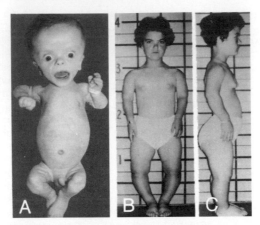

Figure 8.10 A. Three-month-old infant with achondroplasia. Note the large head, short extremities, and protruding abdomen. **B** and **C.** Achondroplasia in a 15-year-old girl. Note dwarfism of the short limb type, the limbs being disproportionately shorter than the trunk. The limbs are bowed; there is an increase in lumbar lordosis; and the face is small relative to the head.

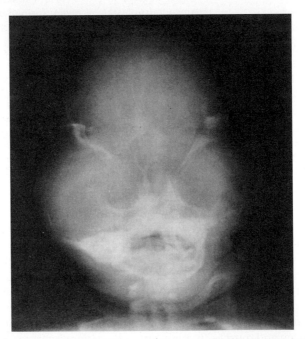

Figure 8.11 Radiograph of a patient with cloverleaf skull characteristic of thanatophoric dwarfism type II. The shape of the skull is due to abnormal growth of the cranial base, caused by a mutation in FGFR-3, followed by craniosynostosis. The sagittal, coronal, and lambdoid sutures are commonly involved.

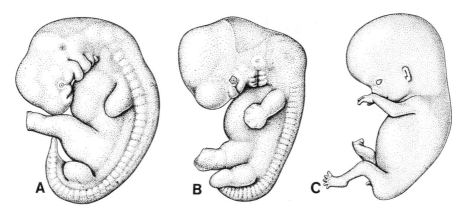

Figure 8.12 Development of the limb buds in human embryos. **A.** At 5 weeks. **B.** At 6 weeks. **C.** At 8 weeks. The hindlimb buds are less well developed than those of the forelimbs.

proximal segment by a circular constriction (Fig. 8.12*B*). Later a second constriction divides the proximal portion into two segments, and the main parts of the extremities can be recognized (Fig. 8.12*C*). Fingers and toes are formed when **cell death** in the AER separates this ridge into five parts (Fig. 8.14*A*). Further formation of the digits depends on their continued outgrowth under the influence of the five segments of ridge ectoderm, condensation of the mesenchyme to form cartilaginous digital rays, and the death of intervening tissue between the rays (Fig. 8.14, *B* and *C*).

Development of the upper and lower limbs is similar except that morphogenesis of the lower limb is approximately 1 to 2 days behind that of the upper limb. Also, during the seventh week of gestation the limbs rotate in opposite directions. The upper limb rotates 90° laterally, so that the extensor muscles lie on the lateral and posterior surface and the thumbs lie laterally, whereas the lower limb rotates approximately 90° medially, placing the extensor muscles on the anterior surface and the big toe medially.

While the external shape is being established, mesenchyme in the buds begins to condense and these cells differentiate into chondrocytes (Fig. 8.13). By the 6th week of development the first **hyaline cartilage models**, foreshadowing the bones of the extremities, are formed by these chondrocytes (Figs. 8.13 and 8.15). Joints are formed in the cartilaginous condensations when chondrogenesis is arrested and a joint **interzone** is induced. Cells in this region increase in number and density and then a joint cavity is formed by cell death. Surrounding cells differentiate into a joint capsule. Factors regulating the positioning of joints are not clear, but the secreted molecule WNT14 appears to be the inductive signal.

Ossification of the bones of the extremities, **endochondral ossification**, begins by the end of the embryonic period. Primary **ossification centers** are

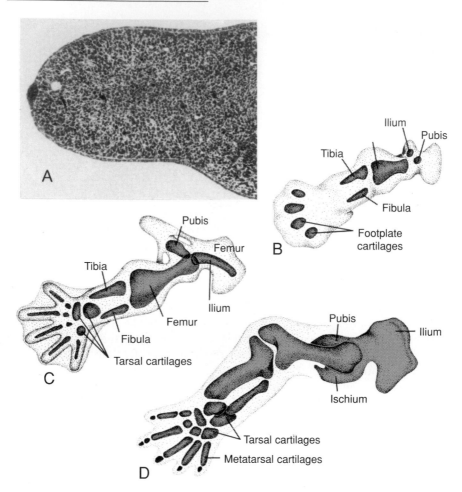

Figure 8.13 A. Longitudinal section through the limb bud of a mouse embryo, showing a core of mesenchyme covered by a layer of ectoderm that thickens at the distal border of the limb to form the AER. In humans this occurs during the fifth week of development. **B.** Lower extremity of an early 6-week embryo, illustrating the first hyaline cartilage models. **C** and **D.** Complete set of cartilage models at the end of the sixth and the beginning of the eighth week, respectively.

present in all long bones of the limbs by the 12th week of development. From the primary center in the shaft or **diaphysis** of the bone, endochondral ossification gradually progresses toward the ends of the cartilaginous model (Fig. 8.15).

   At birth the diaphysis of the bone is usually completely ossified, but the two ends, the **epiphyses**, are still cartilaginous. Shortly thereafter, however, ossification centers arise in the epiphyses. Temporarily a cartilage plate remains between the diaphyseal and epiphyseal ossification centers. This plate, the

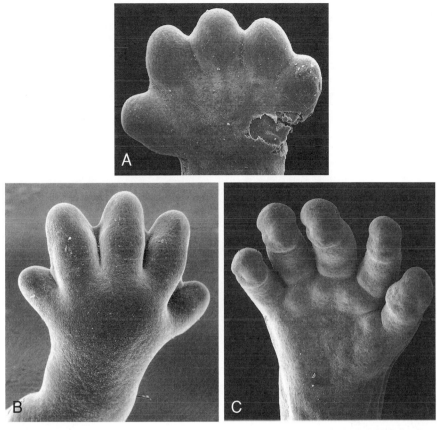

Figure 8.14 Scanning electron micrographs of human hands. **A.** At 48 days. Cell death in the apical ectodermal ridge creates a separate ridge for each digit. **B.** At 51 days. Cell death in the interdigital spaces produces separation of the digits. **C.** At 56 days. Digit separation is complete. The finger pads will create patterns for fingerprints.

epiphyseal plate, plays an important role in growth in the length of the bones. Endochondral ossification proceeds on both sides of the plate (Fig. 8.15). When the bone has acquired its full length, the epiphyseal plates disappear and the epiphyses unite with the shaft of the bone.

In long bones an epiphyseal plate is found on each extremity; in smaller bones, such as the phalanges, it is found only at one extremity; and in irregular bones, such as the vertebrae, one or more primary centers of ossification and usually several secondary centers are present.

MOLECULAR REGULATION OF LIMB DEVELOPMENT

Positioning of the limbs along the craniocaudal axis in the flank regions of the embryo is regulated by the **HOX** genes expressed along this axis. These

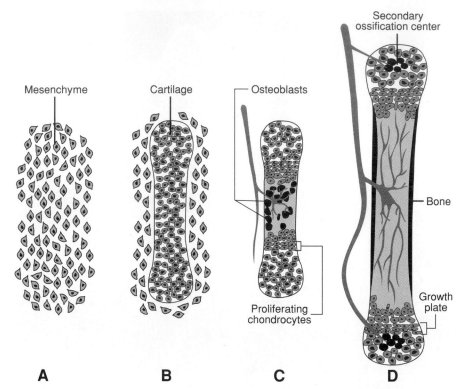

Figure 8.15 Endochondral bone formation. **A.** Mesenchyme cells begin to condense and differentiate into chondrocytes. **B.** Chondrocytes form a cartilaginous model of the prospective bone. **C** and **D.** Blood vessels invade the center of the cartilaginous model bringing oteoblasts (black cells) and restricting proliferating chondrocytic cells to the ends (epiphyses) of the bones. Chondrocytes toward the shaft side (diaphysis) undergo hypertrophy and apoptosis as they mineralize the surrounding matrix. Osteoblasts bind to the mineralized matrix and deposit bone matrices. Later, as blood vessels invade the epiphyses, secondary ossification centers form. Growth of the bones is maintained by proliferation of chondrocytes in the growth plates (**D**).

homeobox genes are expressed in overlapping patterns from head to tail (see Chapter 5), with some having more cranial limits than others. For example, the cranial limit of expression of *HOXB8* is at the cranial border of the forelimb, and misexpression of this gene alters the position of these limbs.

Once positioning along the craniocaudal axis is determined, growth must be regulated along the proximodistal, anteroposterior, and dorsoventral axes (Fig. 8.16). Limb outgrowth, which occurs first, is initiated by FGF-10 secreted by lateral plate mesoderm cells (Fig. 8.16*A*). Once outgrowth is initiated, *bone morphogenetic proteins (BMPs),* expressed in ventral ectoderm, induce formation of the AER by signaling through the homeobox gene *MSX2*. Expression of

Radical fringe (a homologue of *Drosophila fringe*), in the dorsal half of the limb ectoderm, restricts the location of the AER to the distal tip of the limbs. This gene induces expression of ***Ser-2,*** a homologue of *Drosophila serrate,* at the border between cells expressing *Radical fringe* and those that are not. It is at this border that the AER is established. Formation of the border itself is assisted by expression of ***Engrailed-1*** in ventral ectoderm cells, since this gene represses expression of *Radical fringe.* After the ridge is established, it expresses **FGF-4** and **FGF-8**, which maintain the **progress zone**, the rapidly proliferating population of mesenchyme cells adjacent to the ridge (Fig. 8.16*A*). Distal growth of the limb is then effected by these rapidly proliferating cells under the influence of the FGFs. As growth occurs, mesenchymal cells at the proximal end of the progress zone become farther away from the ridge and its influence and begin to slow their division rates and to differentiate.

Patterning of the anteroposterior axis of the limb is regulated by the **zone of polarizing activity (ZPA),** a cluster of cells at the posterior border of the limb near the flank (Fig. 8.16*B*). These cells produce **retinoic acid (vitamin A)**, which initiates expression of ***sonic hedgehog (SHH),*** a secreted factor that regulates the anteroposterior axis. Thus, for example, digits appear in the proper order, with the thumb on the radial (anterior) side. As the limb grows, the ZPA moves distalward to remain in proximity to the posterior border of the AER. Misexpression of retinoic acid or *SHH* in the anterior margin of a limb containing a normally expressing ZPA in the posterior border results in a mirror image duplication of limb structures (Fig. 8.17).

The dorsoventral axis is also regulated by BMPs in the ventral ectoderm which induce expression of the transcription factor *EN1*. In turn, *EN1* represses *WNT7a* expression restricting it to the dorsal limb ectoderm. *WNT7a* is secreted factor that induces expression of *LMX1,* a transcription factor containing a homeodomain, in the dorsal mesenchyme (Fig. 8.16*C*). *LMX1* specifies cells to be dorsal, establishing the dorsoventral components. In addition, *WNT7a* maintains *SHH* expression in the ZPA and therefore indirectly affects anteroposterior patterning as well. These two genes are also intimately linked in signaling pathways in *Drosophila*, and this interaction is conserved in vertebrates. In fact, all of the patterning genes in the limb have feedback loops. Thus, FGFs in the AER activate *SHH* in the ZPA, while *WNT7a* maintains the *SHH* signal.

Although patterning genes for the limb axes have been determined, it is the **HOX genes** that regulate the types and shapes of the bones of the limb (Fig. 8.15*D*). Thus, *HOX* gene expression, which results from the combinatorial expression of *SHH, FGFs,* and *WNT7a,* occurs in phases in three places in the limb that correspond to formation of the proximal (stylopod), middle (zeugopod), and distal (autopod) parts. Genes of the *HOXA* and *HOXD* clusters are the primary determinants in the limb, and variations in their combinatorial patterns of expression may account for differences in forelimb and hindlimb structures. Just as in the craniocaudal axis of the embryo, *HOX* genes are nested in overlapping patterns of expression that somehow regulate patterning

Proximodistal

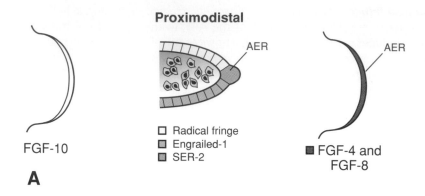

FGF-10

□ Radical fringe
▨ Engrailed-1
▨ SER-2

AER

▨ FGF-4 and
FGF-8

A

Anteroposterior

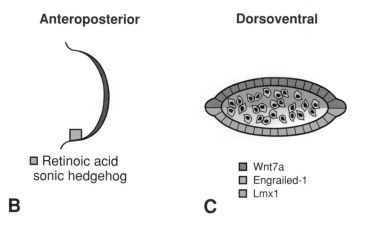

▨ Retinoic acid
sonic hedgehog

B

Dorsoventral

▨ Wnt7a
▨ Engrailed-1
▨ Lmx1

C

HOX Expression

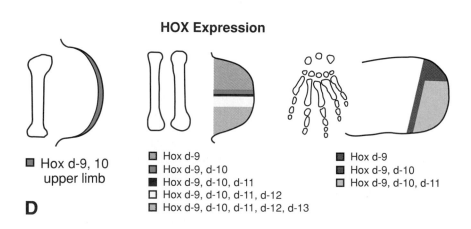

■ Hox d-9, 10
  upper limb

D

■ Hox d-9
■ Hox d-9, d-10
■ Hox d-9, d-10, d-11
□ Hox d-9, d-10, d-11, d-12
▨ Hox d-9, d-10, d-11, d-12, d-13

■ Hox d-9
■ Hox d-9, d-10
▨ Hox d-9, d-10, d-11

(Fig. 8.16*D*). Factors determining forelimb versus hindlimb are the transcription factors *TBX5* (forelimbs) and *TBX4* (hindlimbs).

CLINICAL CORRELATES

Bone Age

Radiologists use the appearance of various ossification centers to determine whether a child has reached his or her proper maturation age. Useful information about **bone age** is obtained from ossification studies in the hands and wrists of children. Prenatal analysis of fetal bones by ultrasonography provides information about fetal growth and gestational age.

Limb Defects

Limb malformations occur in approximately 6/10,000 live births, with 3.4/10,000 affecting the upper limb and 1.1/10,000, the lower. These defects are often associated with other birth defects involving the craniofacial, cardiac, and genitourinary systems. Abnormalities of the limbs vary greatly, and they may be represented by partial (**meromelia**) or complete absence (**amelia**) of one or more of the extremities. Sometimes the long bones are absent, and rudimentary hands and feet are attached to the trunk by small, irregularly shaped bones (**phocomelia,** a form of meromelia) (Fig. 8.18, *A* and *B*). Sometimes all segments of the extremities are present but abnormally short (**micromelia**).

Although these abnormalities are rare and mainly hereditary, cases of teratogen-induced limb defects have been documented. For example, many children with limb malformations were born between 1957 and 1962. Many mothers of these infants had taken **thalidomide,** a drug widely used as a sleeping pill and antinauseant. It was subsequently established that thalidomide causes a characteristic syndrome of malformations consisting of absence or gross deformities of the long bones, intestinal atresia, and cardiac anomalies.

Figure 8.16 (opposite page). Molecular regulation of patterning and growth in the limb. **A.** Limb outgrowth is initiated by *FGF-10* secreted by lateral plate mesoderm in the limb forming regions. Once outgrowth is initiated, the AER is induced by BMPs and restricted in its location by the gene *radical fringe* expressed in dorsal ectoderm. In turn, this expression induces that of *SER2* in cells destined to form the AER. After the ridge is established, it expresses FGF-4 and FGF-8 to maintain the progress zone, the rapidly proliferating mesenchyme cells adjacent to the ridge. **B.** Anteroposterior patterning of the limb is controlled by cells in the ZPA at the posterior border. These cells produce retinoic acid (vitamin A), which initiates expression of *sonic hedgehog*, regulating patterning. **C.** The dorsoventral limb axis is directed by *WNT7a*, which is expressed in the dorsal ectoderm. This gene induces expression of the transcription factor *LMX1* in the dorsal mesenchyme, specifying these cells as dorsal. **D.** Bone type and shape are regulated by *HOX* genes whose expression is determined by the combinatorial expression of *SHH, FGFs,* and *WNT7a. HOXA* and *HOXD* clusters are the primary determinants of bone morphology.

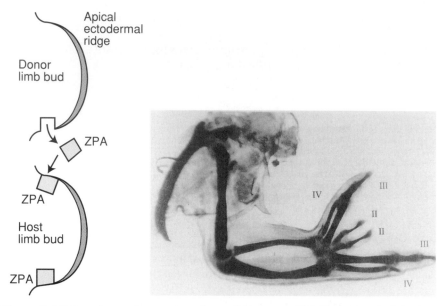

Figure 8.17 Experimental procedure for grafting a new ZPA from one limb bud into another using chick embryos. The result is the production of a limb with mirror image duplication of the digits (chicks have only three digits, numbered II, III, and IV), indicating the role of the ZPA in regulating anteroposterior patterning of the limb. Sonic hedgehog protein is the molecule secreted by the ZPA responsible for this regulation.

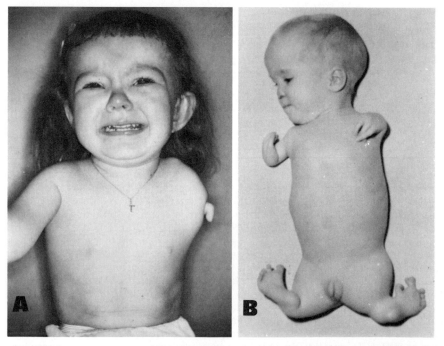

Figure 8.18 A. Child with unilateral amelia. **B.** Patient with a form of meromelia called phocomelia. The hands and feet are attached to the trunk by irregularly shaped bones.

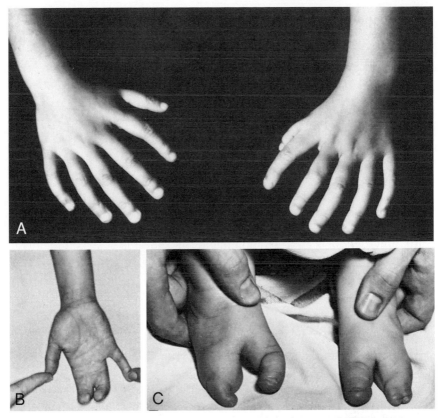

Figure 8.19 Digital defects. **A.** Polydactyly, extra digits. **B.** Syndactyly, fused digits. **C.** Cleft foot, lobster claw deformity.

Since the drug is now being used to treat AIDS and cancer patients, there is concern that its return will result in a new wave of limb defects. Studies indicate that the most sensitive period for teratogen-induced limb malformations is the **fourth and fifth weeks** of development.

A different category of limb abnormalities consists of extra fingers or toes (**polydactyly**) (Fig. 8.19*A*). The extra digits frequently lack proper muscle connections. Abnormalities with an excessive number of bones are mostly bilateral, while the absence of a digit such as a thumb (**ectrodactyly**) is usually unilateral. Polydactyly can be inherited as a dominant trait but may also be induced by teratogens. Abnormal fusion is usually restricted to the fingers or toes (**syndactyly**). Normally mesenchyme between prospective digits in the handplates and footplates breaks down. In 1/2000 births this fails to occur, and the result is fusion of one or more fingers and toes (Fig. 8.19*B*). In some cases the bones actually fuse.

Cleft hand and foot (lobster claw deformity) consists of an abnormal cleft between the second and fourth metacarpal bones and soft tissues. The third metacarpal and phalangeal bones are almost always absent, and the thumb and index finger and the fourth and fifth fingers may be fused (Fig. 8.19C). The two parts of the hand are somewhat opposed to each other and act like a lobster claw.

The role of the *HOX* genes in limb development is illustrated by two abnormal phenotypes produced by mutations in these genes: Mutations in *HOXA13* result in **hand-foot-genital syndrome,** characterized by fusion of the carpal bones and small short digits. Females often have a partially (bicornuate) or completely (didelphic) divided uterus and abnormal positioning of the urethral orifice. Males may have hypospadias. Mutations in *HOXD13* result in a combination of syndactyly and polydactyly (**synpolydactyly**).

Clubfoot usually accompanies syndactyly. The sole of the foot is turned inward, and the foot is adducted and plantar flexed. It is observed mainly in males and in some cases is hereditary. Abnormal positioning of the legs in utero may also cause clubfoot.

Congenital absence or deficiency of the radius is usually a genetic abnormality observed with malformations in other structures, such as **craniosynostosis–radial aplasia syndrome.** Associated digital defects, which may include absent thumbs and a short curved ulna, are usually present.

Amniotic bands may cause ring constrictions and amputations of the limbs or digits (Fig. 8.20). The origin of bands is not clear, but they may represent adhesions between the amnion and affected structures in the fetus. Other investigators believe that bands originate from tears in the amnion that detach and surround part of the fetus.

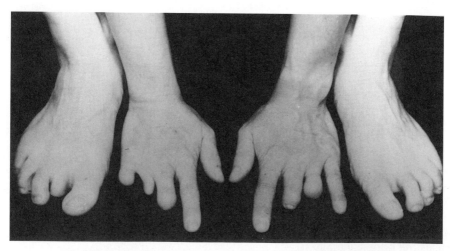

Figure 8.20 Digit amputations resulting from amniotic bands.

Congenital hip dislocation consists of underdevelopment of the acetabulum and head of the femur. It is rather common and occurs mostly in females. Although dislocation usually occurs after birth, the abnormality of the bones develops prenatally. Since many babies with congenital hip dislocation are breech deliveries, it has been thought that breech posture may interfere with development of the hip joint. It is frequently associated with laxity of the joint capsule.

Vertebral Column

During the fourth week of development, cells of the sclerotomes shift their position to surround both the spinal cord and the notochord (Fig. 8.1). This mesenchymal column retains traces of its segmental origin, as the sclerotomic blocks are separated by less dense areas containing **intersegmental arteries** (Fig. 8.21 A).

During further development the caudal portion of each sclerotome segment proliferates extensively and condenses (Fig. 8.21 B). This proliferation is so extensive that it proceeds into the subjacent intersegmental tissue and binds the caudal half of one sclerotome to the cephalic half of the subjacent sclerotome (*arrows* in Fig. 8.21, A and B). Hence, by incorporation of the intersegmental tissue into the **precartilaginous vertebral body** (Fig. 8.21 B), the body of the

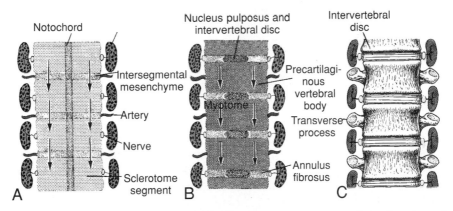

Figure 8.21 Formation of the vertebral column at various stages of development. **A.** At the fourth week of development, sclerotomic segments are separated by less dense intersegmental tissue. Note the position of the myotomes, intersegmental arteries, and segmental nerves. **B.** Condensation and proliferation of the caudal half of one sclerotome proceed into the intersegmental mesenchyme and cranial half of the subjacent sclerotome (*arrows* in **A** and **B**). Note the appearance of the intervertebral discs. **C.** Precartilaginous vertebral bodies are formed by the upper and lower halves of two successive sclerotomes and the intersegmental tissue. Myotomes bridge the intervertebral discs and, therefore, can move the vertebral column.

vertebra becomes intersegmental. Patterning of the shapes of the different vertebra is regulated by *HOX* genes.

Mesenchymal cells between cephalic and caudal parts of the original sclerotome segment do not proliferate but fill the space between two precartilaginous vertebral bodies. In this way they contribute to formation of the **intervertebral disc** (Fig. 8.21*B*). Although the notochord regresses entirely in the region of the vertebral bodies, it persists and enlarges in the region of the intervertebral disc. Here it contributes to the **nucleus pulposus,** which is later surrounded by circular fibers of the **annulus fibrosus.** Combined, these two structures form the **intervertebral disc** (Fig. 8.21*C*).

Rearrangement of sclerotomes into definitive vertebrae causes the myotomes to bridge the intervertebral discs, and this alteration gives them the capacity to move the spine (Fig. 8.21*C*). For the same reason, intersegmental arteries, at first lying between the sclerotomes, now pass midway over the vertebral bodies. Spinal nerves, however, come to lie near the intervertebral discs and leave the vertebral column through the intervertebral foramina.

CLINICAL CORRELATES

Vertebral Defects

The process of formation and rearrangement of segmental sclerotomes into definitive vertebrae is complicated, and it is fairly common to have two successive vertebrae fuse asymmetrically or have half a vertebra missing, a cause of **scoliosis (lateral curving of the spine).** Also, the number of vertebrae is frequently more or less than the norm. A typical example of these abnormalities is found in patients with **Klippel-Feil anomaly.** These patients have fewer than normal cervical vertebrae, and often other vertebrae are fused or abnormal in shape. This anomaly is usually associated with other abnormalities.

One of the most serious vertebral defects is the result of imperfect fusion or nonunion of the vertebral arches. Such an abnormality, known as **cleft vertebra (spina bifida)**, may involve only the bony vertebral arches, leaving the spinal cord intact. In these cases the bony defect is covered by skin, and no neurological deficits occur (**spina bifida occulta**). A more severe abnormality is **spina bifida cystica,** in which the neural tube fails to close, vertebral arches fail to form, and neural tissue is exposed. Any neurological deficits depend on the level and extent of the lesion. This defect, which occurs in 1/1000 births, may be prevented, in many cases, by providing mothers with folic acid prior to conception. Spina bifida can be detected prenatally by ultrasound (Fig. 8.22), and if neural tissue is exposed, amniocentesis can detect elevated levels of α-fetoprotein in the amniotic fluid. (For the various types of spina bifida, see Figs. 19.15 and 19.16.)

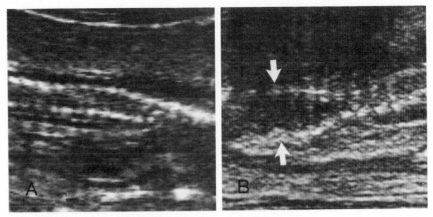

Figure 8.22 Ultrasound scans of the vertebral columns in a normal infant (**A**) and one with spina bifida (**B**) aged 4 months. The cleft vertebrae are readily apparent (*arrows*).

Ribs and Sternum

Ribs form from costal processes of thoracic vertebrae and thus are derived from the sclerotome portion of paraxial mesoderm. The sternum develops independently in somatic mesoderm in the ventral body wall. Two sternal bands are formed on either side of the midline, and these later fuse to form cartilaginous models of the manubrium, sternebrae, and xiphoid process.

Summary

The skeletal system develops from mesenchyme, which is derived from the mesodermal germ layer and from neural crest. Some bones, such as the flat bones of the skull, undergo **membranous ossification;** that is, mesenchyme cells are directly transformed into osteoblasts (Fig. 8.2). In most bones, such as the long bones of the limbs, mesenchyme condenses and forms hyaline cartilage models of bones (Fig. 8.15). Ossification centers appear in these cartilage models, and the bone gradually ossifies by **endochondral ossification.**

The **skull** consists of the **neurocranium** and **viscerocranium** (face). The neurocranium includes a **membranous portion,** which forms the cranial vault, and a cartilaginous portion (**chondrocranium**), which forms the base of the skull. Neural crest cells form the face, most of the cranial vault, and the prechordal part of the chondrocranium (the part that lies rostral to the notochord). Paraxial mesoderm forms the remainder of the skull.

Limbs form as buds along the body wall that appear in the fourth week. Lateral plate mesoderm forms the bones and connective tissue, while muscle cells migrate to the limbs from the somites. The **AER** regulates limb outgrowth, and the **ZPA** controls anteroposterior patterning. Many of the genes that regulate limb growth and patterning have been defined (see Fig. 8.16).

The **vertebral column** and **ribs** develop from the **sclerotome** compartments of the **somites,** and the **sternum is derived from mesoderm in the ventral body wall.** A definitive vertebra is formed by condensation of the caudal half of one sclerotome and fusion with the cranial half of the subjacent sclerotome (Fig. 8.21).

The many abnormalities of the skeletal system include vertebral (spina bifida), cranial (cranioschisis and craniosynostosis), and facial (cleft palate) defects. Major malformations of the limbs are rare, but defects of the radius and digits are often associated with other abnormalities (**syndromes**).

Problems to Solve

1. *Why are cranial sutures important? Are they involved in any abnormalities?*

2. *If you observe congenital absence of the radius or digital defects, such as absent thumb or polydactyly, would you consider examining the infant for other malformations? Why?*

3. *Explain the origin of scoliosis as a vertebral anomaly. What genes might be involved in this abnormality?*

SUGGESTED READING

Cohen MM, MacLean: *Craniosynostosis: Diagnosis, Evaluation and Management.* 2nd ed. New York, Oxford University Press, 2000.

Filly RA: Sonographic anatomy of the normal fetus. *In* Harrison MR, Golbus MS, Filly RA (eds): *The Unborn Patient. Prenatal Diagnosis and Treatment.* 2nd ed. Philadelphia, WB Saunders, 1991.

Filly RA, Golbus MS: Ultrasonography of the normal and pathologic fetal skeleton. *Radiol Clin North Am* 20:311, 1982.

Gorlin RJ: *Syndromes of the Head and Neck.* 2nd ed. New York, McGraw-Hill, 1976.

Hartman C, Tabin CJ: Wnt14 plays a pivotal role in inducing synovial joint formation in the developing appendicular skeleton. *Cell* 104:341, 2001.

Hehr U, Muenke M: Craniosynostosis syndromes: from genes to premature fusion of skull bones. *Mol Gen Metab* 68:139, 1999.

Jiang X, Iseki S, Maxson RE, Sucov HM, Morriss-Kay GM: Tissue origins and interactions in the mammalian skull vault. *Dev Biol* 241:106, 2002.

Laufer E, et al.: Expression of *Radical fringe* in limb bud ectoderm regulates apical ectodermal ridge formation. *Nature* 386:366, 1997.

Lenz W: Thalidomide and congenital abnormalities. *Lancet* 1:1219, 1962.

Mortlock D, Innis JW: Mutation of *HOXA13* in hand-foot-genital syndrome. *Nature Genet* 15:179, 1997.

Muenke M, Schell U: Fibroblast growth factor receptor mutations in human skeletal disorders. *Trends Genet* 11:308, 1995.

Muragaki Y, Mundlos S, Upton J, Olsen BR: Altered growth and branching patterns in synpolydactyly caused by mutations in *HOXD13*. *Science* 272:548, 1996.

Pizette S, Abate-Shen C, Niswander L: BMP controls proximodistal outgrowth, via induction of the apical ectodermal ridge, and dorsoventral patterning in the vertebrate limb. *Development* 128:4463, 2001.

Riddle RD, et al.: Induction of the LIM homeobox gene *Lmx1* by Wnt7a establishes dorsoventral pattern in the vertebrate limb. *Cell* 83:631, 1995.

Rodriguez-Estaban C, et al.: *Radical fringe* positions the apical ectodermal ridge at the dorsoventral boundary of the vertebrate limb. *Nature* 386:360, 1997.

Shubin N, Tabin C, Carroll S: Fossils, genes and the evolution of animal limbs. *Nature* 388:639, 1997.

chapter 9

Muscular System

With the exception of some smooth muscle tissue (see below), the muscular system develops from the mesodermal germ layer and consists of **skeletal, smooth,** and **cardiac muscle.** Skeletal muscle is derived from **paraxial mesoderm,** which forms somites from the occipital to the sacral regions and somitomeres in the head. Smooth muscle differentiates from **splanchnic mesoderm** surrounding the gut and its derivatives and from ectoderm (pupillary, mammary gland, and sweat gland muscles). Cardiac muscle is derived from **splanchnic mesoderm** surrounding the heart tube.

Striated Skeletal Musculature

Somites and **somitomeres** form the musculature of the axial skeleton, body wall, limbs, and head. From the occipital region caudally, somites form and differentiate into the sclerotome, dermatome, and two muscle-forming regions (Fig. 9.1). One of these is in the dorsolateral region of the somite. It expresses the muscle-specific gene *MYO-D* and migrates to provide progenitor cells for limb and body wall (hypomeric) musculature (Figs. 9.1 and 9.2). The other region lies dorsomedially, migrates ventral to cells that form the dermatome, and forms the **myotome.** This region, which expresses the muscle-specific gene *MYF5,* forms epimeric musculature (Figs. 9.1 and 9.2). During differentiation, precursor cells, the **myoblasts,** fuse and form long, multinucleated

199

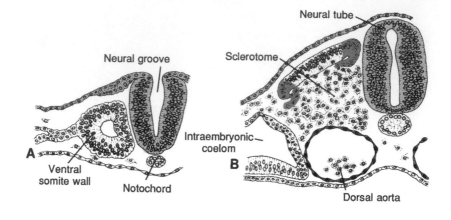

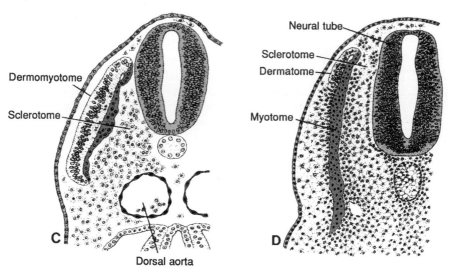

Figure 9.1 Stages in the development of a somite. **A.** Mesoderm cells are arranged around a small cavity. **B.** Cells of the ventral and medial walls of the somite lose their epithelial arrangement and migrate in the direction of the notochord. These cells collectively constitute the sclerotome. Cells at the dorsolateral portion of the somite migrate as precursors to limb and body wall musculature. Dorsomedial cells migrate beneath the remaining dorsal epithelium of the somite to form the myotome. **C.** Cells forming the myotome continue to extend beneath the dorsal epithelium. **D.** After ventral extension of the myotome, dermatome cells lose their epithelial configuration and spread out under the overlying ectoderm to form dermis.

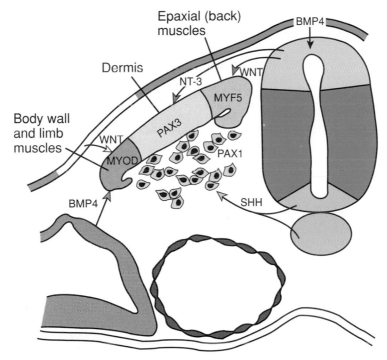

Figure 9.2 Expression patterns of genes that regulate somite differentiation. Sonic hedgehog (SHH), secreted by the notochord and floor plate of the neural tube, causes the ventral part of the somite to form sclerotome and to express *PAX1*, which in turn controls chondrogenesis and vertebral formation. WNT proteins from the dorsal neural tube activate *PAX3*, which demarcates the dermomyotome. WNT proteins also direct the dorsomedial portion of the somite to form epaxial (back) muscles and to express the muscle-specific gene *MYF5*. The middorsal portion of the somite is directed to become dermis by *neurotrophin 3* (*NT-3*) expressed by the dorsal neural tube. Hypaxial (limb and body wall) musculature is derived from the dorsolateral portion of the somite under the combined influence of activating WNT proteins and inhibitory BMP-4 protein, which together activate *MYO-D* expression.

muscle fibers. Myofibrils soon appear in the cytoplasm, and by the end of the third month, cross-striations typical of skeletal muscle appear. A similar process occurs in the seven somitomeres in the head region rostral to the occipital somites. Somitomeres remain loosely organized structures, however, never segregating into sclerotome and dermomyotome segments.

Molecular Regulation of Muscle Development

Genes regulating muscle development have recently been identified. **BMP4** and probably **FGFs** from lateral plate mesoderm, together with **WNT proteins** from adjacent ectoderm, signal the dorsolateral cells of the somite to express the muscle-specific gene **MYO-D.** BMP4 secreted by overlying ectoderm induces production of WNT proteins by the dorsal neural tube, and these proteins cause dorsomedial cells of the somite to activate **MYF5,** another muscle-specific gene (Fig. 9.2). Both of these genes are members of the *MYO-D* muscle-specific family, which also includes the *myogenin* and *MRF4* genes. MYO-D and MYF5 proteins activate the genes for myogenin and MRF5, which in turn promote formation of **myotubes** and **myofibers.** All *MYO-D* family members have DNA binding sites and act as transcription factors to regulate downstream genes in the muscle differentiation pathway.

Patterning of Muscles

Patterns of muscle formation are controlled by connective tissue into which myoblasts migrate. In the head region these connective tissues are derived from neural crest cells; in cervical and occipital regions they differentiate from somitic mesoderm; and in the body wall and limbs they originate from somatic mesoderm.

Derivatives of Precursor Muscle Cells

By the end of the fifth week prospective muscle cells are collected into two parts: a small dorsal portion, the **epimere,** formed from the dorsomedial cells of the somite that reorganized as myotomes; and a larger ventral part, the **hypomere,** formed by migration of dorsolateral cells of the somite (Figs. 9.1*B* and 9.3*A*). Nerves innervating segmental muscles are also divided into a **dorsal primary ramus** for the epimere and a **ventral primary ramus** for the hypomere (Fig. 9.3*B*) and these nerves will remain with their original muscle segment throughout its migration.

Myoblasts of the epimeres form the **extensor muscles** of the vertebral column, and those of the hypomeres give rise to muscles of the limbs and body wall (Fig. 9.3*B*). Myoblasts from cervical hypomeres form the **scalene, geniohyoid,** and **prevertebral muscles.** Those from thoracic segments split into three layers, which in the thorax are represented by the **external intercostal, internal intercostal,** and **innermost intercostal** or **transversus thoracis muscle** (Fig. 9.3*B*). In the abdominal wall these three muscle layers consist of the **external oblique,** the **internal oblique,** and the **transversus abdominis muscles.** The ribs cause the muscles in the wall of the thorax to maintain their segmental character, whereas muscles in the various segments of the abdominal wall fuse to form large sheets of muscle tissue. Myoblasts from the hypoblast of lumbar segments form the **quadratus lumborum muscle,** and those from sacral and coccygeal regions form the **pelvic diaphragm** and **striated muscles of the anus.**

In addition to the three ventrolateral muscle layers, a ventral longitudinal column arises at the ventral tip of the hypomeres (Fig. 9.3*B*). This column

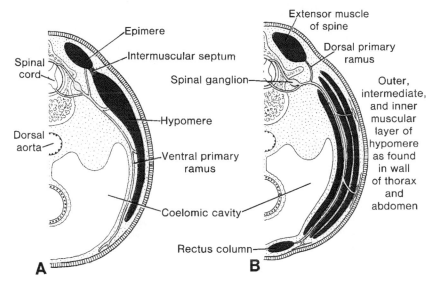

Figure 9.3 A. Transverse section through the thoracic region of a 5-week embryo. The dorsal portion of the body wall musculature (epimere) and the ventral portion (hypomere) are innervated by a dorsal primary ramus and a ventral primary ramus, respectively. **B.** Similar to **A** later in development. The hypomere has formed three muscle layers and a ventral longitudinal muscle column.

is represented by the **rectus abdominis muscle** in the abdominal region and by the **infrahyoid musculature** in the cervical region. In the thorax, the longitudinal muscle normally disappears, but is occasionally represented by the **sternalis muscle.**

Head Musculature

All voluntary muscles of the head region are derived from paraxial mesoderm (somitomeres and somites), including musculature of the tongue, eye (except that of the iris, which is derived from optic cup ectoderm), and that associated with the pharyngeal (visceral) arches (Table 9.1). Patterns of muscle formation in the head are directed by connective tissue elements derived from neural crest cells.

Limb Musculature

The first indication of limb musculature is observed in the seventh week of development as a condensation of mesenchyme near the base of the limb buds (Fig. 9.4*A*). The mesenchyme is derived from dorsolateral cells of the

TABLE 9.1 Origins of the Craniofacial Muscles

Mesodermal Origin	Muscles	Innervation
Somitomeres 1,2	Superior, medial, ventral recti	Oculomotor (III)
Somitomere 3	Superior oblique	Trochlear (IV)
Somitomere 4	Jaw closing	Trigeminal (V)
Somitomere 5	Lateral rectus	Abducens (VI)
Somitomere 6	Jaw opening, other 2nd arch	Facial (VII)
Somitomere 7	Stylopharyngeus	Glossopharyngeal (IX)
Somites 1,2	Intrinsic laryngeals	Vagus (X)
Somites 2–5[a]	Tongue	Hypoglossal (XII)

[a]Somites 2–5 constitute the occipital group (somite 1 degenerates for the most part).

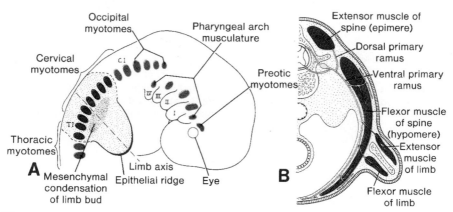

Figure 9.4 A. Myotomes in the head, neck, and thoracic region of a 7-week embryo. Note the location of the preotic and occipital myotomes and condensation of mesenchyme at the base of the limb bud, **B.** Transverse section through the region of attachment of the limb bud. Note the dorsal (extensor) and ventral (flexor) muscular components of the limb.

somites that migrate into the limb bud to form the muscles. As in other regions, connective tissue dictates the pattern of muscle formation, and this tissue is derived from somatic mesoderm, which also gives rise to the bones of the limb.

With elongation of the limb buds, the muscle tissue splits into flexor and extensor components (Fig. 9.4B). Although muscles of the limbs are segmental initially, with time they fuse and are then composed of tissue derived from several segments.

The upper limb buds lie opposite the lower five cervical and upper two thoracic segments (Fig. 9.5, *A* and *B*), and the lower limb buds lie opposite the lower four lumbar and upper two sacral segments (Fig. 9.5C). As soon as the buds form, ventral primary rami from the appropriate spinal nerves penetrate into the mesenchyme (Fig. 9.6). At first each ventral ramus enters with isolated dorsal and ventral branches, but soon these branches unite to form large dorsal

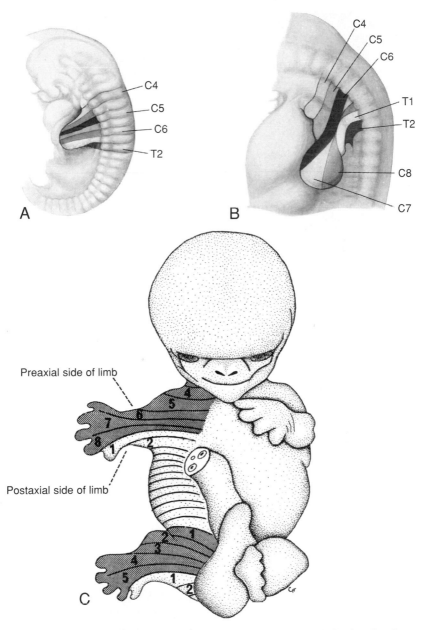

Figure 9.5 Limb buds with their segments of origin indicated. With further development the segmental pattern disappears; however, an orderly sequence in the dermatome pattern can still be recognized in the adult. **A.** Upper limb bud at 5 weeks. **B.** Upper limb bud at 6 weeks. **C.** Limb buds at 7 weeks.

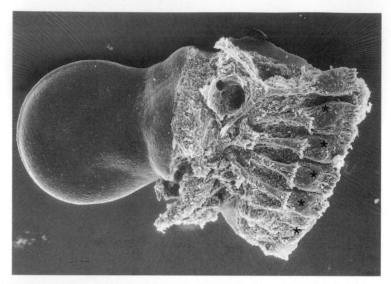

Figure 9.6 Scanning electron micrograph of a mouse upper limb bud, showing spinal nerves entering the limb. *Asterisks,* spinal ganglia.

and ventral nerves. Thus the **radial nerve,** which supplies the extensor musculature, is formed by a combination of the dorsal segmental branches, whereas the **ulnar** and **median nerves,** which supply the flexor musculature, are formed by a combination of the ventral branches. Immediately after the nerves have entered the limb buds, they establish an intimate contact with the differentiating mesodermal condensations, and the early contact between the nerve and muscle cells is a prerequisite for their complete functional differentiation.

Spinal nerves not only play an important role in differentiation and motor innervation of the limb musculature, but also provide sensory innervation for the dermatomes. Although the original dermatomal pattern changes with growth of the extremities, an orderly sequence can still be recognized in the adult (Fig. 9.5).

Clinical Correlates

Partial or complete absence of one or more muscles is rather common. One of the best known examples is total or partial absence of the pectoralis major muscle (**Poland anomaly**). Similarly, the palmaris longus, the serratus anterior, and the quadratus femoris muscles may be partially or entirely absent.

Partial or complete absence of abdominal musculature results in prune-belly syndrome (Fig. 9.7). Usually the abdominal wall is so thin that organs are visible and easily palpated. This defect is often associated with malformations of the urinary tract and bladder.

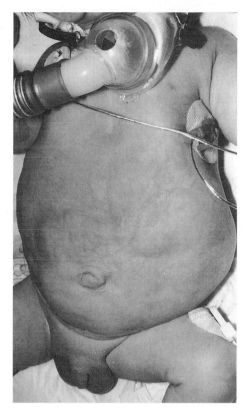

Figure 9.7 Prune belly syndrome: a distended abdomen from aplasia of abdominal wall musculature.

Cardiac Muscle

Cardiac muscle develops from splanchnic mesoderm surrounding the endothelial heart tube. Myoblasts adhere to one another by special attachments that later develop into **intercalated discs.** Myofibrils develop as in skeletal muscle, but myoblasts do not fuse. During later development, a few special bundles of muscle cells with irregularly distributed myofibrils become visible. These bundles, the **Purkinje fibers,** form the conducting system of the heart.

Smooth Muscle

Smooth muscle in the wall of the gut and gut derivatives is derived from splanchnic mesoderm surrounding the endoderm of these structures. Vascular smooth muscle differentiates from mesoderm adjacent to vascular endothelium. Sphincter and dilator muscles of the pupil and muscle tissue in the mammary gland and sweat glands originate from ectoderm.

Summary

Most muscles arise from the **mesoderm.** Skeletal muscles are derived from paraxial mesoderm, including (*a*) somites, which give rise to muscles of the axial skeleton, body wall, and limbs, and (*b*) somitomeres, which give rise to muscles of the head. Progenitor cells for muscle tissues are derived from the dorsolateral and dorsomedial portions of the somites. Cells in the dorsolateral portion express **MYO-D** and migrate to form hypomeric muscle; cells in the dorsomedial portion express **MYF5,** migrate ventral to the dermatome to form the **myotome,** and ultimately form epimeric musculature. By the fifth week muscle precursor cells are divided into a small dorsal portion, the **epimere,** innervated by a **dorsal primary ramus,** and a larger ventral portion, the **hypomere,** innervated by a **ventral primary ramus.** Myoblasts from epimeres form extensor muscles of the vertebral column, while those of the hypomere form limb and body wall musculature. **Connective tissue** derived from somites, somatic mesoderm, and neural crest (head region) provide a template for establishment of muscle patterns. **Most smooth muscles** and **cardiac muscle fibers** are derived from **splanchnic mesoderm.** Smooth muscles of the pupil, mammary gland, and sweat glands differentiate from ectoderm.

Problems to Solve

1. *Muscle cells are derived from what two regions of the somite? Which region forms the epimere and which the hypomere? What muscles form from each of these regions?*

2. *In examining a newborn female, you note that her right nipple is lower than the left and that the right anterior axillary fold is nearly absent. What is your diagnosis?*

3. *Patterning of muscles is dependent on what type of tissue?*

4. *How do you explain the fact that the phrenic nerve, which originates from cervical segments 3, 4, and 5, innervates the diaphragm in the thoracic region?*

SUGGESTED READING

Blagden CS, Hughes SM: Extrinsic influences on limb muscle organization. *Cell Tiss Res* 296:141, 1999.

Brand-Seberi B, Christ B: Genetic and epigenetic control of muscle development in vertebrates. *Cell Tiss Res* 296:199, 1999

Braun T, Arnold HH: *Myf5* and *MyoD* genes are activated in distinct mesenchymal stem cells and determine different skeletal muscle lineages. *EMBO J* 15:310, 1996.

Chevallier A, Kieny M, Mauger A: Limb-somite relationship: origin of the limb musculature. *J Embryol Exp Morphol* 41:245, 1977.

Christ B, Jacob M, Jacob HJ: On the origin and development of the ventrolateral abdominal muscles in the avian embryo. *Anat Embryol* 166:87, 1983.

Cossu G, et al.: Activation of different myogenic pathways: *myf5* is induced by the neural tube and *MyoD* by the dorsal ectoderm in mouse paraxial mesoderm. *Development* 122:429, 1996.

Levi AC, Borghi F, Garavoglia M: Development of the anal canal muscles. *Dis Colon Rectum* 34:262, 1991.

Noden DM: The embryonic origins of avian cephalic and cervical muscles and associated connective tissues. *Am J Anat* 168:257, 1983.

Noden DM: Craniofacial development: new views on old problems. *Anat Rec* 208:113, 1984.

Noden DM: Interactions and fates of avian craniofacial mesenchyme. *Development* 103:121, 1988.

chapter 10

Body Cavities

Formation of the Intraembryonic Cavity

At the end of the third week, intraembryonic mesoderm on each side of the midline differentiates into a paraxial portion, an intermediate portion, and a lateral plate (Fig. 10.1A). When intercellular clefts appear in the lateral mesoderm, the plates are divided into two layers: the **somatic mesoderm layer** and the **splanchnic mesoderm layer.** The latter is continuous with mesoderm of the wall of the yolk sac (Fig. 10.1B). The space bordered by these layers forms the **intraembryonic cavity (body cavity).**

At first the right and left sides of the intraembryonic cavity are in open connection with the extraembryonic cavity, but when the body of the embryo folds cephalocaudally and laterally, this connection is lost (Figs. 10.2, A–E). In this manner a large intraembryonic cavity extending from the thoracic to the pelvic region forms.

CLINICAL CORRELATES

Body Wall Defects

Ventral body wall defects in the thorax or abdomen may involve the heart, abdominal viscera, and urogenital organs. They may be due to a failure of body folding, in which case one or more of the four folds

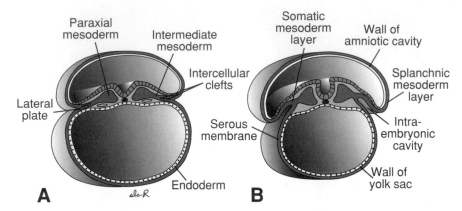

Figure 10.1 A. Transverse section through an embryo of approximately 19 days. Intercellular clefts are visible in the lateral plate mesoderm. **B.** Section through an embryo of approximately 20 days. The lateral plate is divided into somatic and splanchnic mesoderm layers that line the intraembryonic coelom. Tissue bordering the intraembryonic coelom differentiates into serous membranes.

(cephalic, caudal, and two lateral) responsible for closing the ventral body wall at the umbilicus fail to progress to that region. Another cause of these defects is incomplete development of body wall structures, including muscle, bone, and skin.

Cleft sternum is a ventral body wall defect that results from lack of fusion of the bilateral bars of mesoderm responsible for formation of this structure. In some cases the heart protrudes through a sternal defect (either cleft sternum or absence of the lower third of the sternum) and lies outside the body (**ectopia cordis**) (Fig. 10.3*A*). Sometimes the defect involves both the thorax and abdomen, creating a spectrum of abnormalities known as **Cantrell pentalogy**, which includes cleft sternum, ectopia cordis, omphalocele, diaphragmatic hernia (anterior portion), and congenital heart defects (ventricular septal defect, tetralogy of Fallot). Ectopia cordis defects appear to be due to a failure of progression of cephalic and lateral folds.

Omphalocele (Fig. 10.3*B*) is herniation of abdominal viscera through an enlarged umbilical ring. The viscera, which may include liver, small and large intestines, stomach, spleen, or bladder, are covered by amnion. The origin of omphalocele is a failure of the bowel to return to the body cavity from its physiological herniation during the 6th to 10th weeks. Omphalocele, which occurs in 2.5/10,000 births, is associated with a high rate of mortality (25%) and severe malformations, such as cardiac anomalies (50%) and neural tube defects (40%). Chromosomal abnormalities are present in approximately 50% of liveborn infants with omphalocele.

Gastroschisis (Fig. 10.3*C*) is a herniation of abdominal contents through the body wall directly into the amniotic cavity. It occurs lateral to the umbilicus,

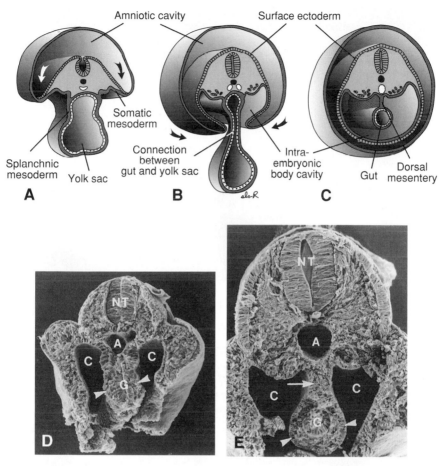

Figure 10.2 Transverse sections through embryos at various stages of development. **A.** The intraembryonic cavity is in open communication with the extraembryonic cavity. **B.** The intraembryonic cavity is about to lose contact with the extraembryonic cavity. **C.** At the end of the fourth week, splanchnic mesoderm layers are continuous with somatic layers as a double-layered membrane, the dorsal mesentery. Dorsal mesentery extends from the caudal limit of the foregut to the end of the hindgut. **D** and **E.** Scanning electron micrographs of sections through mouse embryos showing details similar to those in **B** and **C**, respectively. *G*, Gut tube; *arrowheads*, splanchnic mesoderm; *C*, body cavity; *arrow*, dorsal mesentery; *A*, dorsal aorta; *NT*, neural tube.

usually on the right, through a region weakened by regression of the right umbilical vein, which normally disappears. Viscera are not covered by peritoneum or amnion, and the bowel may be damaged by exposure to amniotic fluid. Both omphalocele and gastroschisis result in elevated levels of α-fetoprotein in the amniotic fluid, which can be detected prenatally.

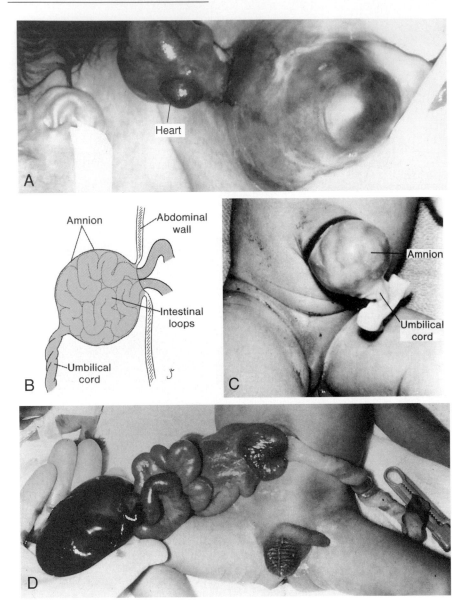

Figure 10.3 Ventral body wall defects. **A.** Infant with ectopia cordis. Mesoderm of the sternum has failed to fuse, and the heart lies outside of the body. **B.** Omphalocele with failure of the intestinal loops to return to the body cavity following physiological herniation. The herniated loops are covered by amnion. **C.** Omphalocele in a newborn. **D.** A newborn with gastroschisis. Loops of bowel return to the body cavity but herniate again through the body wall, usually to the right of the umbilicus in the region of the regressing right umbilical vein. Unlike omphalocele, the defect is not covered by amnion.

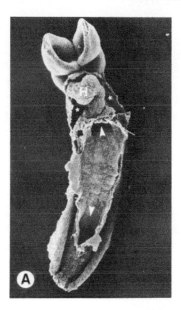

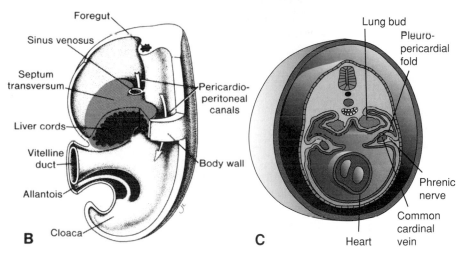

Figure 10.4 A. Scanning electron micrograph showing the ventral view of a mouse embryo (equivalent to approximately the fourth week in human development). The gut tube is closing, the anterior and posterior intestinal portals are visible (*arrowheads*), and the heart (*H*) lies in the primitive pleuropericardial cavity (*asterisks*), which is partially separated from the abdominal cavity by the septum transversum (*arrow*). **B.** Portion of an embryo at approximately 5 weeks with parts of the body wall and septum transversum removed to show the pericardioperitoneal canals. Note the size and thickness of the septum transversum and liver cords penetrating the septum. **C.** Growth of the lung buds into the pericardioperitoneal canals. Note the pleuropericardial folds.

Gastroschisis occurs in 1/10,000 births but is increasing in frequency, especially among young women, and this increase may be related to cocaine use. Unlike omphalocele, gastroschisis is not associated with chromosome abnormalities or other severe defects. Therefore the survival rate is excellent, although volvulus (rotation of the bowel) resulting in a compromised blood supply may kill large regions of the intestine and lead to fetal death.

Serous Membranes

Cells of the somatic mesoderm lining the intraembryonic cavity become mesothelial and form the **parietal layer of the serous membranes** lining the outside of the peritoneal, pleural, and pericardial cavities. In a similar manner, cells of the splanchnic mesoderm layer form the **visceral layer of the serous membranes** covering the abdominal organs, lungs, and heart (Fig. 10.1). Visceral and parietal layers are continuous with each other as the **dorsal mesentery** (Fig. 10.2, *C* and *E*), which suspends the gut tube in the peritoneal cavity. Initially this dorsal mesentery is a thick band of mesoderm running continuously from the caudal limit of the foregut to the end of the hindgut. **Ventral mesentery** exists only from the caudal foregut to the upper portion of the duodenum and results from thinning of mesoderm of the **septum transversum** (see Chapter 13). These mesenteries are double layers of peritoneum that provide a pathway for blood vessels, nerves, and lymphatics to the organs.

Diaphragm and Thoracic Cavity

The **septum transversum** is a thick plate of mesodermal tissue occupying the space between the thoracic cavity and the stalk of the yolk sac (Fig. 10.4, *A* and *B*). This septum does not separate the thoracic and abdominal cavities completely but leaves large openings, the **pericardioperitoneal canals,** on each side of the foregut (Fig. 10.4*B*).

When lung buds begin to grow, they expand caudolaterally within the pericardioperitoneal canals (Fig. 10.4*C*). As a result of the rapid growth of the lungs, the pericardioperitoneal canals become too small, and the lungs begin to expand into the mesenchyme of the body wall dorsally, laterally, and ventrally (Fig. 10.4*C*). Ventral and lateral expansion is posterior to the **pleuropericardial folds.** At first these folds appear as small ridges projecting into the primitive undivided thoracic cavity (Fig. 10.4*C*). With expansion of the lungs, mesoderm of the body wall splits into two components (Fig. 10.5): (*a*) the definitive wall of the thorax and (*b*) the **pleuropericardial membranes,** which are extensions of the pleuropericardial folds that contain the **common cardinal veins** and **phrenic nerves.** Subsequently, descent of the heart and positional changes of

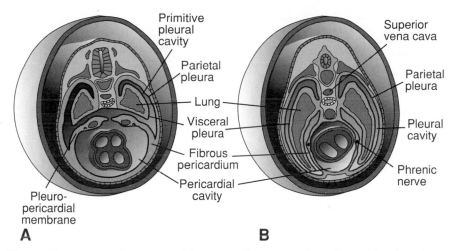

Primitive pleural cavity
Parietal pleura
Lung
Visceral pleura
Fibrous pericardium
Pericardial cavity
Pleuro-pericardial membrane

A

Superior vena cava
Parietal pleura
Pleural cavity
Phrenic nerve

B

Figure 10.5 A. Transformation of the pericardioperitoneal canals into the pleural cavities and formation of the pleuropericardial membranes. Note the pleuropericardial folds containing the common cardinal vein and phrenic nerve. Mesenchyme of the body wall splits into the pleuropericardial membranes and definitive body wall. **B.** The thorax after fusion of the pleuropericardial folds with each other and with the root of the lungs. Note the position of the phrenic nerve, now in the fibrous pericardium. The right common cardinal vein has developed into the superior vena cava.

the sinus venosus shift the common cardinal veins toward the midline, and the pleuropericardial membranes are drawn out in mesentery-like fashion (Fig. 10.5*A*). Finally, they fuse with each other and with the root of the lungs, and the thoracic cavity is divided into the definitive **pericardial cavity** and two **pleural cavities** (Fig. 10.5*B*). In the adult, the pleuropericardial membranes form the **fibrous pericardium.**

Formation of the Diaphragm

Although the pleural cavities are separate from the pericardial cavity, they remain in open communication with the abdominal (peritoneal) cavity, since the diaphragm is incomplete. During further development, the opening between the prospective pleural and peritoneal cavities is closed by crescent-shaped folds, the **pleuroperitoneal folds,** which project into the caudal end of the pericardioperitoneal canals (Fig. 10.6*A*). Gradually the folds extend medially and ventrally so that by the seventh week they fuse with the mesentery of the esophagus and with the septum transversum (Fig. 10.6*B*). Hence the connection between the pleural and peritoneal portions of the body cavity is closed by the pleuroperitoneal membranes. Further expansion of the pleural cavities relative to mesenchyme of the body wall adds a peripheral rim to the pleuroperitoneal membranes (Fig. 10.6*C*). Once this rim is established, myoblasts

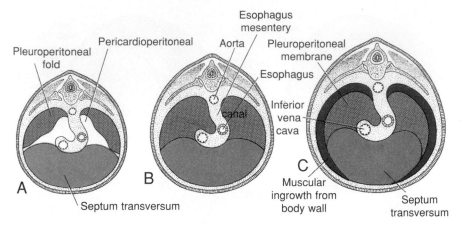

Figure 10.6 Development of the diaphragm. **A.** Pleuroperitoneal folds appear at the beginning of the fifth week. **B.** Pleuroperitoneal folds fuse with the septum transversum and mesentery of the esophagus in the seventh week, separating the thoracic cavity from the abdominal cavity. **C.** Transverse section at the fourth month of development. An additional rim derived from the body wall forms the most peripheral part of the diaphragm.

originating in the body wall penetrate the membranes to form the muscular part of the diaphragm.

Thus the diaphragm is derived from the following structures: (*a*) the septum transversum, which forms the central tendon of the diaphragm; (*b*) the two pleuroperitoneal membranes; (*c*) muscular components from the lateral and dorsal body walls; and (*d*) the mesentery of the esophagus, in which the **crura** of the diaphragm develop (Fig. 10.6*C*).

Initially the septum transversum lies opposite cervical somites, and nerve components of the **third, fourth, and fifth cervical segments** of the spinal cord grow into the septum. At first the nerves, known as **phrenic nerves,** pass into the septum through the pleuropericardial folds (Fig. 10.4*B*). This explains why further expansion of the lungs and descent of the septum shift the phrenic nerves that innervate the diaphragm into the fibrous pericardium (Fig. 10.5).

Although the septum transversum lies opposite cervical segments during the fourth week, by the sixth week the developing diaphragm is at the level of thoracic somites. The repositioning of the diaphragm is caused by rapid growth of the dorsal part of the embryo (vertebral column), compared with that of the ventral part. By the beginning of the third month some of the dorsal bands of the diaphragm originate at the level of the first lumbar vertebra.

The phrenic nerves supply the diaphragm with its motor and sensory innervation. Since the most peripheral part of the diaphragm is derived from mesenchyme of the thoracic wall, it is generally accepted that some of the lower intercostal (thoracic) nerves contribute sensory fibers to the peripheral part of the diaphragm.

CLINICAL CORRELATES

Diaphragmatic Hernias

A congenital diaphragmatic hernia, one of the more common malformations in the newborn (1/2000), is most frequently caused by failure of one or both of the pleuroperitoneal membranes to close the pericardioperitoneal canals. In that case the peritoneal and pleural cavities are continuous with one another along the posterior body wall. This hernia allows abdominal viscera to enter the pleural cavity. In 85 to 90% of cases the hernia is on the left side, and intestinal loops, stomach, spleen, and part of the liver may enter the thoracic cavity (Fig. 10.7). The abdominal viscera in the chest push the heart anteriorly and compress the lungs, which are commonly hypoplastic. A large defect is associated with a high rate of mortality (75%) from pulmonary hypoplasia and dysfunction.

Occasionally a small part of the muscular fibers of the diaphragm fails to develop, and a hernia may remain undiscovered until the child is several years old. Such a defect, frequently seen in the anterior portion of the diaphragm, is a parasternal hernia. A small peritoneal sac containing intestinal loops may enter the chest between the sternal and costal portions of the diaphragm (Fig. 10.7A).

Another type of diaphragmatic hernia, **esophageal hernia,** is thought to be due to congenital shortness of the esophagus. Upper portions of the stomach are retained in the thorax, and the stomach is constricted at the level of the diaphragm.

SUMMARY

At the end of the third week, intercellular clefts appear in the mesoderm on each side of the midline. When these spaces fuse, the **intraembryonic cavity (body cavity)**, bordered by a **somatic mesoderm** and a **splanchnic mesoderm layer**, is formed (Figs. 10.1 and 10.2). With cephalocaudal and lateral folding of the embryo, the intraembryonic cavity extends from the thoracic to the pelvic region. Somatic mesoderm will form the **parietal layer** of the **serous membranes** lining the outside **of the peritoneal, pleural,** and **pericardial cavities**. The **splanchnic layer** will form the **visceral layer of the serous membranes** covering the lungs, heart, and abdominal organs. These layers are continuous at the root of these organs in their cavities (as if a finger were stuck into a balloon, with the layer surrounding the finger being the splanchnic or visceral layer and the rest of the balloon, the somatic or parietal layer surrounding the body cavity). The serous membranes in the abdomen are called **peritoneum**.

The **diaphragm** divides the body cavity into the **thoracic** and **peritoneal cavities**. It develops from four components: (*a*) **septum transversum (central tendon)**; (*b*) **pleuroperitoneal membranes**; (*c*) **dorsal mesentery of the esophagus**; and (*d*) **muscular components of the body wall** (Fig. 10.6). Congenital

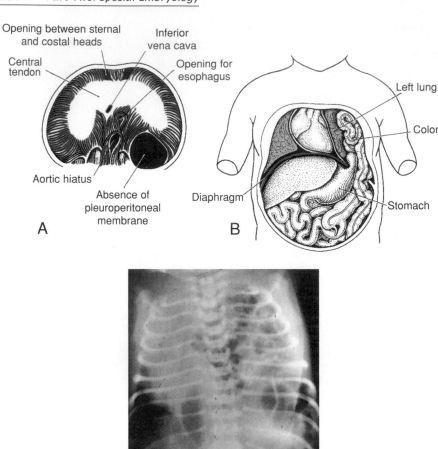

Figure 10.7 Congenital diaphragmatic hernia. **A**. Abdominal surface of the diaphragm showing a large defect of the pleuroperitoneal membrane. **B**. Hernia of the intestinal loops and part of the stomach into the left pleural cavity. The heart and mediastinum are frequently pushed to the right and the left lung compressed. **C**. Radiograph of a newborn with a large defect in the left side of the diaphragm. Abdominal viscera have entered the thorax through the defect.

diaphragmatic hernias involving a defect of the pleuroperitoneal membrane on the left side occur frequently.

The **thoracic cavity** is divided into the **pericardial cavity** and two **pleural cavities** for the lungs by the **pleuropericardial membranes** (Fig. 10.5).

Double layers of peritoneum form **mesenteries** that suspend the gut tube and provide a pathway for vessels, nerves, and lymphatics to the organs. Initially, the gut tube from the caudal end of the foregut to the end of the hindgut is suspended from the dorsal body wall by **dorsal mesentery** (Fig. 10.2, *C* and *E*). **Ventral mesentery** derived from the septum transversum exists only in the region of the terminal part of the esophagus, the stomach, and upper portion of the duodenum (see Chapter 13).

Problems to Solve

1. *A newborn infant cannot breathe and soon dies. An autopsy reveals a large diaphragmatic defect on the left side, with the stomach and intestines occupying the left side of the thorax. Both lungs are severely hypoplastic. What is the embryological basis for this defect?*

2. *A child is born with a large defect lateral to the umbilicus. Most of the large and the small bowel protrude through the defect and are not covered by amnion. What is the embryological basis for this abnormality, and should you be concerned that other malformations may be present?*

SUGGESTED READING

Cunniff C, Jones KL, Jones MC: Patterns of malformations in children with congenital diaphragmatic defects. *J Pediatr* 116:258, 1990.

Puri P, Gormak F: Lethal nonpulmonary anomalies associated with congenital diaphragmatic hernia: implications for early intrauterine surgery. *J Pediatr Surg* 35:29, 1984.

Skandalakis JE, Gray SW: *Embryology for Surgeons: The Embryological Basis for the Treatment of Congenital Anomalies.* 2nd ed. Baltimore, Williams & Wilkins, 1994.

chapter 11

Cardiovascular System

Establishment of the Cardiogenic Field

The vascular system appears in the middle of the third week, when the embryo is no longer able to satisfy its nutritional requirements by diffusion alone. Cardiac progenitor cells lie in the epiblast, immediately lateral to the primitive streak. From there they migrate through the streak. Cells destined to form cranial segments of the heart, the outflow tract, migrate first, and cells forming more caudal portions, right ventricle, left ventricle, and sinus venosus, respectively, migrate in sequential order. The cells proceed toward the cranium and position themselves rostral to the buccopharyngeal membrane and neural folds (Fig. 11.1). Here they reside in the splanchnic layer of the lateral plate mesoderm. At this time, late in the presomite stage of development, they are induced by the underlying pharyngeal endoderm to form cardiac myoblasts. Blood islands also appear in this mesoderm, where they will form blood cells and vessels by the process of vasculogenesis (see Chapter 5; p 103) (Fig. 11.1). With time, the islands unite and form a **horseshoe-shaped** endothelial-lined tube surrounded by myoblasts. This region is known as the **cardiogenic field;** the intraembryonic cavity over it later develops into the **pericardial cavity** (Fig. 11.1*D*).

In addition to the cardiogenic region, other blood islands appear bilaterally, parallel and close to the midline of the embryonic shield. These islands form a pair of longitudinal vessels, the **dorsal aortae.**

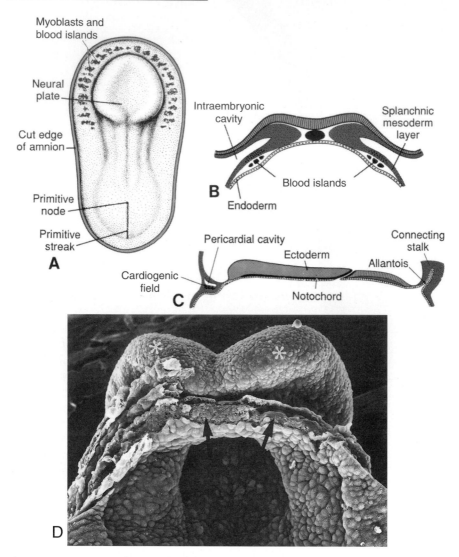

Figure 11.1 A. Dorsal view of a late presomite embryo (approximately 18 days) after removal of the amnion. Prospective myoblasts and hemangioblasts reside in the splanchnic mesoderm in front of the neural plate and on each side of the embryo. **B.** Transverse section through a similar-staged embryo to show the position of the blood islands in the splanchnic mesoderm layer. **C.** Cephalocaudal section through a similar-staged embryo showing the position of the pericardial cavity and cardiogenic field. **D.** Scanning electron micrograph of a mouse embryo equivalent to 19 days in the human, showing coalescence of the blood islands by vasculogenesis into a horseshoe-shaped heart tube (*arrows*) lying in the primitive pericardial cavity under the cranial neural folds (*asterisks*).

Formation and Position of the Heart Tube

Initially, the central portion of the cardiogenic area is anterior to the buccopharyngeal membrane and the neural plate (Fig. 11.2*A*). With closure of the neural tube and formation of the brain vesicles, however, the central nervous system grows cephalad so rapidly that it extends over the central cardiogenic area and the future pericardial cavity (Fig. 11.2). As a result of growth of the brain and cephalic folding of the embryo, the **buccopharyngeal membrane** is pulled forward, while the heart and pericardial cavity move first to the cervical region and finally to the thorax (Fig. 11.2).

As the embryo folds cephalocaudally, it also folds laterally (Fig. 11.3). As a result, the caudal regions of the paired cardiac primordia merge except at their caudalmost ends. Simultaneously, the crescent part of the horseshoe-shaped area expands to form the future outflow tract and ventricular regions. Thus, the heart becomes a continuous expanded tube consisting of an inner endothelial lining and an outer myocardial layer. It receives venous drainage at its caudal pole and begins to pump blood out of the first aortic arch into the dorsal aorta at its cranial pole (Figs. 11.4 and 11.5).

The developing heart tube bulges more and more into the pericardial cavity. Initially, however, the tube remains attached to the dorsal side of the pericardial cavity by a fold of mesodermal tissue, the **dorsal mesocardium** (Figs. 11.3 and 11.5). No ventral mesocardium is ever formed. With further development, the dorsal mesocardium disappears, creating the **transverse pericardial sinus,** which connects both sides of the pericardial cavity. The heart is now suspended in the cavity by blood vessels at its cranial and caudal poles (Fig. 11.5).

During these events, the myocardium thickens and secretes a thick layer of extracellular matrix, rich in hyaluronic acid, that separates it from the endothelium (Figs. 11.3 and 11.5). In addition, mesothelial cells from the region of the sinus venosus migrate over the heart to form the **epicardium.** Thus the heart tube consists of three layers: (*a*) the **endocardium,** forming the internal endothelial lining of the heart; (*b*) the **myocardium,** forming the muscular wall; and (*c*) the **epicardium** or **visceral pericardium,** covering the outside of the tube. This outer layer is responsible for formation of the coronary arteries, including their endothelial lining and smooth muscle.

Formation of the Cardiac Loop

The heart tube continues to elongate and bend on day 23. The cephalic portion of the tube bends ventrally, caudally, and to the right (Fig. 11.6, *B* and *C*), and the atrial (caudal) portion shifts dorsocranially and to the left (Figs. 11.6 and 11.7*A*). This bending, which may be due to cell shape changes, creates the **cardiac loop.** It is complete by day 28.

While the cardiac loop is forming, local expansions become visible throughout the length of the tube. The **atrial portion,** initially a paired structure outside

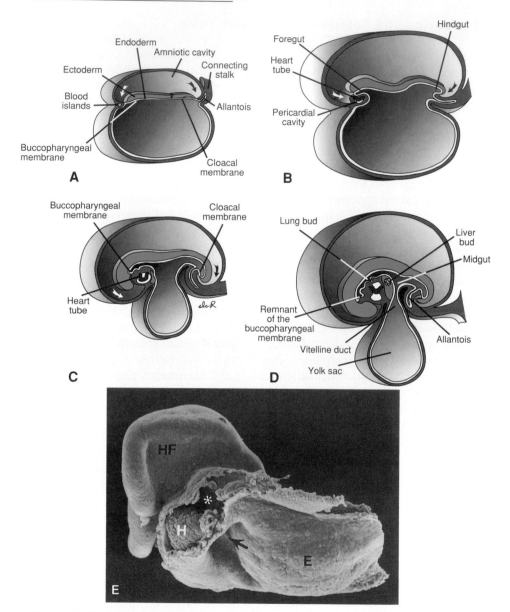

Figure 11.2 Figures showing effects of the rapid growth of the brain on positioning of the heart. Initially the cardiogenic area and the pericardial cavity are in front of the buccopharyngeal membrane. **A.** 18 days. **B.** 20 days. **C.** 21 days. **D.** 22 days. **E.** Scanning electron micrograph of a mouse embryo at a stage similar to that shown in **C.** The amnion, yolk sac, and caudal half of the embryo have been removed. The head folds (*HF*) are expanding and curving over the heart (*H*) and pericardial cavity (*asterisk*). The intestinal opening (*arrow*) of the gut into the primitive pharynx and the endoderm (*E*) of the open region of the gut tube are shown.

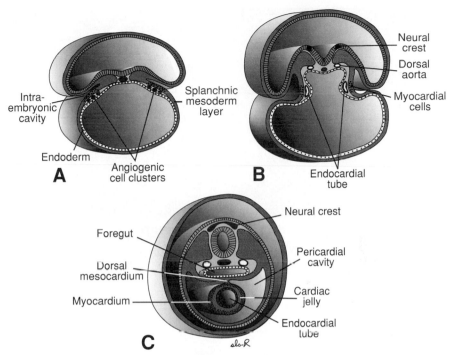

Intra-embryonic cavity

Splanchnic mesoderm layer

Endoderm

Angiogenic cell clusters

A

Neural crest

Dorsal aorta

Myocardial cells

Endocardial tube

B

Foregut

Dorsal mesocardium

Myocardium

Neural crest

Pericardial cavity

Cardiac jelly

Endocardial tube

C

Figure 11.3 Transverse sections through embryos at different stages of development, showing formation of a single heart tube from paired primordia. **A.** Early presomite embryo (17 days). **B.** Late presomite embryo (18 days). **C.** Eight-somite stage (22 days). Fusion occurs only in the caudal region of the horseshoe-shaped tube (see Fig. 12.4). The outflow tract and most of the ventricular region form by expansion and growth of the crescent portion of the horseshoe.

the pericardial cavity, forms a common atrium and is incorporated into the pericardial cavity (Figs. 11.7*A*). The **atrioventricular junction** remains narrow and forms the **atrioventricular canal,** which connects the common atrium and the early embryonic ventricle (Fig. 11.8). The **bulbus cordis** is narrow except for its proximal third. This portion will form the **trabeculated part of the right ventricle** (Figs. 11.7*B* and 11.8). The midportion, the **conus cordis,** will form the outflow tracts of both ventricles. The distal part of the bulbus, the **truncus arteriosus,** will form the roots and proximal portion of the aorta and pulmonary artery (Fig. 11.8). The junction between the ventricle and the bulbus cordis, externally indicated by the **bulboventricular sulcus** (Fig. 11.6*C*), remains narrow. It is called the **primary interventricular foramen** (Fig. 11.8). Thus, the cardiac tube is organized by regions along its craniocaudal axis from the conotruncus to the right ventricle to the left ventricle to the atrial region, respectively (Fig. 11.6, *A – C*). Evidence suggests that organization of these segments is regulated by homeobox genes in a manner similar to that for the craniocaudal axis of the embryo (see Chapter 5).

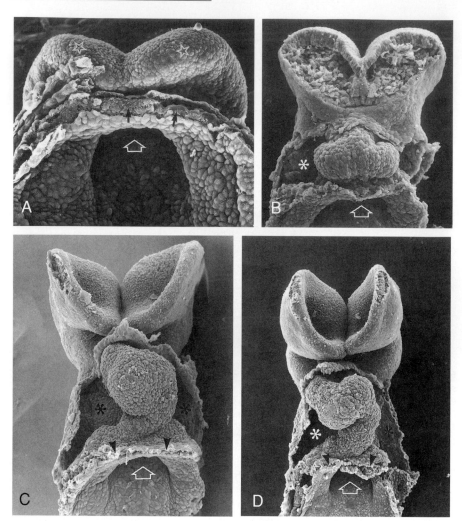

Figure 11.4 Formation of the heart tube on days 19, 20, 21, and 22 in scanning electron micrographs of mouse embryos at equivalent stages of human development. **A.** The heart tube (*arrows*) is horseshoe shaped in the pericardial cavity beneath the neural folds (*stars*). **B.** The crescent portion of the horseshoe expands to form the ventricular and outflow tract regions, while lateral folding brings the caudal (venous) poles of the horseshoe together (see Fig. 12.3). **C.** The caudal regions begin to fuse. **D.** Fusion of the caudal regions is complete, leaving the caudal poles embedded in the septum transversum (*arrowheads*). Cardiac looping has also been initiated. *Asterisk,* pericardial cavity; *large arrow,* anterior intestinal portal.

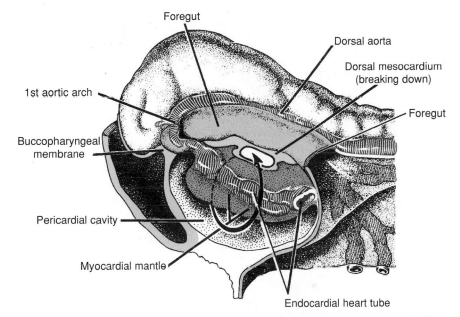

Figure 11.5 Cephalic end of an early somite embryo. The developing endocardial heart tube and its investing layer bulge into the pericardial cavity. The dorsal mesocardium is breaking down.

At the end of loop formation, the smooth-walled heart tube begins to form primitive trabeculae in two sharply defined areas just proximal and distal to the **primary interventricular foramen** (Fig. 11.8). The bulbus temporarily remains smooth walled. The primitive ventricle, which is now trabeculated, is called the **primitive left ventricle.** Likewise, the trabeculated proximal third of the bulbus cordis may be called the **primitive right ventricle** (Fig. 11.8).

The conotruncal portion of the heart tube, initially on the right side of the pericardial cavity, shifts gradually to a more medial position. This change in position is the result of formation of two transverse dilations of the atrium, bulging on each side of the bulbus cordis (Figs. 11.7*B* and 11.8).

CLINICAL CORRELATES

Abnormalities of Cardiac Looping

Dextrocardia, in which the heart lies on the right side of the thorax instead of the left, is caused because the heart loops to the left instead of the right. Dextrocardia may coincide with **situs inversus,** a complete reversal of asymmetry in all organs. Situs inversus, which occurs in 1/7000 individuals, usually is associated with normal physiology, although there is a slight risk of heart defects. In other cases sidedness is random, such that some organs are reversed and others are not; this is **heterotaxy.** These cases are classified as

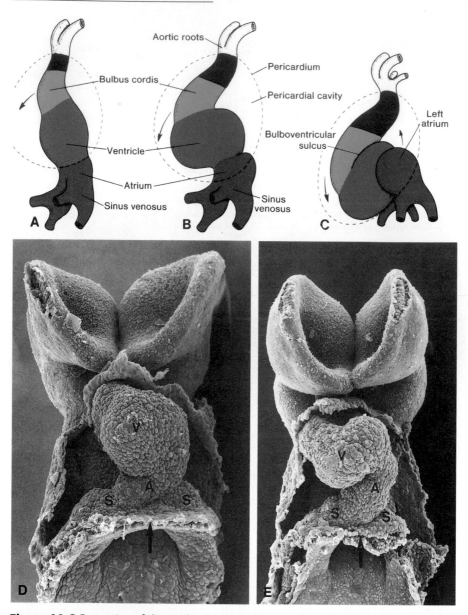

Figure 11.6 Formation of the cardiac loop. **A.** 22 days. **B.** 23 days. **C.** 24 days. *Broken line,* pericardium. **D** and **E.** Scanning electron micrographs of mouse embryos showing frontal views of the process shown in the diagrams. Initially the cardiac tube is short and relatively straight **(D),** but as it lengthens, it bends (loops), bringing the atrial region cranial and dorsal to the ventricular region **(E).** The tube is organized in segments, illustrated by the different colors, from the outflow region to the right ventricle to the left ventricle to the atrial region. These segments represent a craniocaudal axis that appears to be regulated by homeobox gene expression. *A,* primitive atrium; *arrow,* septum transversum; *S,* sinus venosus; *V,* ventricle.

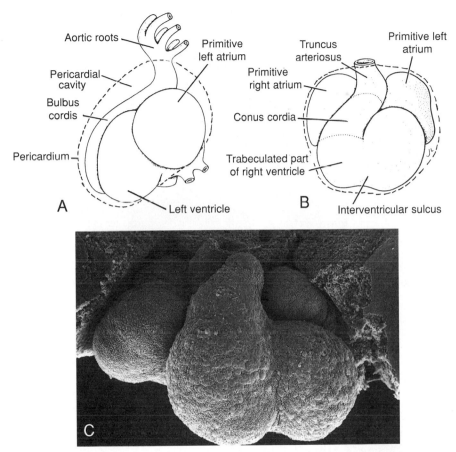

Figure 11.7 Heart of a 5-mm embryo (28 days). **A.** Viewed from the left. **B.** Frontal view. The bulbus cordis is divided into the truncus arteriosus, conus cordis, and trabeculated part of the right ventricle. *Broken line,* pericardium. **C.** Scanning electron micrograph of the heart of a similar-staged mouse embryo showing a view similar to **B.**

laterality sequences. Patients with these conditions appear to be predominantly left sided bilaterally or right sided bilaterally. The spleen reflects the differences: those with left-sided bilaterality have polysplenia; those with right-sided bilaterality have asplenia or hypoplastic spleen. Patients with laterality sequences also have increased incidences of other malformations, especially heart defects. Genes regulating sidedness are expressed during gastrulation (see Chapter 4).

Molecular Regulation of Cardiac Development

Signals from anterior (cranial) endoderm induce a heart-forming region in overlying splanchnic mesoderm by turning on the transcription factor **NKX2.5**. The

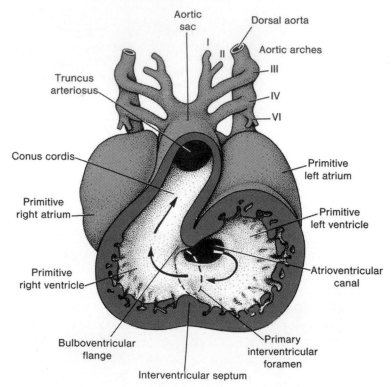

Figure 11.8 Frontal section through the heart of a 30-day embryo showing the primary interventricular foramen and entrance of the atrium into the primitive left ventricle. Note the bulboventricular flange. *Arrows*, direction of blood flow.

signals require secretion of bone morphogenetic proteins (BMPs) 2 and 4 and inhibitors (crescent) of WNT genes in the endoderm and lateral plate mesoderm (Fig. 11.9). This combination is responsible for inducing expression of *NKX2.5* that specifies the cardiogenic field and, later, plays a role in septation and in development of the conduction system.

NKX2.5 contains a homeodomain and is a homologue of the gene *tinman*, which regulates heart development in *Drosophila*. *TBX5* is another transcription factor that contains a DNA-binding motif known as the T-box. Expressed later than *NKX2.5*, it plays a role in septation.

Development of the Sinus Venosus

In the middle of the fourth week, the **sinus venosus** receives venous blood from the **right** and **left sinus horns** (Fig. 11.10*A*). Each horn receives blood

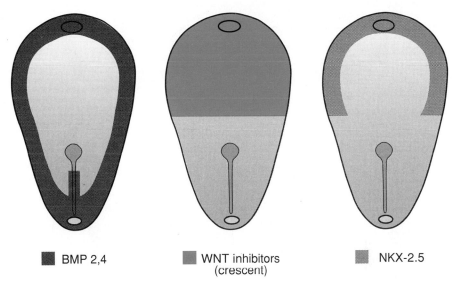

■ BMP 2,4   ■ WNT inhibitors   ■ NKX-2.5
                     (crescent)

Figure 11.9 Heart induction. BMPs secreted in the posterior portion of the primitive streak and periphery of the embryo, in combination with inhibition of *WNT* expression by *crescent* in the anterior half of the embryo, induce expression of *NKX2.5* in the heart forming region of the lateral plate mesoderm (splanchnic layer). *NKX2.5* is then responsible for heart induction.

from three important veins: (*a*) the **vitelline** or **omphalomesenteric vein,** (*b*) the **umbilical vein,** and (*c*) the **common cardinal vein.** At first communication between the sinus and the atrium is wide. Soon, however, the entrance of the sinus shifts to the right (Fig. 11.10*B*). This shift is caused primarily by left-to-right shunts of blood, which occur in the venous system during the fourth and fifth weeks of development.

With obliteration of the right umbilical vein and the left vitelline vein during the fifth week, the left sinus horn rapidly loses its importance (Fig. 11.10*B*). When the left common cardinal vein is obliterated at 10 weeks, all that remains of the left sinus horn is the **oblique vein of the left atrium** and the **coronary sinus** (Fig. 11.11).

As a result of left-to-right shunts of blood, the right sinus horn and veins enlarge greatly. The right horn, which now forms the only communication between the original sinus venosus and the atrium, is incorporated into the right atrium to form the smooth-walled part of the right atrium (Fig. 11.12). Its entrance, the **sinuatrial orifice,** is flanked on each side by a valvular fold, the **right** and **left venous valves** (Fig. 11.12*A*). Dorsocranially the valves fuse, forming a ridge known as the **septum spurium** (Fig. 11.12*A*). Initially the valves are large, but when the right sinus horn is incorporated into the wall of the atrium, the left venous valve and the septum spurium fuse with the developing atrial septum (Fig. 11.12*C*). The superior portion of the right venous valve disappears entirely. The inferior portion develops into two parts: (*a*) the **valve of**

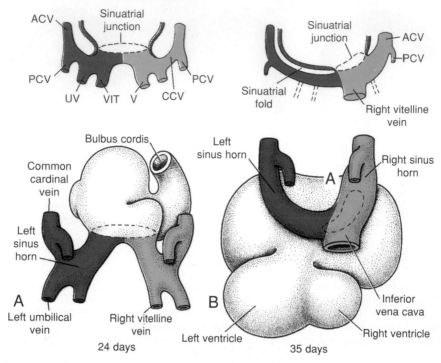

Figure 11.10 Dorsal view of two stages in the development of the sinus venosus at approximately 24 days **(A)** and 35 days **(B)**. *Broken line*, the entrance of the sinus venosus into the atrial cavity. Each drawing is accompanied by a scheme to show in transverse section the great veins and their relation to the atrial cavity. *ACV*, anterior cardinal vein; *PCV*, posterior cardinal vein; *UV*, umbilical vein; *VIT V*, vitelline vein; *CCV*, common cardinal vein. (See also Fig. 11.41.)

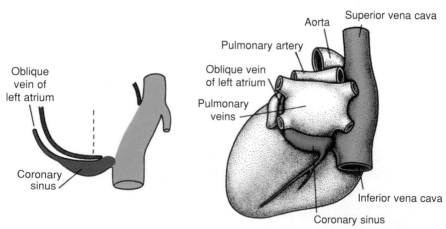

Figure 11.11 Final stage in development of the sinus venosus and great veins.

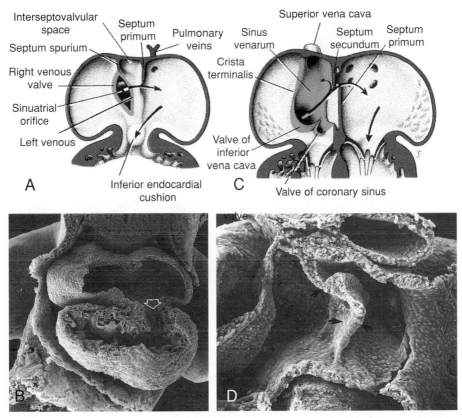

Figure 11.12 Ventral view of coronal sections through the heart at the level of the atrioventricular canal to show development of the venous valves. **A.** 5 weeks. **B.** Scanning electron micrograph of a similar-staged mouse heart showing initial formation of the septum primum; septum spurium is not visible. Note the atrioventricular canal (*arrow*). **C.** Fetal stage. The sinus venarum (*blue*) is smooth walled; it derives from the right sinus horn. *Arrows*, blood flow. **D.** High magnification of the interatrial septum (*arrows*) of a mouse embryo at a stage similar to **C.** The foramen ovale is not visible.

the inferior vena cava, and (*b*) the **valve of the coronary sinus** (Fig. 11.12*C*). The **crista terminalis** forms the dividing line between the original trabeculated part of the right atrium and the smooth-walled part (**sinus venarum**), which originates from the right sinus horn (Fig. 11.12*C*).

Formation of the Cardiac Septa

The major septa of the heart are formed between the 27th and 37th days of development, when the embryo grows in length from 5 mm to approximately 16 to 17 mm. One method by which a septum may be formed involves two

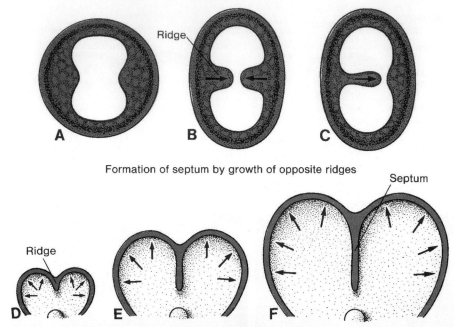

Formation of septum by growth of opposite ridges

Figure 11.13 A and **B.** Septum formation by two actively growing ridges that approach each other until they fuse. **C.** Septum formed by a single actively growing cell mass. **D, E,** and **F.** Septum formation by merging of two expanding portions of the wall of the heart. Such a septum never completely separates two cavities.

actively growing masses of tissue that approach each other until they fuse, dividing the lumen into two separate canals (Fig. 11.13, *A* and *B*). Such a septum may also be formed by active growth of a single tissue mass that continues to expand until it reaches the opposite side of the lumen (Fig. 11.13*C*). Formation of such tissue masses depends on synthesis and deposition of extracellular matrices and cell proliferation. The masses, known as **endocardial cushions,** develop in the **atrioventricular** and **conotruncal** regions. In these locations they assist in formation of the **atrial and ventricular (membranous portion) septa,** the **atrioventricular canals and valves,** and the **aortic and pulmonary channels.**

The other manner in which a septum is formed does not involve endocardial cushions. If, for example, a narrow strip of tissue in the wall of the atrium or ventricle should fail to grow while areas on each side of it expand rapidly, a narrow ridge forms between the two expanding portions (Fig. 11.13,*D* and *E*). When growth of the expanding portions continues on either side of the narrow portion, the two walls approach each other and eventually merge, forming a septum (Fig. 11.13 *F*). Such a septum never completely divides the original lumen but leaves a narrow communicating canal between the two expanded sections. It is usually closed secondarily by tissue contributed by neighboring proliferating tissues. Such a septum partially divides the atria and ventricles.

CLINICAL CORRELATES

Endocardial Cushions and Heart Defects

Because of their key location, abnormalities in endocardial cushion formation contribute to many cardiac malformations, including **atrial** and **ventricular septal defects** and defects involving the **great vessels** (i.e., **transposition of the great vessels** and **tetralogy of Fallot**). Since cells populating the conotruncal cushions include **neural crest cells** and since crest cells also contribute extensively to development of the head and neck, abnormalities in these cells, produced by teratogenic agents or genetic causes, often produce both heart and craniofacial defects in the same individual.

SEPTUM FORMATION IN THE COMMON ATRIUM

At the end of the fourth week, a sickle-shaped crest grows from the roof of the common atrium into the lumen. This crest is the first portion of the **septum primum** (Figs. 11.12 *A* and 11.14, *A* and *B*). The two limbs of this septum extend toward the endocardial cushions in the atrioventricular canal. The opening between the lower rim of the septum primum and the endocardial cushions is the **ostium primum** (Fig. 11.14, *A* and *B*). With further development, extensions of the superior and inferior endocardial cushions grow along the edge of the septum primum, closing the ostium primum (Fig. 11.14, *C* and *D*). Before closure is complete, however, **cell death** produces perforations in the upper portion of the septum primum. Coalescence of these perforations forms the **ostium secundum,** ensuring free blood flow from the right to the left primitive atrium (Fig. 11.14, *B* and *D*).

When the lumen of the right atrium expands as a result of incorporation of the sinus horn, a new crescent-shaped fold appears. This new fold, the **septum secundum** (Fig. 11.14, *C* and *D*), never forms a complete partition in the atrial cavity (Fig. 11.14*G*). Its anterior limb extends downward to the septum in the atrioventricular canal. When the left venous valve and the septum spurium fuse with the right side of the septum secundum, the free concave edge of the septum secundum begins to overlap the ostium secundum (Fig. 11.12, *A* and *B*). The opening left by the septum secundum is called the **oval foramen (foramen ovale)**. When the upper part of the septum primum gradually disappears, the remaining part becomes the **valve of the oval foramen.** The passage between the two atrial cavities consists of an obliquely elongated cleft (Fig. 11.14, *E–G*) through which blood from the right atrium flows to the left side (*arrows* in Figs. 11.12*B* and 11.14*E*).

After birth, when lung circulation begins and pressure in the left atrium increases, the valve of the oval foramen is pressed against the septum secundum,

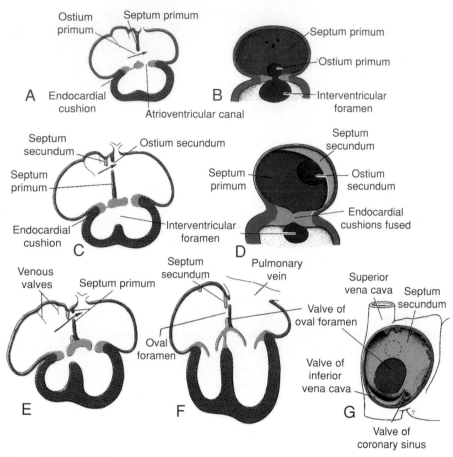

Figure 11.14 Atrial septa at various stages of development. **A.** 30 days (6 mm). **B.** Same stage as **A,** viewed from the right. **C.** 33 days (9 mm). **D.** Same stage as **C,** viewed from the right **E.** 37 days (14 mm). **F.** Newborn. **G.** The atrial septum from the right; same stage as **F.**

obliterating the oval foramen and separating the right and left atria. In about 20% of cases, fusion of the septum primum and septum secundum is incomplete, and a narrow oblique cleft remains between the two atria. This condition is called **probe patency** of the oval foramen; it does not allow intracardiac shunting of blood.

Further Differentiation of the Atria

While the primitive right atrium enlarges by incorporation of the right sinus horn, the primitive left atrium is likewise expanding. Initially, a single embryonic **pulmonary vein** develops as an outgrowth of the posterior left atrial wall, just to the left of the septum primum (Fig. 11.15A). This vein gains connection with

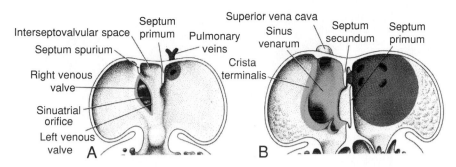

Figure 11.15 Coronal sections through the heart to show development of the smooth-walled portions of the right and left atrium. Both the wall of the right sinus horn (*blue*) and the pulmonary veins (*red*) are incorporated into the heart to form the smooth-walled parts of the atria.

veins of the developing lung buds. During further development, the pulmonary vein and its branches are incorporated into the left atrium, forming the large **smooth-walled** part of the adult atrium. Although initially one vein enters the left atrium, ultimately four pulmonary veins enter (Fig. 11.15*B*) as the branches are incorporated into the expanding atrial wall.

In the fully developed heart, the original embryonic left atrium is represented by little more than the **trabeculated atrial appendage,** while the smooth-walled part originates from the pulmonary veins (Fig. 11.15). On the right side the original embryonic right atrium becomes the trabeculated **right atrial appendage** containing the pectinate muscles, and the smooth-walled **sinus venarum** originates from the right horn of the sinus venosus.

SEPTUM FORMATION IN THE ATRIOVENTRICULAR CANAL

At the end of the fourth week, two mesenchymal cushions, the **atrioventricular endocardial cushions,** appear at the superior and inferior borders of the atrioventricular canal (Figs. 11.16 and 11.17). Initially the atrioventricular canal gives access only to the primitive left ventricle and is separated from the bulbus cordis by the **bulbo(cono)ventricular flange** (Fig. 11.8). Near the end of the fifth week, however, the posterior extremity of the flange terminates almost midway along the base of the superior endocardial cushion and is much less prominent than before (Fig. 11.17). Since the atrioventricular canal enlarges to the right, blood passing through the atrioventricular orifice now has direct access to the primitive left as well as the primitive right ventricle.

In addition to the superior and inferior endocardial cushions, the two **lateral atrioventricular cushions** appear on the right and left borders of the canal (Figs. 11.16 and 11.17). The superior and inferior cushions, in the meantime, project further into the lumen and fuse, resulting in a complete division of the canal into right and left atrioventricular orifices by the end of the fifth week (Fig. 11.16).

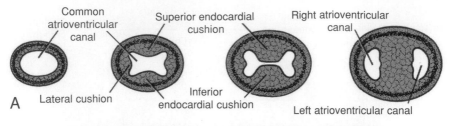

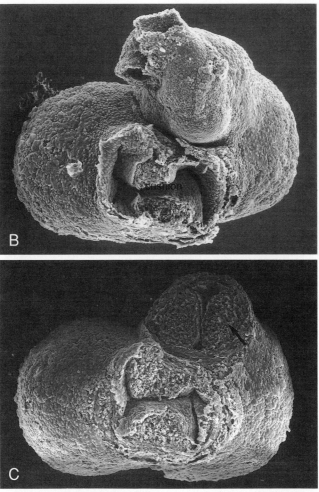

Figure 11.16 Formation of the septum in the atrioventricular canal. **A.** From left to right, days 23, 26, 31, and 35. The initial circular opening widens transversely. **B** and **C.** Scanning electron micrographs of hearts from mouse embryos, showing growth and fusion of the superior and inferior endocardial cushions in the atrioventricular canal. In **C,** cushions of the ouflow tract (*arrow*) are also fusing.

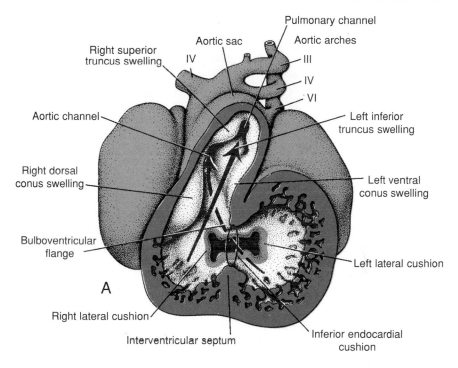

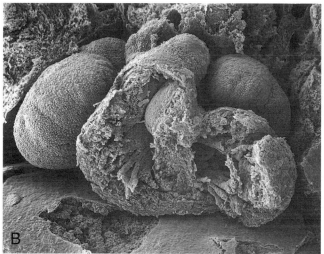

Figure 11.17 A. Frontal section through the heart of a day 35 embryo. At this stage of development blood from the atrial cavity enters the primitive left ventricle as well as the primitive right ventricle. Note development of the cushions in the atrioventricular canal. Cushions in the truncus and conus are also visible. *Ring*, primitive interventricular foramen. *Arrows*, blood flow. **B.** Scanning electron micrograph of a mouse embryo at a slightly later stage showing fusion of the atrioventricular cushions and contact between those in the outflow tract.

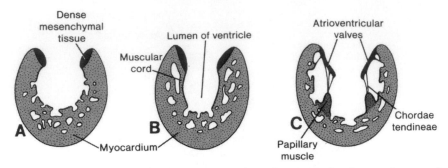

Figure 11.18 Formation of the atrioventricular valves and chordae tendineae. The valves are hollowed out from the ventricular side but remain attached to the ventricular wall by the chordae tendineae.

Atrioventricular Valves

After the atrioventricular endocardial cushions fuse, each atrioventricular orifice is surrounded by local proliferations of mesenchymal tissue (Fig. 11.18*A*). When the bloodstream hollows out and thins tissue on the ventricular surface of these proliferations, valves form and remain attached to the ventricular wall by muscular cords (Fig. 11.18*B*). Finally, muscular tissue in the cords degenerates and is replaced by dense connective tissue. The valves then consist of connective tissue covered by endocardium. They are connected to thick trabeculae in the wall of the ventricle, the **papillary muscles,** by means of **chordae tendineae** (Fig.11.18*C*). In this manner two valve leaflets, constituting the **bicuspid,** or **mitral, valve,** form in the left atrioventricular canal, and three, constituting the **tricuspid valve,** form on the right side.

CLINICAL CORRELATES

Heart Defects

Heart and vascular abnormalities make up the largest category of human birth defects, accounting for 1% of malformations among live-born infants. The incidence among stillborns is 10 times as high. It is estimated that 8% of cardiac malformations are due to genetic factors, 2% are due to environmental agents, and most are due to a complex interplay between genetic and environmental influences (**multifactorial** causes). Classic examples of cardiovacular teratogens include **rubella virus** and **thalidomide.** Others include **isotretinoin (vitamin A), alcohol,** and many other compounds. Maternal diseases, such as insulin-dependent **diabetes** and **hypertension,** have also been linked to cardiac defects. Chromosomal abnormalities are associated with heart malformations, with 6 to 10% of newborns with cardiac defects having an unbalanced chromosomal abnormality. Furthermore, 33% of children with chromosomal abnormalities have a congenital heart defect, with an incidence of nearly 100% in children with trisomy 18. Finally, cardiac malformations are associated with a number of genetic syndromes, including craniofacial abnormalities, such as **DiGeorge, Goldenhar,** and **Down** syndromes (see Chapter 15).

Genes regulating cardiac development are being identified and mapped and mutations that result in heart defects are being discovered. For example, mutations in the heart-specifying gene *NKX2.5,* on chromosome 5q35, produce atrial septal defects (secundum type) and atrioventricular conduction delays in an autosomal dominant fashion. Mutations in the *TBX5* gene result in **Holt-Oram syndrome,** characterized by preaxial (radial) limb abnormalities and atrial septal defects. Defects in the muscular portion of the interventricular septum may also occur. Holt-Oram syndrome is one of a group of **heart-hand syndromes** illustrating that the same genes may participate in multiple developmental processes. For example, *TBX5* is expressed in distal segments of the limb bud and in the heart primordia. Holt-Oram syndrome is inherited as an autosomal dominant trait with a frequency of 1/100,000 live births.

Atrial septal defect (ASD) is a congenital heart abnormality with an incidence of 6.4/10,000 births and with a 2:1 prevalence in female versus male infants. One of the most significant defects is the **ostium secundum** defect, characterized by a large opening between the left and right atria. It is caused either by excessive cell death and resorption of the septum primum (Fig. 11.19, *B* and *C*) or by inadequate development of the septum secundum (Fig. 11.19, *D* and *E*). Depending on the size of the opening, considerable intracardiac shunting may occur from left to right.

The most serious abnormality in this group is complete absence of the atrial septum (Fig. 11.19*F*). This condition, known as common atrium or **cor triloculare biventriculare,** is always associated with serious defects elsewhere in the heart.

Occasionally, the oval foramen closes during prenatal life. This abnormality, **premature closure of the oval foramen,** leads to massive hypertrophy of the right atrium and ventricle and underdevelopment of the left side of the heart. Death usually occurs shortly after birth.

Endocardial cushions of the atrioventricular canal not only divide this canal into a right and left orifice, but also participate in formation of the membranous portion of the interventricular septum and in closure of the ostium primum. This region has the appearance of a cross, with the atrial and ventricular septa forming the post and the atrioventricular cushions the crossbar. The integrity of this cross is an important sign in ultrasound scans of the heart (Fig. 11.29*C*). Whenever the cushions fail to fuse, the result is a **persistent atrioventricular canal,** combined with a defect in the cardiac septum (Fig.11.20*A*). This septal defect has an atrial and a ventricular

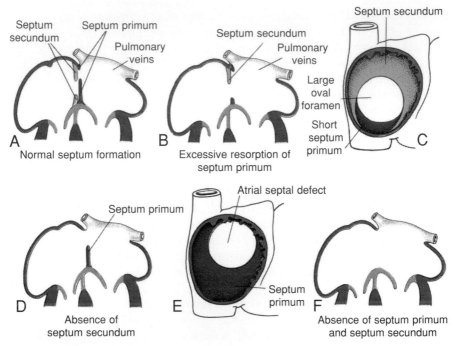

Figure 11.19 A. Normal atrial septum formation. **B** and **C.** Ostium secundum defect caused by excessive resorption of the septum primum. **D** and **E.** Similar defect caused by failure of development of the septum secundum. **F.** Common atrium, or cor triloculare biventriculare, resulting from complete failure of the septum primum and septum secundum to form.

component, separated by abnormal valve leaflets in the single atrioventricular orifice (Fig.11.20*C*).

Occasionally, endocardial cushions in the atrioventricular canal partially fuse. The result is a defect in the atrial septum, but the interventricular septum is closed (Fig. 11.20, *D* and *E*). This defect, the **ostium primum defect,** is usually combined with a cleft in the anterior leaflet of the tricuspid valve (Fig. 11.20*C*).

Tricuspid atresia, which involves obliteration of the right atrioventricular orifice (Fig. 11.21*B*), is characterized by the absence or fusion of the tricuspid valves. Tricuspid atresia is always associated with (*a*) patency of the oval foramen, (*b*) ventricular septal defect, (*c*) underdevelopment of the right ventricle, and (*d*) hypertrophy of the left ventricle.

SEPTUM FORMATION IN THE TRUNCUS ARTERIOSUS AND CONUS CORDIS

During the fifth week, pairs of opposing ridges appear in the truncus. These ridges, the **truncus swellings,** or **cushions,** lie on the right superior wall (**right**

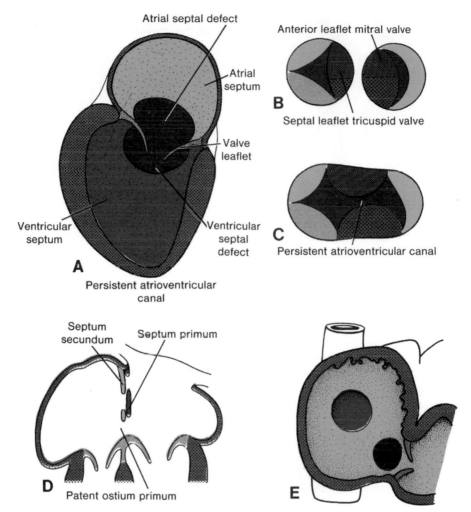

Figure 11.20 A. Persistent common atrioventricular canal. This abnormality is always accompanied by a septum defect in the atrial as well as in the ventricular portion of the cardiac partitions. **B.** Valves in the atrioventricular orifices under normal conditions. **C.** Split valves in a persistent atrioventricular canal. **D** and **E.** Ostium primum defect caused by incomplete fusion of the atrioventricular endocardial cushions.

superior truncus swelling) and on the left inferior wall (**left inferior truncus swelling**) (Fig. 11.17). The right superior truncus swelling grows distally and to the left, and the left inferior truncus swelling grows distally and to the right. Hence, while growing toward the aortic sac, the swellings twist around each other, foreshadowing the spiral course of the future septum (Figs. 11.22 and 11.23). After complete fusion, the ridges form the **aorticopulmonary septum,** dividing the truncus into an **aortic** and a **pulmonary channel.**

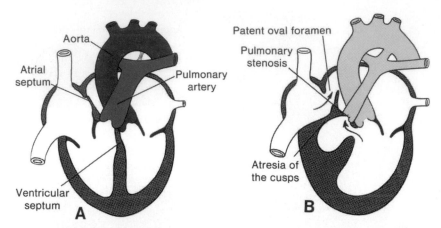

Figure 11.21 A. Normal heart. **B.** Tricuspid atresia. Note the small right ventricle and the large left ventricle.

When the truncus swellings appear, similar swellings (cushions) develop along the right dorsal and left ventral walls of the **conus cordis** (Figs. 11.17 and 11.23). The conus swellings grow toward each other and distally to unite with the truncus septum. When the two conus swellings have fused, the septum divides the conus into an anterolateral portion (the ouflow tract of the right ventricle) (Fig. 11.24) and a posteromedial portion (the outflow tract of the left ventricle) (Fig. 11.25).

Neural crest cells, migrating from the edges of the neural folds in the hindbrain region, contribute to endocardial cushion formation in both the conus cordis and truncus arteriosus. Abnormal migration, proliferation, or differentiation of these cells results in congenital malformations in this region, such as tetralogy of Fallot (Fig 11.29), pulmonary stenoses, transposition of the great vessels and persistent truncus arteriosus (Fig 11.30). Since neural crest cells also contribute to craniofacial development, it is not uncommon to see facial and cardiac abnormalities in the same individual.

SEPTUM FORMATION IN THE VENTRICLES

By the end of the fourth week, the two primitive ventricles begin to expand. This is accomplished by continuous growth of the myocardium on the outside and continuous diverticulation and trabecula formation on the inside (Figs. 11.8, 11.17, and 11.25).

The medial walls of the expanding ventricles become apposed and gradually merge, forming the **muscular interventricular septum** (Fig. 11.25). Sometimes the two walls do not merge completely, and a more or less deep apical cleft between the two ventricles appears. The space between the free rim of

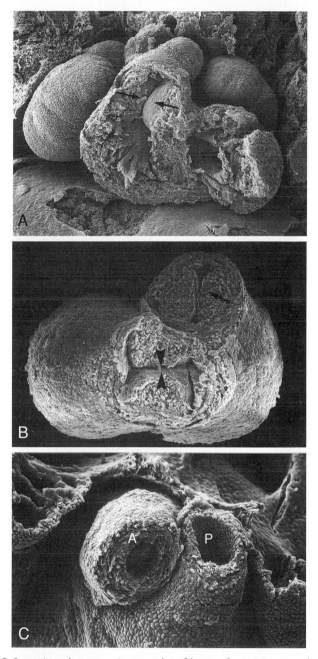

Figure 11.22 Scanning electron micrographs of hearts from mouse embryos showing formation of the conotruncal ridges (cushions) that form a septum in the outflow tract to divide this region into aortic and pulmonary channels. **A.** Frontal section showing cushion contact (*arrows*) in the outflow tract. **B.** Cross section through the atrioventricular canal (*arrowheads*) and outflow tract (*arrow*). Cushions in both regions have made initial contact. **C.** Cross section through the aortic (*A*) and pulmonary (*P*) vessels showing their entwined course caused by spiraling of the conotruncal ridges (see Fig. 11.23). Note the thickness of the aorta.

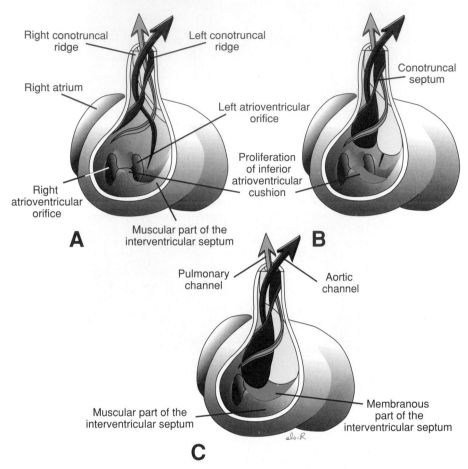

Right conotruncal ridge

Left conotruncal ridge

Right atrium

Conotruncal septum

Left atrioventricular orifice

Proliferation of inferior atrioventricular cushion

Right atrioventricular orifice

A   Muscular part of the interventricular septum

B

Pulmonary channel

Aortic channel

Muscular part of the interventricular septum

Membranous part of the interventricular septum

C

Figure 11.23 Development of the conotruncal ridges (cushions) and closure of the interventricular foramen. Proliferations of the right and left conus cushions, combined with proliferation of the inferior endocardial cushion, close the interventricular foramen and form the membranous portion of the interventricular septum. **A.** 6 weeks (12 mm). **B.** Beginning of the seventh week (14.5 mm). **C.** End of the seventh week (20 mm).

the muscular ventricular septum and the fused endocardial cushions permits communication between the two ventricles.

The **interventricular foramen,** above the muscular portion of the interventricular septum, shrinks on completion of the **conus septum** (Fig. 11.23). During further development, outgrowth of tissue from the inferior endocardial cushion along the top of the muscular interventricular septum closes the foramen (Fig. 11.23). This tissue fuses with the abutting parts of the conus septum. Complete closure of the interventricular foramen forms **the membranous part of the interventricular septum.**

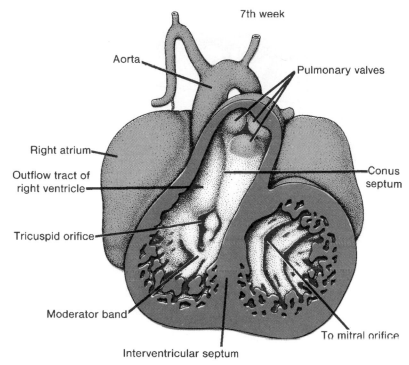

Figure 11.24 Frontal section through the heart of a 7-week embryo. Note the conus septum and position of the pulmonary valves.

Semilunar Valves

When partitioning of the truncus is almost complete, primordia of the semilunar valves become visible as small tubercles found on the main truncus swellings. One of each pair is assigned to the pulmonary and aortic channels, respectively (Fig. 11.26). A third tubercle appears in both channels opposite the fused truncus swellings. Gradually the tubercles hollow out at their upper surface, forming the **semilunar valves** (Fig. 11.27). Recent evidence shows that neural crest cells contribute to formation of these valves.

CLINICAL CORRELATES

Heart Defects

Ventricular septal defect (**VSD**) involving the membranous portion of the septum (Fig. 11.28) is the most common congenital cardiac malformation, occurring as an isolated condition in 12/10,000 births. Although it may be found as an isolated lesion, VSD is often associated with abnormalities in partitioning of the conotruncal region. Depending on the size of the opening, blood carried

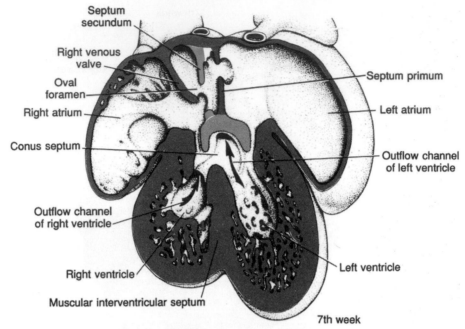

Figure 11.25 Frontal section through the heart of an embryo at the end of the seventh week. The conus septum is complete, and blood from the left ventricle enters the aorta. Note the septum in the atrial region.

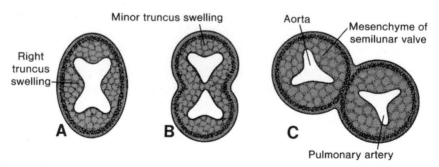

Figure 11.26 Transverse sections through the truncus arteriosus at the level of the semilunar valves at 5 (**A**), 6 (**B**), and 7 (**C**) weeks of development.

by the pulmonary artery may be 1.2 to 1.7 times as abundant as that carried by the aorta. Occasionally the defect is not restricted to the membranous part but also involves the muscular part of the septum.

Tetralogy of Fallot, the most frequently occurring abnormality of the **conotruncal** region (Fig. 11.29), is due to an unequal division of the conus resulting from anterior displacement of the conotruncal septum. Displacement of the septum produces four cardiovascular alterations: (*a*) a narrow right

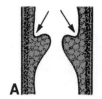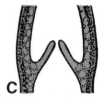

Figure 11.27 Longitudinal sections through the semilunar valves at the sixth (**A**), seventh (**B**), and ninth (**C**) weeks of development. The upper surface is hollowed (*arrows*) to form the valves.

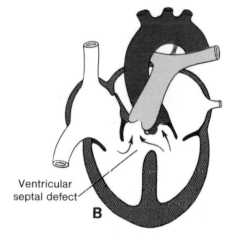

Ventricular septal defect

Figure 11.28 A. Normal heart. **B.** Isolated defect in the membranous portion of the interventricular septum. Blood from the left ventricle flows to the right through the interventricular foramen (*arrows*).

ventricular outflow region, a **pulmonary infundibular stenosis;** (*b*) a large defect of the interventricular septum; (*c*) an overriding aorta that arises directly above the septal defect; and (*d*) hypertrophy of the right ventricular wall because of higher pressure on the right side. Tetralogy of Fallot, which is not fatal, occurs in 9.6/10,000 births.

Persistent truncus arteriosus results when the conotruncal ridges fail to fuse and to descend toward the ventricles (Fig. 11.30). In such a case, which occurs in 0.8/10,000 births, the pulmonary artery arises some distance above the origin of the undivided truncus. Since the ridges also participate in formation of the interventricular septum, the persistent truncus is always accompanied by a defective interventricular septum. The undivided truncus thus overrides both ventricles and receives blood from both sides.

Transposition of the great vessels occurs when the conotruncal septum fails to follow its normal spiral course and runs straight down (Fig. 11.31*A*). As a consequence, the aorta originates from the right ventricle, and the

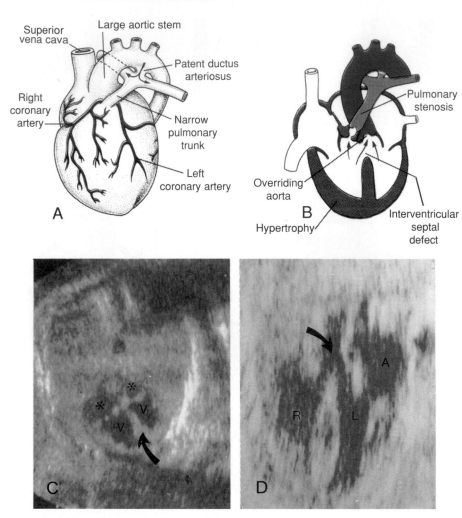

Figure 11.29 Tetralogy of Fallot. **A.** Surface view. **B.** The four components of the defect: pulmonary stenosis, overriding aorta, interventricular septal defect, and hypertrophy of the right ventricle. **C.** Ultrasound scan showing a normal heart with atria (*asterisks*), ventricles (V), and interventricular septum (*arrow*). **D.** Scan of a heart showing the characteristic features of the tetralogy, including hypertrophy of the right ventricle (*R*) and overriding aorta (*arrow*). *A*, atrium; *L*, left ventricle.

pulmonary artery originates from the left ventricle. This condition, which occurs in 4.8/10,000 births, sometimes is associated with a defect in the membranous part of the interventricular septum. It is usually accompanied by an open ductus arteriosus. Since neural crest cells contribute to the formation of the truncal cushions, insults to these cells contribute to cardiac-defects involving the outflow tract.

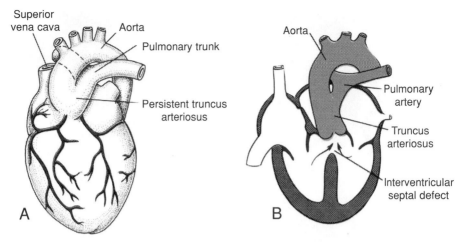

Figure 11.30 Persistent truncus arteriosus. The pulmonary artery originates from a common truncus (**A**). The septum in the truncus and conus has failed to form (**B**). This abnormality is always accompanied by an interventricular septal defect.

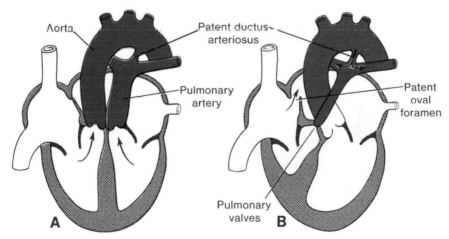

Figure 11.31 A. Transposition of the great vessels. **B.** Pulmonary valvular atresia with a normal aortic root. The only access route to the lungs is by way of a patent ductus arteriosus.

Valvular stenosis of the pulmonary artery or aorta occurs when the semilunar valves are fused for a variable distance. The incidence of the abnormality is similar for both regions, being approximately 3 to 4 per 10,000 births. In the case of a **valvular stenosis of the pulmonary artery,** the trunk of the pulmonary artery is narrow or even atretic (Fig. 11.31 *B*). The patent oval foramen then forms the only outlet for blood from the right side of the

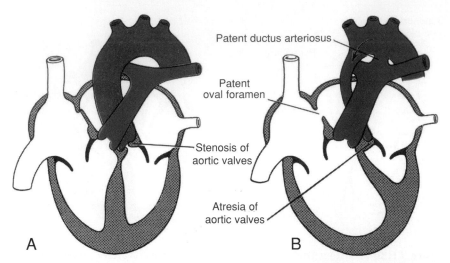

Figure 11.32 A. Aortic valvular stenosis. **B.** Aortic valvular atresia. *Arrow* in the arch of the aorta indicates direction of blood flow. The coronary arteries are supplied by this retroflux. Note the small left ventricle and the large right ventricle.

heart. The ductus arteriosus, always patent, is the only access route to the pulmonary circulation.

In **aortic valvular stenosis** (Fig. 11.32), fusion of the thickened valves may be so complete that only a pinhole opening remains. The size of the aorta itself is usually normal.

When fusion of the semilunar aortic valves is complete–**aortic valvular atresia** (Fig. 11.32*B*)–the aorta, left ventricle, and left atrium are markedly underdeveloped. The abnormality is usually accompanied by an open ductus arteriosus, which delivers blood into the aorta.

Ectopia cordis is a rare anomaly in which the heart lies on the surface of the chest. It is caused by failure of the embryo to close the ventral body wall (see Chapter 10).

Formation of the Conducting System of the Heart

Initially the **pacemaker** for the heart lies in the caudal part of the left cardiac tube. Later the sinus venosus assumes this function, and as the sinus is incorporated into the right atrium, pacemaker tissue lies near the opening of the superior vena cava. Thus, the **sinuatrial node** is formed.

The **atrioventricular node and bundle (bundle of His)** are derived from two sources: (*a*) cells in the left wall of the sinus venosus, and (*b*) cells from the atrioventricular canal. Once the sinus venosus is incorporated into the right atrium, these cells lie in their final position at the base of the interatrial septum.

Vascular Development

ARTERIAL SYSTEM

Aortic Arches

When pharyngeal arches form during the fourth and fifth weeks of development, each arch receives its own cranial nerve and its own artery (see Chapter 15). These arteries, the **aortic arches,** arise from the **aortic sac,** the most distal part of the truncus arteriosus (Figs. 11.8 and 11.33). The aortic arches are embedded in mesenchyme of the pharyngeal arches and terminate in the right and left dorsal aortae. (In the region of the arches the dorsal aortae remain paired, but caudal to this region they fuse to form a single vessel.) The pharyngeal arches and their vessels appear in a cranial to caudal sequence, so that they are not all present simultaneously. The aortic sac contributes a branch to each new arch as it forms, giving rise to a total of five pairs of arteries. (The fifth arch either never forms or forms incompletely and then regresses. Consequently the five arches are numbered I, II, III, IV, and VI [Fig. 11.34].) During further development, this arterial pattern becomes modified, and some vessels regress completely.

Division of the truncus arteriosus by the aorticopulmonary septum divides the outflow channel of the heart into the **ventral aorta** and the **pulmonary artery.** The aortic sac then forms right and left horns, which subsequently give rise to the **brachiocephalic artery** and the proximal segment of the **aortic arch,** respectively (Fig. 11.35, *B* and *C*).

By day 27, most of the **first aortic arch** has disappeared (Fig. 11.34), although a small portion persists to form the **maxillary artery.** Similarly, the

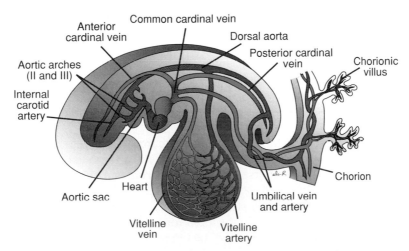

Figure 11.33 Main intraembryonic and extraembryonic arteries (*red*) and veins (*blue*) in a 4-mm embryo (end of the fourth week). Only the vessels on the left side of the embryo are shown.

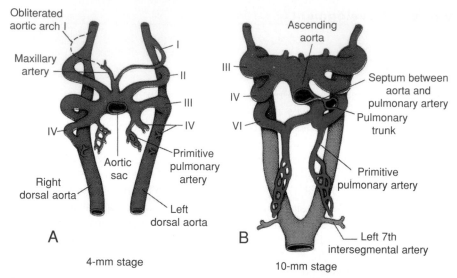

Obliterated aortic arch I

Maxillary artery

I

II

III

IV

IV

Aortic sac

Right dorsal aorta

Primitive pulmonary artery

Left dorsal aorta

A

4-mm stage

Ascending aorta

III

IV

VI

Septum between aorta and pulmonary artery

Pulmonary trunk

Primitive pulmonary artery

B

Left 7th intersegmental artery

10-mm stage

Figure 11.34 A. Aortic arches at the end of the fourth week. The first arch is obliterated before the sixth is formed. **B.** Aortic arch system at the beginning of the sixth week. Note the aorticopulmonary septum and the large pulmonary arteries.

second aortic arch soon disappears. The remaining portions of this arch are the **hyoid** and **stapedial arteries.** The third arch is large; the fourth and sixth arches are in the process of formation. Even though the sixth arch is not completed, the **primitive pulmonary artery** is already present as a major branch (Fig. 11.34*A*).

In a 29-day embryo, the first and second aortic arches have disappeared (Fig. 11.34*B*). The third, fourth, and sixth arches are large. The truncoaortic sac has divided so that the sixth arches are now continuous with the pulmonary trunk.

With further development, the aortic arch system loses its original symmetrical form, as shown in Figure 11.35*A*, and establishes the definitive pattern illustrated in Figure 11.35, *B* and *C*. This representation may clarify the transformation from the embryonic to the adult arterial system. The following changes occur:

The **third aortic arch** forms the **common carotid artery** and the first part of the **internal carotid artery.** The remainder of the internal carotid is formed by the cranial portion of the dorsal aorta. The **external carotid artery** is a sprout of the third aortic arch.

The **fourth aortic arch** persists on both sides, but its ultimate fate is different on the right and left sides. On the left it forms part of the arch of the aorta, between the left common carotid and the left subclavian arteries. On the right it forms the most proximal segment of the right subclavian artery, the distal part of which is formed by a portion of the right dorsal aorta and the seventh intersegmental artery (Fig. 11.35*B*).

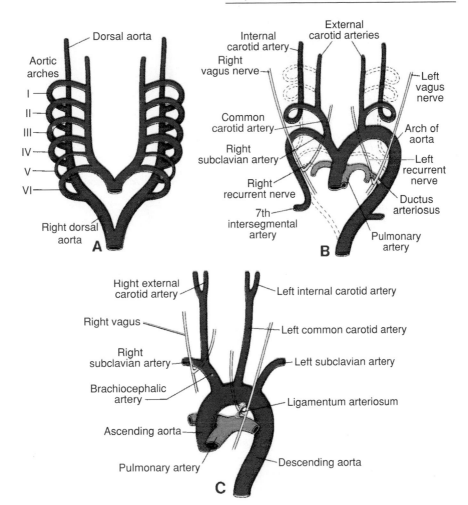

Figure 11.35 A. Aortic arches and dorsal aortae before transformation into the definitive vascular pattern. **B.** Aortic arches and dorsal aortae after the transformation. *Broken lines,* obliterated components. Note the patent ductus arteriosus and position of the seventh intersegmental artery on the left. **C.** The great arteries in the adult. Compare the distance between the place of origin of the left common carotid artery and the left subclavian in **B** and **C.** After disappearance of the distal part of the sixth aortic arch (the fifth arches never form completely), the right recurrent laryngeal nerve hooks around the right subclavian artery. On the left the nerve remains in place and hooks around the ligamentum arteriosum.

The **fifth aortic arch** either never forms or forms incompletely and then regresses.

The **sixth aortic arch,** also known as the **pulmonary arch,** gives off an important branch that grows toward the developing lung bud (Fig. 11.34B). On the right side the proximal part becomes the proximal segment of the right pulmonary artery. The distal portion of this arch loses its connection

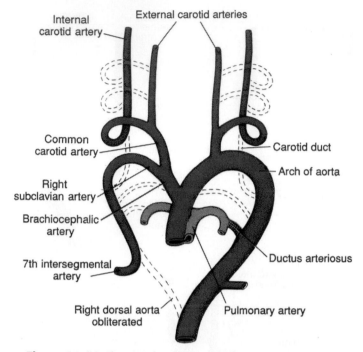

Internal carotid artery

External carotid arteries

Common carotid artery

Carotid duct

Arch of aorta

Right subclavian artery

Brachiocephalic artery

7th intersegmental artery

Ductus arteriosus

Right dorsal aorta obliterated

Pulmonary artery

Figure 11.36 Changes from the original aortic arch system.

with the dorsal aorta and disappears. On the left the distal part persists during intrauterine life as the **ductus arteriosus.**

A number of other changes occur along with alterations in the aortic arch system: (*a*) The dorsal aorta between the entrance of the third and fourth arches, known as the **carotid duct,** is obliterated (Fig. 11.36). (*b*) The right dorsal aorta disappears between the origin of the seventh intersegmental artery and the junction with the left dorsal aorta (Fig. 11.36). (*c*) Cephalic folding, growth of the forebrain, and elongation of the neck push the heart into the thoracic cavity. Hence the carotid and brachiocephalic arteries elongate considerably (Fig. 11.35*C*). As a further result of this caudal shift, the left subclavian artery, distally fixed in the arm bud, shifts its point of origin from the aorta at the level of the seventh intersegmental artery (Fig. 11.35*B*) to an increasingly higher point until it comes close to the origin of the left common carotid artery (Fig. 11.35*C*). (*d*) As a result of the caudal shift of the heart and the disappearance of various portions of the aortic arches, the course of the **recurrent laryngeal nerves** becomes different on the right and left sides. Initially these nerves, branches of the vagus, supply the sixth pharyngeal arches. When the heart descends, they hook around the sixth aortic arches and ascend again to the larynx, which accounts for their recurrent course. On the right, when the distal part of the sixth aortic arch and the fifth aortic arch disappear, the recurrent laryngeal nerve moves up and hooks around the right subclavian artery. On the left the nerve does not move up, since the distal part of the sixth aortic arch persists as the **ductus arteriosus,** which later forms the **ligamentum arteriosum** (Fig. 11.35).

Vitelline and Umbilical Arteries

The **vitelline arteries,** initially a number of paired vessels supplying the yolk sac (Fig. 11.33), gradually fuse and form the arteries in the dorsal mesentery of the gut. In the adult they are represented by the **celiac, superior mesenteric,** and **inferior mesenteric arteries.** These vessels supply derivatives of the **foregut, midgut,** and **hindgut,** respectively.

The **umbilical arteries,** initially paired ventral branches of the dorsal aorta, course to the placenta in close association with the allantois (Fig. 11.33). During the fourth week, however, each artery acquires a secondary connection with the dorsal branch of the aorta, the **common iliac artery,** and loses its earliest origin. After birth the proximal portions of the umbilical arteries persist as the **internal iliac** and **superior vesical arteries,** and the distal parts are obliterated to form the **medial umbilical ligaments.**

CLINICAL CORRELATES

Arterial System Defects

Under normal conditions the **ductus arteriosus** is functionally closed through contraction of its muscular wall shortly after birth to form the **ligamentum arteriosum.** Anatomical closure by means of intima proliferation takes 1 to 3 months. A **patent ductus arteriosus,** one of the most frequently occurring abnormalities of the great vessels (8/10,000 births), especially in premature infants, either may be an isolated abnormality or may accompany other heart defects (Figs. 11.29A and 11.31). In particular, defects that cause large differences between aortic and pulmonary pressures may cause increased blood flow through the ductus, preventing its normal closure.

In **coarctation of the aorta** (Fig. 11.37, *A* and *B*), which occurs in 3.2/10,000 births, the aortic lumen below the origin of the left subclavian artery is significantly narrowed. Since the constriction may be above or below the entrance of the ductus arteriosus, two types, **preductal** and **postductal,** may be distinguished. The cause of aortic narrowing is primarily an abnormality in the media of the aorta, followed by intima proliferations. In the preductal type the ductus arteriosus persists, whereas in the postductal type, which is more common, this channel is usually obliterated. In the latter case collateral circulation between the proximal and distal parts of the aorta is established by way of large intercostal and internal thoracic arteries. In this manner the lower part of the body is supplied with blood.

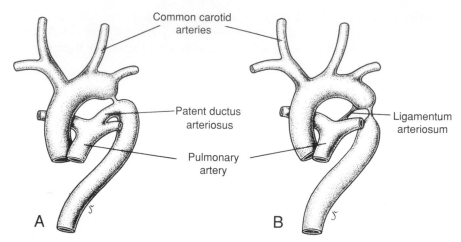

Figure 11.37 Coarctation of the aorta. **A.** Preductal type. **B.** Postductal type. The caudal part of the body is supplied by large hypertrophied intercostal and internal thoracic arteries.

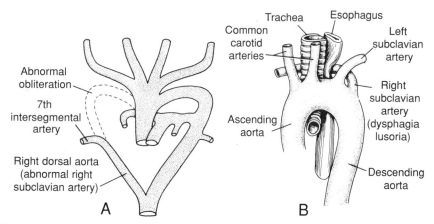

Figure 11.38 Abnormal origin of the right subclavian artery. **A.** Obliteration of the right fourth aortic arch and the proximal portion of the right dorsal aorta with persistence of the distal portion of the right dorsal aorta. **B.** The abnormal right subclavian artery crosses the midline behind the esophagus and may compress it.

Abnormal origin of the right subclavian artery (Fig. 11.38, *A* and *B*) occurs when the artery is formed by the distal portion of the right dorsal aorta and the seventh intersegmental artery. The right fourth aortic arch and the proximal part of the right dorsal aorta are obliterated. With shortening of the aorta between the left common carotid and left subclavian arteries, the origin of the abnormal right subclavian artery finally settles just below that of the left subclavian artery. Since its stem is derived from the right dorsal aorta,

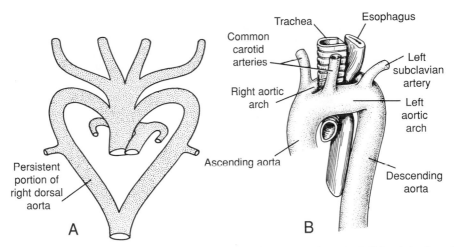

Figure 11.39 Double aortic arch. **A.** Persistence of the distal portion of the right dorsal aorta. **B.** The double aortic arch forms a vascular ring around the trachea and esophagus.

it must cross the midline behind the esophagus to reach the right arm. This location does not usually cause problems with swallowing or breathing, since neither the esophagus nor the trachea is severely compressed.

With a **double aortic arch** the right dorsal aorta persists between the origin of the seventh intersegmental artery and its junction with the left dorsal aorta (Fig. 11.39) A **vascular ring** surrounds the trachea and esophagus and commonly compresses these structures, causing difficulties in breathing and swallowing.

In a **right aortic arch,** the left fourth arch and left dorsal aorta are obliterated and replaced by the corresponding vessels on the right side. Occasionally, when the ligamentum arteriosum lies on the left side and passes behind the esophagus, it causes complaints with swallowing.

An **interrupted aortic arch** is caused by obliteration of the fourth aortic arch on the left side (Fig. 11.40, *A* and *B*). It is frequently combined with an abnormal origin of the right subclavian artery. The ductus arteriosus remains open, and the descending aorta and subclavian arteries are supplied with blood of low oxygen content. The aortic trunk supplies the two common carotid arteries.

VENOUS SYSTEM

In the fifth week, three pairs of major veins can be distinguished: (*a*) the **vitelline veins, or omphalomesenteric veins,** carrying blood from the yolk sac to the sinus venosus; (*b*) the **umbilical veins,** originating in the chorionic villi and carrying oxygenated blood to the embryo; and (*c*) the **cardinal veins,** draining the body of the embryo proper (Fig. 11.41).

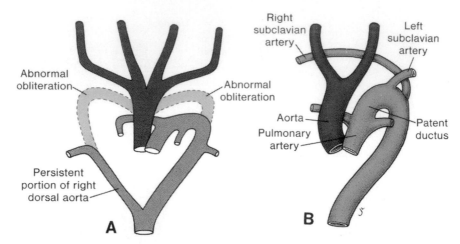

Figure 11.40 A. Obliteration of the fourth aortic arch on the right and left and persistence of the distal portion of the right dorsal aorta. **B.** Case of interrupted aortic arch. The aorta supplies the head; the pulmonary artery, by way of the ductus arteriosus, supplies the rest of the body.

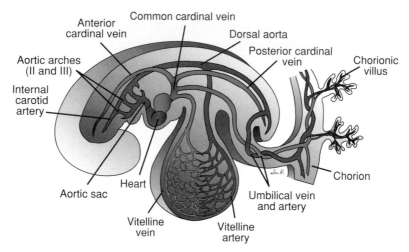

Figure 11.41 Main components of the venous and arterial systems in a 4-mm embryo (end of the fourth week).

Vitelline Veins

Before entering the sinus venosus, the vitelline veins form a plexus around the duodenum and pass through the septum transversum. The liver cords growing into the septum interrupt the course of the veins, and an extensive vascular network, the **hepatic sinusoids,** forms (Fig. 11.42).

With reduction of the left sinus horn, blood from the left side of the liver is rechanneled toward the right, resulting in an enlargement of the right vitelline

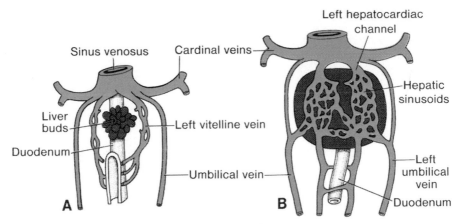

Figure 11.42 Development of the vitelline and umbilical veins during the (**A**) fourth and (**B**) fifth weeks. Note the plexus around the duodenum, formation of the hepatic sinusoids, and initiation of left-to-right shunts between the vitelline veins.

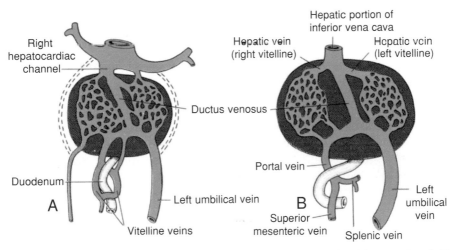

Figure 11.43 Development of vitelline and umbilical veins in the (**A**) second and (**B**) third months. Note formation of the ductus venosus, portal vein, and hepatic portion of the inferior vena cava. The splenic and superior mesenteric veins enter the portal vein.

vein (right hepatocardiac channel). Ultimately the right hepatocardiac channel forms the **hepatocardiac portion of the inferior vena cava.** The proximal part of the left vitelline vein disappears (Fig. 11.43, *A* and *B*). The anastomotic network around the duodenum develops into a single vessel, the **portal vein** (Fig. 11.43*B*). The **superior mesenteric vein,** which drains the primary intestinal loop, derives from the right vitelline vein. The distal portion of the left vitelline vein also disappears (Fig. 11.43, *A* and *B*).

Umbilical Veins

Initially the umbilical veins pass on each side of the liver, but some connect to the hepatic sinusoids (Fig. 11.42, *A* and *B*). The proximal part of both umbilical veins and the remainder of the right umbilical vein then disappear, so that the left vein is the only one to carry blood from the placenta to the liver (Fig. 11.43). With the increase of the placental circulation, a direct communication forms between the left umbilical vein and the right hepatocardiac channel, the **ductus venosus** (Fig. 11.43, *A* and *B*). This vessel bypasses the sinusoidal plexus of the liver. After birth the left umbilical vein and ductus venosus are obliterated and form the **ligamentum teres hepatis** and **ligamentum venosum,** respectively.

Cardinal Veins

Initially the cardinal veins form the main venous drainage system of the embryo. This system consists of the **anterior cardinal veins,** which drain the cephalic part of the embryo, and the **posterior cardinal veins,** which drain the rest of the embryo. The anterior and posterior veins join before entering the sinus horn and form the short **common cardinal veins.** During the fourth week, the cardinal veins form a symmetrical system (Fig. 11.44*A*).

During the fifth to the seventh week a number of additional veins are formed: (*a*) the **subcardinal veins,** which mainly drain the kidneys; (*b*) the **sacrocardinal veins,** which drain the lower extremities; and (*c*) the **supracardinal veins,** which drain the body wall by way of the intercostal veins, taking over the functions of the posterior cardinal veins (Fig. 11.44).

Formation of the vena cava system is characterized by the appearance of anastomoses between left and right in such a manner that the blood from the left is channeled to the right side.

The **anastomosis between the anterior cardinal veins** develops into the **left brachiocephalic vein** (Fig. 11.44, *A* and *B*). Most of the blood from the left side of the head and the left upper extremity is then channeled to the right. The terminal portion of the left posterior cardinal vein entering into the left brachiocephalic vein is retained as a small vessel, the **left superior intercostal vein** (Fig. 11.44*B*). This vessel receives blood from the second and third intercostal spaces. The **superior vena cava** is formed by the right common cardinal vein and the proximal portion of the right anterior cardinal vein.

The **anastomosis between the subcardinal veins** forms the **left renal vein.** When this communication has been established, the left subcardinal vein disappears, and only its distal portion remains as the **left gonadal vein.** Hence the right subcardinal vein becomes the main drainage channel and develops into the **renal segment of the inferior vena cava** (Fig. 11.44*B*).

The **anastomosis between the sacrocardinal veins** forms the **left common iliac vein** (Fig. 11.44*B*). The right sacrocardinal vein becomes the

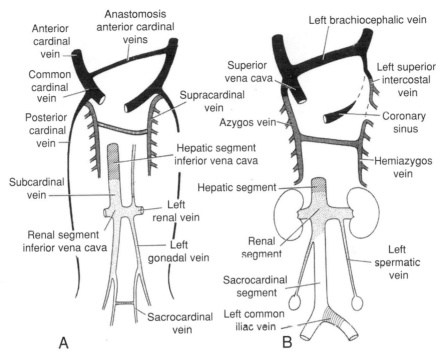

Figure 11.44 Development of the inferior vena cava, azygos vein, and superior vena cava. **A.** Seventh week. The anastomosis lies between the subcardinals, supracardinals, sacrocardinals, and anterior cardinals. **B.** The venous system at birth showing the three components of the inferior vena cava.

sacrocardinal segment of the inferior vena cava. When the renal segment of the inferior vena cava connects with the hepatic segment, which is derived from the right vitelline vein, the inferior vena cava, consisting of hepatic, renal, and sacrocardinal segments, is complete.

With obliteration of the major portion of the posterior cardinal veins, the supracardinal veins assume a greater role in draining the body wall. The 4th to 11th right intercostal veins empty into the right supracardinal vein, which together with a portion of the posterior cardinal vein forms the **azygos vein** (Fig. 11.44). On the left the 4th to 7th intercostal veins enter into the left supracardinal vein, and the left supracardinal vein, then known as the **hemiazygos vein,** empties into the azygos vein (Fig. 11.44*B*).

CLINICAL CORRELATES

Venous System Defects

The complicated development of the vena cava accounts for the fact that deviations from the normal pattern are common.

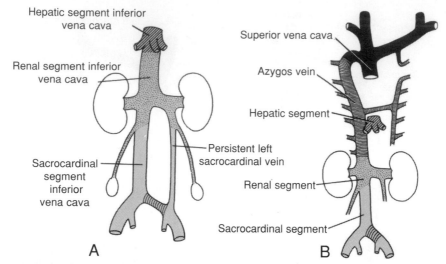

Figure 11.45 A. Double inferior vena cava at the lumbar level arising from the persistence of the left sacrocardinal vein. **B.** Absent inferior vena cava. The lower half of the body is drained by the azygos vein, which enters the superior vena cava. The hepatic vein enters the heart at the site of the inferior vena cava.

A **double inferior vena cava** occurs when the left sacrocardinal vein fails to lose its connection with the left subcardinal vein (Fig. 11.45*A*). The left common iliac vein may or may not be present, but the left gonadal vein remains as in normal conditions.

Absence of the inferior vena cava arises when the right subcardinal vein fails to make its connection with the liver and shunts its blood directly into the right supracardinal vein (Figs. 11.44 and 11.45*B*). Hence the bloodstream from the caudal part of the body reaches the heart by way of the azygos vein and superior vena cava. The hepatic vein enters into the right atrium at the site of the inferior vena cava. Usually this abnormality is associated with other heart malformations.

Left superior vena cava is caused by persistence of the left anterior cardinal vein and obliteration of the common cardinal and proximal part of the anterior cardinal veins on the right (Fig. 11.46*A*). In such a case, blood from the right is channeled toward the left by way of the brachiocephalic vein. The left superior vena cava drains into the right atrium by way of the left sinus horn, that is, the coronary sinus.

A **double superior vena cava** is characterized by the persistence of the left anterior cardinal vein and failure of the left brachiocephalic vein to form (Fig. 11.46*B*). The persistent left anterior cardinal vein, the **left superior vena cava,** drains into the right atrium by way of the coronary sinus.

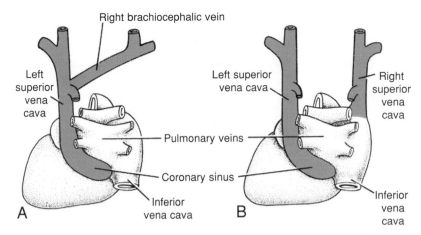

Figure 11.46 A. Left superior vena cava draining into the right atrium by way of the coronary sinus (dorsal view). **B.** Double superior vena cava. The communicating (brachiocephalic) vein between the two anterior cardinals has failed to develop (dorsal view).

Circulation Before and After Birth

FETAL CIRCULATION

Before birth, blood from the placenta, about 80 % saturated with oxygen, returns to the fetus by way of the umbilical vein. On approaching the liver, most of this blood flows through the ductus venosus directly into the inferior vena cava, short-circuiting the liver. A smaller amount enters the liver sinusoids and mixes with blood from the portal circulation (Fig. 11.47). A **sphincter mechanism** in the **ductus venosus,** close to the entrance of the umbilical vein, regulates flow of umbilical blood through the liver sinusoids. This sphincter closes when a uterine contraction renders the venous return too high, preventing a sudden overloading of the heart.

After a short course in the inferior vena cava, where placental blood mixes with deoxygenated blood returning from the lower limbs, it enters the right atrium. Here it is guided toward the oval foramen by the valve of the inferior vena cava, and most of the blood passes directly into the left atrium. A small amount is prevented from doing so by the lower edge of the septum secundum, the **crista dividens,** and remains in the right atrium. Here it mixes with desaturated blood returning from the head and arms by way of the superior vena cava.

From the left atrium, where it mixes with a small amount of desaturated blood returning from the lungs, blood enters the left ventricle and ascending aorta. Since the coronary and carotid arteries are the first branches of the ascending aorta, the heart musculature and the brain are supplied with well-oxygenated blood. Desaturated blood from the superior vena cava flows by way of the right ventricle into the pulmonary trunk. During fetal life, resistance in the

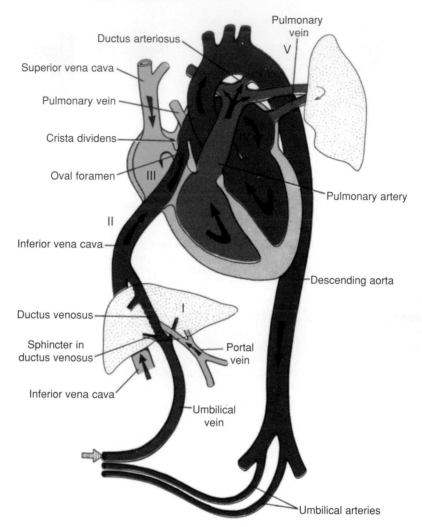

Figure 11.47 Fetal circulation before birth. *Arrows,* direction of blood flow. Note where oxygenated blood mixes with deoxygenated blood: in the liver (*I*), in the inferior vena cava (*II*), in the right atrium (*III*), in the left atrium (*IV*), and at the entrance of the ductus arteriosus into the descending aorta (*V*).

pulmonary vessels is high, such that most of this blood passes directly through the **ductus arteriosus** into the descending aorta, where it mixes with blood from the proximal aorta. After coursing through the descending aorta, blood flows toward the placenta by way of the two umbilical arteries. The oxygen saturation in the umbilical arteries is approximately 58 %.

During its course from the placenta to the organs of the fetus, blood in the umbilical vein gradually loses its high oxygen content as it mixes with desaturated blood. Theoretically, mixing may occur in the following places (Fig. 11.47, *I–V*): in the liver (*I*), by mixture with a small amount of blood returning from the portal system; in the inferior vena cava (*II*), which carries deoxygenated blood returning from the lower extremities, pelvis, and kidneys; in the right atrium (*III*), by mixture with blood returning from the head and limbs; in the left atrium (*IV*), by mixture with blood returning from the lungs; and at the entrance of the ductus arteriosus into the descending aorta (*V*).

CIRCULATORY CHANGES AT BIRTH

Changes in the vascular system at birth are caused by cessation of placental blood flow and the beginning of respiration. Since the ductus arteriosus closes by muscular contraction of its wall, the amount of blood flowing through the lung vessels increases rapidly. This, in turn, raises pressure in the left atrium. Simultaneously, pressure in the right atrium decreases as a result of interruption of placental blood flow. The septum primum is then apposed to the septum secundum, and functionally the oval foramen closes.

To summarize, the following changes occur in the vascular system after birth (Fig. 11.48):

Closure of the umbilical arteries, accomplished by contraction of the smooth musculature in their walls, is probably caused by thermal and mechanical stimuli and a change in oxygen tension. Functionally the arteries close a few minutes after birth, although the actual obliteration of the lumen by fibrous proliferation may take 2 to 3 months. Distal parts of the umbilical arteries form the **medial umbilical ligaments,** and the proximal portions remain open as the **superior vesical arteries** (Fig. 11.48).

Closure of the umbilical vein and ductus venosus occurs shortly after that of the umbilical arteries. Hence blood from the placenta may enter the newborn for some time after birth. After obliteration, the umbilical vein forms the **ligamentum teres hepatis** in the lower margin of the falciform ligament. The ductus venosus, which courses from the ligamentum teres to the inferior vena cava, is also obliterated and forms the **ligamentum venosum.**

Closure of the ductus arteriosus by contraction of its muscular wall occurs almost immediately after birth; it is mediated by **bradykinin,** a substance released from the lungs during initial inflation. Complete anatomical obliteration by proliferation of the intima is thought to take 1 to 3 months. In the adult the obliterated ductus arteriosus forms the **ligamentum arteriosum.**

Closure of the oval foramen is caused by an increased pressure in the left atrium, combined with a decrease in pressure on the right side. The first breath presses the septum primum against the septum secundum. During the first days of life, however, this closure is reversible. Crying by the baby creates a shunt from right to left, which accounts for cyanotic periods in the newborn. Constant apposition gradually leads to fusion of the two septa in about 1 year. In 20% of individuals, however, perfect anatomical closure may never be obtained **(probe patent foramen ovale).**

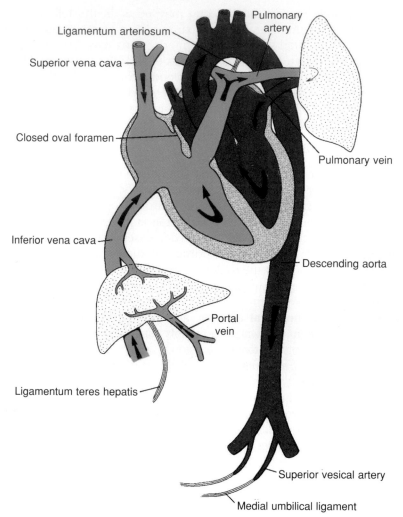

Pulmonary artery

Ligamentum arteriosum

Superior vena cava

Closed oval foramen

Pulmonary vein

Inferior vena cava

Descending aorta

Portal vein

Ligamentum teres hepatis

Superior vesical artery

Medial umbilical ligament

Figure 11.48 Human circulation after birth. Note the changes occurring as a result of the beginning of respiration and interruption of placental blood flow. *Arrows,* direction of blood flow.

Lymphatic System

The lymphatic system begins its development later than the cardiovascular system, not appearing until the fifth week of gestation. The origin of lymphatic vessels is not clear, but they may form from mesenchyme in situ or may arise as saclike outgrowths from the endothelium of veins. Six primary lymph sacs are formed: two **jugular,** at the junction of the subclavian and anterior cardinal veins; two **iliac,** at the junction of the iliac and posterior cardinal veins; one **retroperitoneal,** near the root of the mesentery; and one **cisterna chyli,** dorsal to the retroperitoneal sac. Numerous channels connect the sacs with each other and drain lymph from the limbs, body wall, head, and neck. Two main channels, the right and left thoracic ducts, join the jugular sacs with the cisterna chyli, and soon an anastomosis forms between these ducts. The **thoracic duct** then develops from the distal portion of the right thoracic duct, the anastomosis, and the cranial portion of the left thoracic duct. The **right lymphatic duct** is derived from the cranial portion of the right thoracic duct. Both ducts maintain their original connections with the venous system and empty into the junction of the internal jugular and subclavian veins. Numerous anastomoses produce many variations in the final form of the thoracic duct.

Summary

The entire cardiovascular system—heart, blood vessels, and blood cells—originates from the mesodermal germ layer. Although initially paired, by the 22nd day of development the two tubes (Figs. 11.3 and 11.4) form a single, slightly bent heart tube (Fig. 11.6) consisting of an inner endocardial tube and a surrounding myocardial mantle. During the 4th to 7th weeks the heart divides into a typical four-chambered structure.

Septum formation in the heart in part arises from development of **endocardial cushion** tissue in the atrioventricular canal (**atrioventricular cushions**) and in the conotruncal region (**conotruncal swellings**). Because of the key location of cushion tissue, many cardiac malformations are related to abnormal cushion morphogenesis.

Septum Formation in the Atrium. The **septum primum,** a sickle-shaped crest descending from the roof of the atrium, begins to divide the atrium in two but leaves a lumen, the **ostium primum,** for communication between the two sides (Fig. 11.14). Later, when the ostium primum is obliterated by fusion of the septum primum with the endocardial cushions, the **ostium secundum** is formed by cell death that creates an opening in the septum primum. Finally, a **septum secundum** forms, but an interatrial opening, the **oval foramen,** persists. Only **at birth,** when pressure in the left atrium increases, do the two septa press against each other and close the communication between the two. Abnormalities in the atrial septum may vary from total absence (Fig. 11.19) to a small opening known as **probe patency** of the oval foramen.

Septum Formation in the Atrioventricular Canal. Four **endocardial cushions** surround the atrioventricular canal. Fusion of the opposing superior and inferior cushions divides the orifice into right and left atrioventricular canals. Cushion tissue then becomes fibrous and forms the mitral (bicuspid) valve on the left and the tricuspid valve on the right (Fig. 11.17). Persistence of the common atrioventricular canal (Fig. 11.20) and abnormal division of the canal (Fig. 11.21*B*) are well-known defects.

Septum Formation in the Ventricles. The interventricular septum consists of a thick **muscular** part and a thin **membranous** portion (Fig. 11.25) formed by (*a*) an inferior endocardial atrioventricular cushion, (*b*) the right conus swelling, and (*c*) the left conus swelling (Fig. 11.23). In many cases these three components fail to fuse, resulting in an open interventricular foramen. Although this abnormality may be isolated, it is commonly combined with other compensatory defects (Figs. 11.28 and 11.29).

Septum Formation in the Bulbus. The bulbus is divided into (*a*) the truncus (aorta and pulmonary trunk), (*b*) the conus (outflow tract of the aorta and pulmonary trunk), and (*c*) the trabeculated portion of the right ventricle. The truncus region is divided by the spiral **aorticopulmonary septum** into the two main arteries (Fig. 11.22). The conus swellings divide the outflow tracts of the aortic and pulmonary channels and with tissue from the inferior endocardial cushion close the interventricular foramen (Fig. 11.23). Many vascular abnormalities, such as **transposition of the great vessels** and **pulmonary valvular atresia,** result from abnormal division of the conotruncal region; they may involve neural crest cells that contribute to septum formation in the conotruncal region.

   The aortic arches lie in each of the five pharyngeal arches (Figs. 11.35). Four important derivatives of the original aortic arch system are (*a*) the carotid arteries (third arches); (*b*) the arch of the aorta (left fourth aortic arch); (*c*) the pulmonary artery (sixth aortic arch), which during fetal life is connected to the aorta through the ductus arteriosus; and (*d*) the right subclavian artery formed by the right fourth aortic arch, distal portion of the right dorsal aorta, and the seventh intersegmental artery (Fig. 11.35*B*). The most common vascular aortic arch abnormalities include (*a*) open ductus arteriosus and coarctation of the aorta (Fig. 11.37) and (*b*) persistent right aortic arch and abnormal right subclavian artery (Figs. 11.38 and 11.39), both causing respiratory and swallowing complaints.

   The **vitelline arteries** initially supply the yolk sac but later form the **celiac, superior mesenteric,** and **inferior mesenteric arteries,** which supply the **foregut, midgut,** and **hindgut** regions, respectively.

   The paired **umbilical arteries** arise from the common iliac arteries. After birth the distal portions of these arteries are obliterated to form the **medial umbilical ligaments,** whereas the proximal portions persist as the **internal iliac** and **vesicular arteries.**

Venous System. Three systems can be recognized: (*a*) the **vitelline system,** which develops into the **portal system;** (*b*) the cardinal system, which forms the **caval system;** and (*c*) the **umbilical system,** which disappears after birth. The complicated caval system is characterized by many abnormalities, such as double inferior and superior vena cava and left superior vena cava (Fig. 11.46).

Changes at Birth. During prenatal life the placental circulation provides the fetus with its oxygen, but after birth the lungs take on gas exchange. In the circulatory system the following changes take place at birth and in the first postnatal months: (*a*) the ductus arteriosus closes; (*b*) the oval foramen closes; (*c*) the umbilical vein and ductus venosus close and remain as the **ligamentum teres hepatis** and **ligamentum venosum;** and (*d*) the umbilical arteries form the **medial umbilical ligaments.**

Lymphatic System. The lymphatic system develops later than the cardiovascular system, originating as five sacs: two jugular, two iliac, one retroperitoneal, and one cisterna chyli. Numerous channels form to connect the sacs and provide drainage from other structures. Ultimately the **thoracic duct** forms from anastomosis of the right and left thoracic ducts, the distal part of the right thoracic duct, and the cranial part of the left thoracic duct. The **right lymphatic duct** develops from the cranial part of the right thoracic duct.

Problems to Solve

1. *A prenatal ultrasound of a 35-year-old woman in her 12th week of gestation reveals an abnormal image of the fetal heart. Instead of a four-chambered view provided by the typical cross, a portion just below the crosspiece is missing. What structures constitute the cross, and what defect does this infant probably have?*

2. *A child is born with severe craniofacial defects and transposition of the great vessels. What cell population may play a role in both abnormalities, and what type of insult might have produced this effect?*

3. *What type of tissue is critical for dividing the heart into four chambers and the outflow tract into pulmonary and aortic channels?*

4. *A patient complains about having difficulty swallowing. What vascular abnormality or abnormalities might produce this complaint? What is its embryological origin?*

SUGGESTED READING

Adkins RB, et al.: Dysphagia associated with aortic arch anomaly in adults. *Am Surg* 52:238, 1986.

Basson CT, et al.: Mutations in human *TBX5* cause limb and cardiac malformation in Holt-Oram syndrome. *Nat Genet* 15:30, 1997.

Bruyer HJ, Kargas SA, Levy JM: The causes and underlying developmental mechanisms of congenital cardiovascular malformation: a critical review. *Am J Med Genet* 3:411, 1987.

Clark EB: Cardiac embryology: its relevance to congenital heart disease. *Am J Dis Child* 140:41, 1986.

Coffin D, Poole TJ: Embryonic vascular development: immunohistochemical identification of the origin and subsequent morphogenesis of the major vessel primordia of quail embryos. *Development* 102:735, 1988.

Fishman MC, Chien KR: Fashioning the vertebrate heart: earliest embryonic decisions. *Development* 124:2099, 1997.

Harvey RP: *NK*-2 homeobox genes and heart development. *Dev Biol* 178:203, 1996.

Hirakow R: Development of the cardiac blood vessels in staged human embryos. *Acta Anat* 115:220, 1983.

Ho E, Shimada Y: Formation of the epicardium studied with the scanning electron microscope. *Dev Biol* 66:579, 1978.

Jiang X, Rowitch DH, Soriano P, McMahon AP, Sucov HM: Fate of the mammalian neural crest. *Development* 127:1607, 2000.

Kirklin JW, et al.: Complete transposition of the great arteries: treatment in the current era. *Pediatr Clin North Am* 37:171, 1990.

Li QY, et al.: Holt-Oram syndrome is caused by mutations in *TBX5*, a member of the *Brachyury* (*T*) gene family. *Nat Genet* 15:21, 1997.

Manasek FJ, Burnside MB, Waterman RE: Myocardial cell shape change as a mechanism of embryonic heart looping. *Dev Biol* 29:349, 1972.

Marvin MJ, DiRocco GD, Gardiner A, Bush SA, Lassar AB: Inhibition of Wnt activity induces heart formation from posterior mesoderm. *Genes Dev* 15:316, 2001.

Noden DM: Origins and assembly of avian embryonic blood vessels. *Ann N Y Acad Sci* 588:236, 1990.

Schott JJ, et al.: Congenital heart disease caused by mutations in the transcription factor *NKX2–5*. *Science* 281:108, 1998.

Skandalakis JE, Gray SW: *Embryology for Surgeons: The Embryological Basis for the Treatment of Congenital Anomalies.* 2nd ed. Baltimore, Williams & Wilkins, 1994.

Waldo K, Miyagawa-Tomita S, Kumiski D, Kirby ML: Cardiac neural crest cells provide new insight into septation of the cadiac outflow tract: aortic sac to ventricular septal closure. *Dev Biol* 196:129, 1998.

chapter 12

Respiratory System

Formation of the Lung Buds

When the embryo is approximately 4 weeks old, the **respiratory diverticulum (lung bud)** appears as an outgrowth from the ventral wall of the foregut (Fig. 12.1*A*). The location of the bud along the gut tube is determined by signals from the surrounding mesenchyme, including fibroblast growth factors (FGFs) that "instruct"the endoderm. Hence **epithelium** of the internal lining of the larynx, trachea, and bronchi, as well as that of the lungs, is entirely of **endodermal origin.** The **cartilaginous, muscular,** and **connective tissue** components of the trachea and lungs are derived from **splanchnic mesoderm** surrounding the foregut.

Initially the lung bud is in open communication with the foregut (Fig. 12.1*B*). When the diverticulum expands caudally, however, two longitudinal ridges, the **tracheoesophageal ridges,** separate it from the foregut (Fig. 12.2*A*). Subsequently, when these ridges fuse to form the **tracheoesophageal septum,** the foregut is divided into a dorsal portion, the **esophagus,** and a ventral portion, the **trachea** and **lung buds** (Fig. 12.2, *B* and *C*). The respiratory primordium maintains its communication with the pharynx through the **laryngeal orifice** (Fig. 12.2*D*).

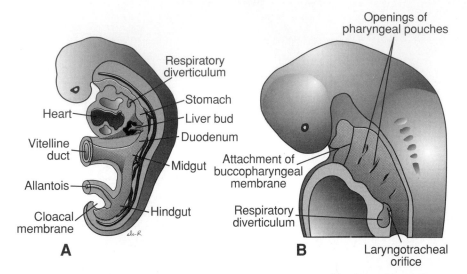

Figure 12.1 A. Embryo of approximately 25 days gestation showing the relation of the respiratory diverticulum to the heart, stomach, and liver. **B.** Sagittal section through the cephalic end of a 5-week embryo showing the openings of the pharyngeal pouches and the laryngotracheal orifice.

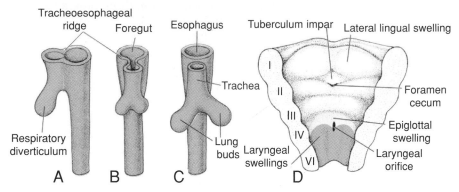

Figure 12.2 A, B, and **C.** Successive stages in development of the respiratory diverticulum showing the tracheoesophageal ridges and formation of the septum, splitting the foregut into esophagus and trachea with lung buds. **D.** The ventral portion of the pharynx seen from above showing the laryngeal orifice and surrounding swelling.

CLINICAL CORRELATES

Abnormalities in partitioning of the esophagus and trachea by the tracheo-esaphageal septum result in **esophageal atresia** with or without **tracheo-esaphageal fistulas (TEFs).** These defects occur in approximately in 1/3000 births, and 90 % result in the upper portion of the esophagus ending in a blind

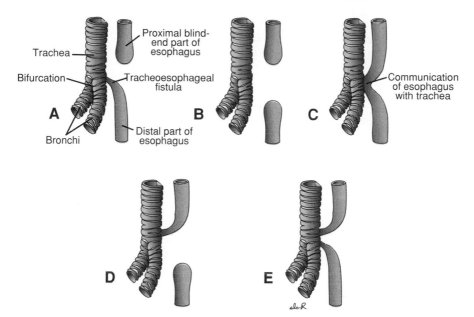

Trachea

Bifurcation

Proximal blind-end part of esophagus

Tracheoesophageal fistula

A   **B**   **C**

Bronchi

Distal part of esophagus

Communication of esophagus with trachea

D   **E**

Figure 12.3 Various types of esophageal atresia and/or tracheoesophageal fistulae. **A.** The most frequent abnormality (90% of cases) occurs with the upper esophagus ending in a blind pouch and the lower segment forming a fistula with the trachea. **B.** Isolated esophageal atresia (4% of cases). **C.** H-type tracheoesophageal fistula (4% of cases). **D** and **E.** Other variations (each 1% of cases).

pouch and the lower segment forming a fistula with the trachea (Fig. 12.3*A*). Isolated esophageal atresia (Fig. 12.3*B*) and H-type TEF without esophageal atresia (Fig. 12.3*C*) each account for 4% of these defects. Other variations (Fig. 12.3, *D* and *E*) each account for approximately 1% of these defects. These abnormalities are associated with other birth defects, including cardiac abnormalities, which occur in 33% of these cases. In this regard TEFs are a component of the **VACTERL** association (**V**ertebral anomalies, **A**nal atresia, **C**ardiac defects, **T**racheoesophageal fistula, **E**sophageal atresia, **R**enal anomalies, and **L**imb defects), a collection of defects of unknown causation, but occurring more frequently than predicted by chance alone.

A complication of some TEFs is polyhydramnios, since in some types of TEF amniotic fluid does not pass to the stomach and intestines. Also, gastric contents and/or amniotic fluid may enter the trachea through a fistula, causing pneumonitis and pneumonia.

Larynx

The internal lining of the larynx originates from endoderm, but the cartilages and muscles originate from mesenchyme of the **fourth** and **sixth pharyngeal**

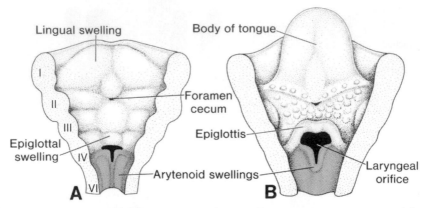

Figure 12.4 Laryngeal orifice and surrounding swellings at successive stages of development. **A.** 6 weeks. **B.** 12 weeks.

arches. As a result of rapid proliferation of this mesenchyme, the laryngeal orifice changes in appearance from a sagittal slit to a T-shaped opening (Fig. 12.4*A*). Subsequently, when mesenchyme of the two arches transforms into the **thyroid, cricoid,** and **arytenoid cartilages,** the characteristic adult shape of the laryngeal orifice can be recognized (Fig. 12.4*B*).

At about the time that the cartilages are formed, the laryngeal epithelium also proliferates rapidly, resulting in a temporary occlusion of the lumen. Subsequently, vacuolization and recanalization produce a pair of lateral recesses, the **laryngeal ventricles.** These recesses are bounded by folds of tissue that differentiate into the **false** and **true vocal cords.**

Since musculature of the larynx is derived from mesenchyme of the fourth and sixth pharyngeal arches, all laryngeal muscles are innervated by branches of the tenth cranial nerve, the **vagus nerve.** The **superior laryngeal** nerve innervates derivatives of the fourth pharyngeal arch, and the **recurrent laryngeal nerve** innervates derivatives of the sixth pharyngeal arch. (For further details on the laryngeal cartilages, see Chapter 15.)

Trachea, Bronchi, and Lungs

During its separation from the foregut, the **lung bud** forms the trachea and two lateral outpocketings, the **bronchial buds** (Fig. 12.2, *B* and *C*). At the beginning of the fifth week, each of these buds enlarges to form right and left main bronchi. The right then forms three secondary bronchi, and the left, two (Fig. 12.5*A*), thus foreshadowing the three lobes on the right side and two on the left (Fig. 12.5, *B* and *C*).

With subsequent growth in caudal and lateral directions, the lung buds expand into the body cavity (Fig. 12.6). **The spaces for the lungs, the pericardioperitoneal canals,** are narrow. They lie on each side of the foregut (Fig. 10.4)

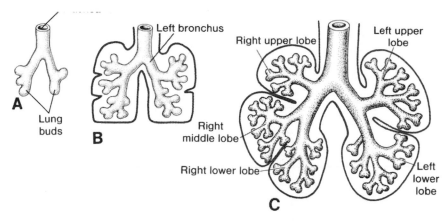

Figure 12.5 Stages in development of the trachea and lungs. **A.** 5 weeks. **B.** 6 weeks. **C.** 8 weeks.

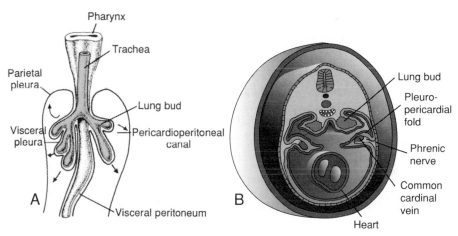

Figure 12.6 Expansion of the lung buds into the pericardioperitoneal canals. At this stage the canals are in communication with the peritoneal and pericardial cavities. **A.** Ventral view of lung buds. **B.** Transverse section through the lung buds showing the pleuropericardial folds that will divide the thoracic portion of the body cavity into the pleural and pericardial cavities.

and are gradually filled by the expanding lung buds. Ultimately the pleuroperitoneal and pleuropericardial folds separate the pericardioperitoneal canals from the peritoneal and pericardial cavities, respectively, and the remaining spaces form the **primitive pleural cavities** (see Chapter 10). The mesoderm, which covers the outside of the lung, develops into the **visceral pleura**. The somatic mesoderm layer, covering the body wall from the inside, becomes the **parietal pleura** (Fig. 12.6*A*). The space between the parietal and visceral pleura is the **pleural cavity** (Fig. 12.7).

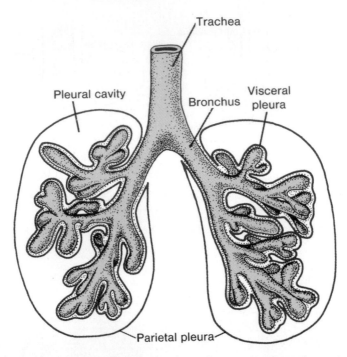

Figure 12.7 Once the pericardioperitoneal canals separate from the pericardial and peritoneal cavities, respectively, the lungs expand in the pleural cavities. Note the visceral and parietal pleura and definitive pleural cavity. The visceral pleura extends between the lobes of the lungs.

During further development, secondary bronchi divide repeatedly in a dichotomous fashion, forming 10 **tertiary (segmental)** bronchi in the right lung and 8 in the left, creating the **bronchopulmonary segments** of the adult lung. By the end of the sixth month, approximately 17 generations of subdivisions have formed. Before the bronchial tree reaches its final shape, however, **an additional 6 divisions form during postnatal life.** Branching is regulated by epithelial-mesenchymal interactions between the endoderm of the lung buds and splanchnic mesoderm that surrounds them. Signals for branching, which emit from the mesoderm, involve members of the fibroblast growth factor (FGF) family. While all of these new subdivisions are occurring and the bronchial tree is developing, the lungs assume a more caudal position, so that by the time of birth the bifurcation of the trachea is opposite the fourth thoracic vertebra.

Maturation of the Lungs (Table 12.1)

Up to the seventh prenatal month, the bronchioles divide continuously into more and smaller canals (canalicular phase) (Fig. 12.8*A*), and the vascular

TABLE 12.1 **Maturation of the Lungs**

Pseudoglandular period	5–16 weeks	Branching has continued to form terminal bronchioles. No respiratory bronchioles or alveoli are present.
Canalicular period	16–26 weeks	Each terminal bronchiole divides into 2 or more respiratory bronchioles, which in turn divide into 3–6 alveolar ducts.
Terminal sac period	26 weeks to birth	Terminal sacs (primitive alveoli) form, and capillaries establish close contact.
Alveolar period	8 months to childhood	Mature alveoli have well-developed epithelial endothelial (capillary) contacts.

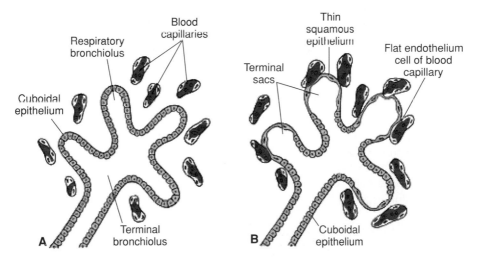

Figure 12.8 Histological and functional development of the lung. **A.** The canalicular period lasts from the 16th to the 26th week. Note the cuboidal cells lining the respiratory bronchioli. **B.** The terminal sac period begins at the end of the sixth and beginning of the seventh prenatal month. Cuboidal cells become very thin and intimately associated with the endothelium of blood and lymph capillaries or form terminal sacs (primitive alveoli).

supply increases steadily. Respiration becomes possible when some of the cells of the cuboidal **respiratory bronchioles** change into thin, flat cells (Fig. 12.8*B*). These cells are intimately associated with numerous blood and lymph capillaries, and the surrounding spaces are now known as **terminal sacs** or **primitive alveoli.** During the seventh month, sufficient numbers of capillaries are present to guarantee adequate gas exchange, and the premature infant is able to survive.

During the last 2 months of prenatal life and for several years thereafter, the number of terminal sacs increases steadily. In addition, cells lining the sacs,

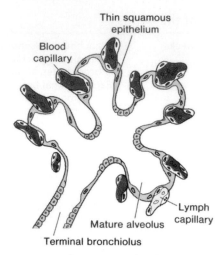

Thin squamous
epithelium

Blood
capillary

Lymph
capillary

Mature alveolus

Terminal bronchiolus

Figure 12.9 Lung tissue in a newborn. Note the thin squamous epithelial cells (also known as alveolar epithelial cells, type I) and surrounding capillaries protruding into mature alveoli.

known as **type I alveolar epithelial cells,** become thinner, so that surrounding capillaries protrude into the alveolar sacs (Fig. 12.9). This intimate contact between epithelial and endothelial cells makes up the **blood-air barrier. Mature alveoli** are not present before birth. In addition to endothelial cells and flat alveolar epithelial cells, another cell type develops at the end of the sixth month. These cells, **type II alveolar epithelial cells,** produce **surfactant,** a phospholipid-rich fluid capable of lowering surface tension at the air-alveolar interface.

Before birth the lungs are full of fluid that contains a high chloride concentration, little protein, some mucus from the bronchial glands, and surfactant from the alveolar epithelial cells (type II). The amount of surfactant in the fluid increases, particularly during the last 2 weeks before birth.

Fetal **breathing movements** begin before birth and cause aspiration of amniotic fluid. These movements are important for stimulating lung development and conditioning respiratory muscles. When respiration begins at birth, most of the lung fluid is rapidly resorbed by the blood and lymph capillaries, and a small amount is probably expelled via the trachea and bronchi during delivery. When the fluid is resorbed from alveolar sacs, surfactant remains deposited as a thin phospholipid coat on alveolar cell membranes. With air entering alveoli during the first breath, the surfactant coat prevents development of an air-water (blood) interface with high surface tension. Without the fatty surfactant layer, the alveoli would collapse during expiration (atelectasis).

Respiratory movements after birth bring air into the lungs, which expand and fill the pleural cavity. Although the alveoli increase somewhat in size, growth of the lungs after birth is due primarily to an increase in the number of respiratory bronchioles and alveoli. It is estimated that only one-sixth of the adult number of alveoli are present at birth. The remaining alveoli are formed during the first 10 years of postnatal life through the continuous formation of new primitive alveoli.

CLINICAL CORRELATES

Surfactant is particularly important for survival of the **premature infant.** When surfactant is insufficient, the air-water (blood) surface membrane tension becomes high, bringing great risk that alveoli will collapse during expiration. As a result, **respiratory distress syndrome** (RDS) develops. This is a common cause of death in the premature infant. In these cases the partially collapsed alveoli contain a fluid with a high protein content, many hyaline membranes, and lamellar bodies, probably derived from the surfactant layer. RDS, which is therefore also known as **hyaline membrane disease,** accounts for approximately 20% of deaths among newborns. Recent development of artificial surfactant and treatment of premature babies with glucocorticoids to stimulate surfactant production have reduced the mortality associated with RDS and allowed survival of some babies as young as 5.5 months of gestation.

Although many abnormalities of the lung and bronchial tree have been described (e.g., blind-ending trachea with absence of lungs and agenesis of one lung), most of these gross abnormalities are rare. Abnormal divisions of the bronchial tree are more common; some result in supernumerary lobules. These variations of the bronchial tree have little functional significance, but they may cause unexpected difficulties during bronchoscopies.

More interesting are **ectopic lung lobes** arising from the trachea or esophagus. It is believed that these lobes are formed from additional respiratory buds of the foregut that develop independently of the main respiratory system.

Most important clinically are **congenital cysts of the lung,** which are formed by dilation of terminal or larger bronchi. These cysts may be small and multiple, giving the lung a honeycomb appearance on radiograph, or they may be restricted to one or more larger ones. Cystic structures of the lung usually drain poorly and frequently cause chronic infections.

Summary

The respiratory system is an outgrowth of the ventral wall of the foregut, and the epithelium of the larynx, trachea, bronchi, and alveoli originates in the endoderm. The cartilaginous, muscular, and connective tissue components arise in the mesoderm. In the fourth week of development, the **tracheoesophageal septum** separates the trachea from the foregut, dividing the foregut into the lung bud anteriorly and the esophagus posteriorly. Contact between the two is maintained through the larynx, which is formed by tissue of the fourth and sixth pharyngeal arches. The lung bud develops into two main bronchi: the right forms three secondary bronchi and three lobes; the left forms two secondary bronchi and two lobes. Faulty partitioning of the foregut by the tracheoesophageal septum causes esophageal atresias and tracheoesophageal fistulas (Fig. 12.3).

After a pseudoglandular (5–16 weeks) and canalicular (16–26 weeks) phase, cells of the cuboidal lined bronchioles change into thin, flat cells, **type I alveolar epithelial cells**, intimately associated with blood and lymph capillaries. In the seventh month, gas exchange between the blood and air in the **primitive alveoli** is possible. Before birth the lungs are filled with fluid with little protein, some mucus, and **surfactant**, which is produced by **type II alveolar epithelial cells** and which forms a phospholipid coat on the alveolar membranes. At the beginning of respiration the lung fluid is resorbed except for the surfactant coat, which prevents the collapse of the alveoli during expiration by reducing the surface tension at the air-blood capillary interface. Absent or insufficient surfactant in the premature baby causes **respiratory distress syndrome (RDS)** because of collapse of the primitive alveoli (**hyaline membrane disease**).

Growth of the lungs after birth is primarily due to an increase in the **number** of respiratory bronchioles and alveoli and not to an increase in the **size** of the alveoli. New alveoli are formed during the first 10 years of postnatal life.

Problems to Solve

1. *A prenatal ultrasound revealed polyhydramnios, and at birth the baby had excessive fluids in its mouth. What type of birth defect might be present, and what is its embryological origin? Would you examine the child carefully for other birth defects? Why?*

2. *A baby born at 6 months gestation is having trouble breathing. Why?*

SUGGESTED READING

Bellusci S, et al.: Fibroblast growth factor 10 (FGF 10) and branching morphogenesis in the embryonic mouse lung. *Development* 124:4867, 1997.

Endo H, Oka T: An immunohistochemical study of bronchial cells producing surfactant protein A in the developing human fetal lung. *Early Hum Dev* 25:149, 1991.

Kozuma S, Nemoto A, Okai T, Mizuno M: Maturational sequence of fetal breathing movements. *Biol Neonate* 60(suppl 1):36, 1991.

Shannon JM, Nielson LD, Gebb SA, Randell SH: Mesenchyme specifies epithelial differentiation in reciprocal recombinants of embryonic lung and trachea. *Dev Dynam* 212:482, 1998.

Whitsett JA: Molecular aspects of the pulmonary surfactant system in the newborn. *In* Chernick V, Mellins RB (eds): *Basic Mechanisms of Pediatric Respiratory Disease: Cellular and Integrative.* Philadelphia, BC Decker, 1991.

chapter 13

Digestive System

Divisions of the Gut Tube

As a result of cephalocaudal and lateral folding of the embryo, a portion of the endoderm-lined yolk sac cavity is incorporated into the embryo to form the **primitive gut.** Two other portions of the endoderm-lined cavity, the **yolk sac** and the **allantois,** remain outside the embryo (Fig. 13.1, A–D).

In the cephalic and caudal parts of the embryo, the primitive gut forms a blind-ending tube, the **foregut** and **hindgut,** respectively. The middle part, the **midgut,** remains temporally connected to the yolk sac by means of the **vitelline duct,** or **yolk stalk** (Fig. 13.1 D).

Development of the primitive gut and its derivatives is usually discussed in four sections: (a) The **pharyngeal gut,** or **pharynx,** extends from the buccopharyngeal membrane to the tracheobronchial diverticulum (Fig. 13.1 D); since this section is particularly important for development of the head and neck, it is discussed in Chapter 15. (b) The **foregut** lies caudal to the pharyngeal tube and extends as far caudally as the liver outgrowth. (c) The **midgut** begins caudal to the liver bud and extends to the junction of the right two-thirds and left third of the transverse colon in the adult. (d) The **hindgut** extends from the left third of the transverse colon to the cloacal membrane (Fig. 13.1). Endoderm forms the epithelial lining of the digestive tract and gives rise to the **parenchyma** of glands, such as the liver and pancreas. Muscle, connective tissue, and peritoneal components of the wall of the gut are derived from splanchnic mesoderm.

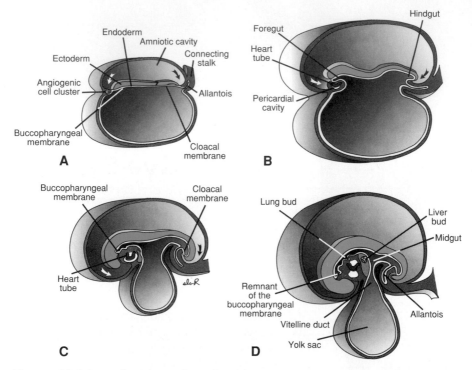

Figure 13.1 Sagittal sections through embryos at various stages of development demonstrating the effect of cephalocaudal and lateral folding on the position of the endoderm-lined cavity. Note formation of the foregut, midgut, and hindgut. **A.** Presomite embryo. **B.** Embryo with 7 somites. **C.** Embryo with 14 somites. **D.** At the end of the first month.

Molecular Regulation of Gut Tube Development

Differentiation of various regions of the gut and its derivatives is dependent upon a reciprocal interaction between the endoderm (epithelium) of the gut tube and surrounding splanchnic mesoderm. Mesoderm dictates the type of structure that will form, for example lungs in the thoracic region and descending colon from the hindgut region, through a **HOX code** similar to the one that establishes the anterior (cranial) posterior (caudal) body axis. Induction of this *HOX* code is a result of **sonic hedgehog (SHH)** expressed throughout the gut endoderm. Thus, in the region of the mid- and hindgut, expression of *SHH* in gut endoderm establishes a nested expression of the *HOX* code in the mesoderm (Fig. 13.2). Once the mesoderm is specified by this code, it instructs the endoderm to form the various components of the mid- and hindgut regions, including the small intestine, cecum, colon, and cloaca (Fig. 13.2). Similar interactions are responsible for partitioning the foregut.

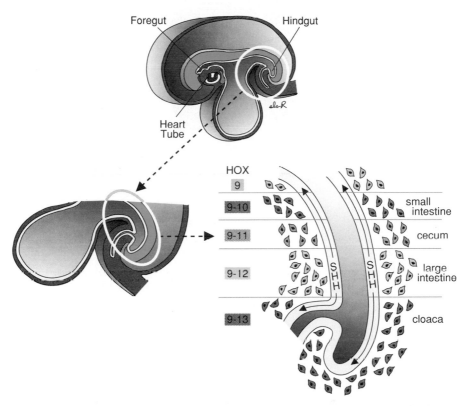

Figure 13.2 Diagrams of the mid- and hindgut regions. The morphogen sonic hedgehog (SHH) is secreted by gut endoderm and induces a nested expression of *HOX* genes in surrounding mesoderm. *HOX* expression then initiates a cascade of genes that "instruct" gut endoderm to differentiate into its regional identities. Signaling between the two tissues is an example of an epithelial-mesenchymal interaction.

Mesenteries

Portions of the gut tube and its derivatives are suspended from the dorsal and ventral body wall by **mesenteries,** double layers of peritoneum that enclose an organ and connect it to the body wall. Such organs are called **intraperitoneal,** whereas organs that lie against the posterior body wall and are covered by peritoneum on their anterior surface only (e.g., the kidneys) are considered **retroperitoneal. Peritoneal ligaments** are double layers of peritoneum (mesenteries) that pass from one organ to another or from an organ to the body wall. Mesenteries and ligaments provide pathways for vessels, nerves, and lymphatics to and from abdominal viscera (Figs. 13.3 and 13.4).

Initially the foregut, midgut, and hindgut are in broad contact with the mesenchyme of the posterior abdominal wall (Fig. 13.3). By the fifth week,

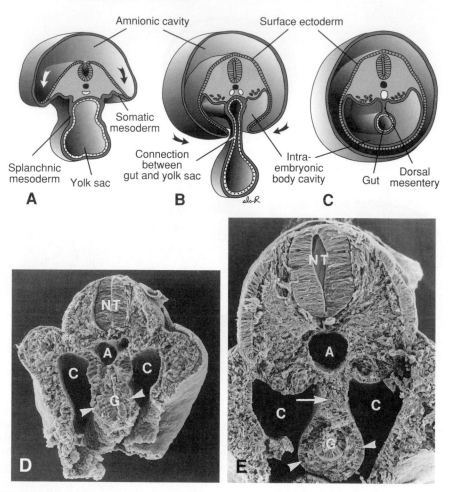

Figure 13.3 Transverse sections through embryos at various stages of development.
A. The intraembryonic cavity, bordered by splanchnic and somatic layers of lateral plate
mesoderm, is in open communication with the extraembryonic cavity. **B.** The intraembryonic cavity is losing its wide connection with the extraembryonic cavity. **C.** At the
end of the fourth week splanchnic mesoderm layers are fused in the midline and form
a double-layered membrane (dorsal mesentery) between right and left halves of the
body cavity. Ventral mesentery exists only in the region of the septum transversum (not
shown). **D.** Scanning electron micrograph of a mouse embryo at approximately the same
stage as in **B.** The mesoderm (*arrowheads*) surrounds the gut tube (*G*) and suspends it
from the posterior body wall into the body cavity (*C*). **E.** Scanning electron micrograph
of a mouse embryo at approximately the same stage as in **C.** Mesoderm suspends the
gut tube from the posterior body wall into the body cavity (*C*) and is thinning to form
the dorsal mesentery (*arrow*). *NT,* neural tube; *A,* dorsal aorta.

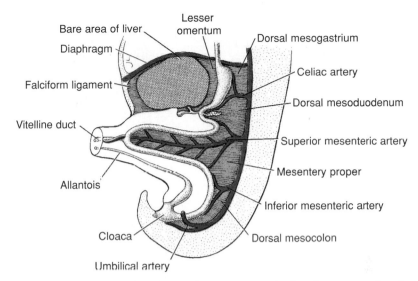

Figure 13.4 Primitive dorsal and ventral mesenteries. The liver is connected to the ventral abdominal wall and to the stomach by the falciform ligament and lesser omentum, respectively. The superior mesenteric artery runs through the mesentery proper and continues toward the yolk sac as the vitelline artery.

however, the connecting tissue bridge has narrowed, and the caudal part of the foregut, the midgut, and a major part of the hindgut are suspended from the abdominal wall by the **dorsal mesentery** (Figs. 13.3C and 13.4), which extends from the lower end of the esophagus to the cloacal region of the hindgut. In the region of the stomach it forms the **dorsal mesogastrium** or **greater omentum;** in the region of the duodenum it forms the dorsal **mesoduodenum;** and in the region of the colon it forms the **dorsal mesocolon.** Dorsal mesentery of the jejunal and ileal loops forms the **mesentery proper.**

Ventral mesentery, which exists only in the region of the terminal part of the esophagus, the stomach, and the upper part of the duodenum (Fig. 13.4), is derived from the **septum transversum.** Growth of the liver into the mesenchyme of the septum transversum divides the ventral mesentery into (*a*) the **lesser omentum,** extending from the lower portion of the esophagus, the stomach, and the upper portion of the duodenum to the liver, and (*b*) the **falciform ligament,** extending from the liver to the ventral body wall (Fig. 13.4).

Foregut

ESOPHAGUS

When the embryo is approximately 4 weeks old, the **respiratory diverticulum (lung bud)** appears at the ventral wall of the foregut at the border with the

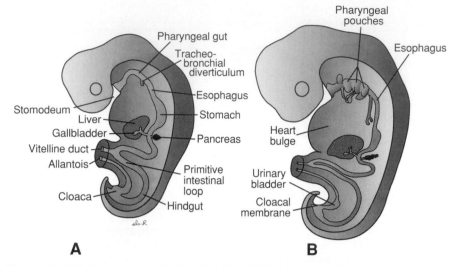

Figure 13.5 Embryos during the fourth **(A)** and fifth **(B)** weeks of development showing formation of the gastrointestinal tract and the various derivatives originating from the endodermal germ layer.

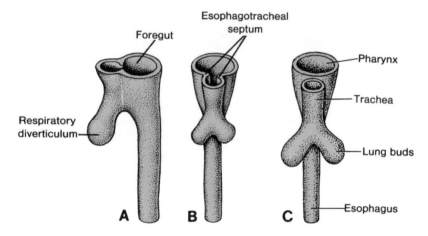

Figure 13.6 Successive stages in development of the respiratory diverticulum and esophagus through partitioning of the foregut. **A.** At the end of the third week (lateral view). **B** and **C.** During the fourth week (ventral view).

pharyngeal gut (Fig. 13.5). The **tracheoesophageal septum** gradually partitions this **diverticulum** from the dorsal part of the foregut (Fig. 13.6). In this manner the foregut divides into a ventral portion, the **respiratory primordium,** and a dorsal portion, the **esophagus** (see Chapter 12).

At first the esophagus is short (Fig. 13.5A), but with descent of the heart and lungs it lengthens rapidly (Fig. 13.5B). The muscular coat, which is formed by surrounding splanchnic mesenchyme, is striated in its upper two-thirds and innervated by the vagus; the muscle coat is smooth in the lower third and is innervated by the splanchnic plexus.

CLINICAL CORRELATES

Esophageal Abnormalities

Esophageal atresia and/or **tracheoesophageal fistula** results either from spontaneous posterior deviation of the **tracheoesophageal septum** or from some mechanical factor pushing the dorsal wall of the foregut anteriorly. In its most common form the proximal part of the esophagus ends as a blind sac, and the distal part is connected to the trachea by a narrow canal just above the bifurcation (Fig. 13.7A). Other types of defects in this region occur much less frequently (Fig. 13.7, B–E) (see Chapter-12).

Atresia of the esophagus prevents normal passage of amniotic fluid into the intestinal tract, resulting in accumulation of excess fluid in the amniotic sac (**polyhydramnios**). In addition to atresias, the lumen of the esophagus may narrow, producing **esophageal stenosis,** usually in the lower third. Stenosis may be caused by incomplete recanalization, vascular abnormalities, or

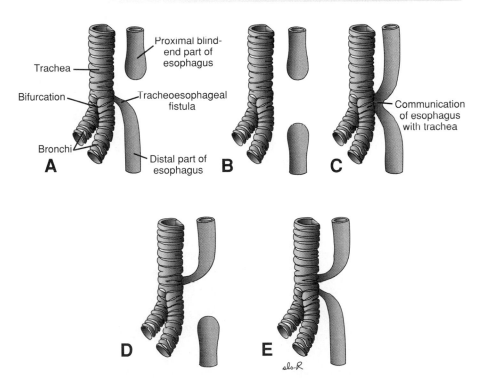

Figure 13.7 Variations of esophageal atresia and/or tracheoesophageal fistula in order of their frequency of appearance: **A,** 90%; **B,** 4%; **C,** 4%; **D,** 1%; and **E,** 1%.

accidents that compromise blood flow. Occasionally the esophagus fails to lengthen sufficiently and the stomach is pulled up into the esophageal hiatus through the diaphragm. The result is a **congenital hiatal hernia.**

STOMACH

The stomach appears as a fusiform dilation of the foregut in the fourth week of development (Fig. 13.8). During the following weeks, its appearance and position change greatly as a result of the different rates of growth in various regions of its wall and the changes in position of surrounding organs. Positional changes of the stomach are most easily explained by assuming that it rotates around a longitudinal and an anteroposterior axis (Fig. 13.8).

The stomach rotates 90° clockwise around its longitudinal axis, causing its left side to face anteriorly and its right side to face posteriorly (Fig. 13.8, *A–C*). Hence the left vagus nerve, initially innervating the left side of the stomach, now innervates the anterior wall; similarly, the right vagus nerve innervates the

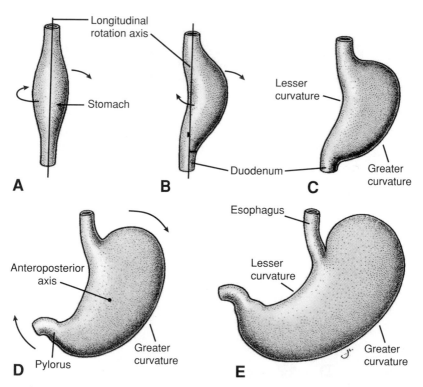

Figure 13.8 A, B, and **C.** Rotation of the stomach along its longitudinal axis as seen anteriorly. **D** and **E.** Rotation of the stomach around the anteroposterior axis. Note the change in position of the pylorus and cardia.

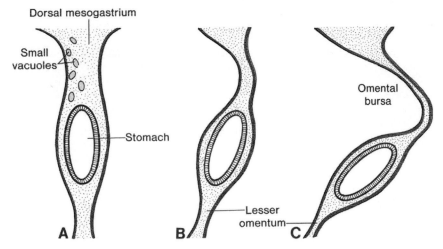

Figure 13.9 A. Transverse section through a 4-week embryo showing intercellular clefts appearing in the dorsal mesogastrium. **B** and **C.** The clefts have fused, and the omental bursa is formed as an extension of the right side of the intraembryonic cavity behind the stomach.

posterior wall. During this rotation the original posterior wall of the stomach grows faster than the anterior portion, forming the **greater** and **lesser curvatures** (Fig. 13.8C).

The cephalic and caudal ends of the stomach originally lie in the midline, but during further growth the stomach rotates around an anteroposterior axis, such that the caudal or **pyloric part** moves to the right and upward and the cephalic or **cardiac portion** moves to the left and slightly downward (Fig. 13.8, D and E). The stomach thus assumes its final position, its axis running from above left to below right.

Since the stomach is attached to the dorsal body wall by the **dorsal mesogastrium** and to the ventral body wall by the **ventral mesogastrium** (Figs. 13.4 and 13.9A), its rotation and disproportionate growth alter the position of these mesenteries. Rotation about the longitudinal axis pulls the dorsal mesogastrium to the left, creating a space behind the stomach called the **omental bursa (lesser peritoneal sac)** (Figs. 13.9 and 13.10). This rotation also pulls the ventral mesogastrium to the right. As this process continues in the fifth week of development, the spleen primordium appears as a mesodermal proliferation between the two leaves of the dorsal mesogastrium (Figs. 13.10 and 13.11). With continued rotation of the stomach, the dorsal mesogastrium lengthens, and the portion between the spleen and dorsal midline swings to the left and fuses with the peritoneum of the posterior abdominal wall (Figs. 13.10 and 13.11). The posterior leaf of the dorsal mesogastrium and the peritoneum along this line of fusion degenerate. The spleen, which remains intraperitoneal, is then connected to the body wall in the region of the left kidney by the **lienorenal ligament** and to the stomach by the **gastrolienal ligament** (Figs. 13.10 and

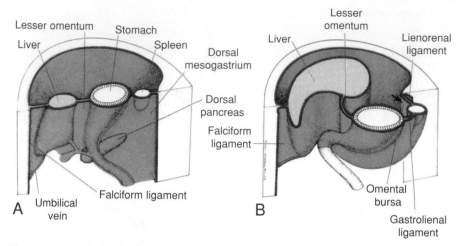

Figure 13.10 A. The positions of the spleen, stomach, and pancreas at the end of the fifth week. Note the position of the spleen and pancreas in the dorsal mesogastrium. **B.** Position of spleen and stomach at the 11th week. Note formation of the omental bursa or lesser peritoneal sac.

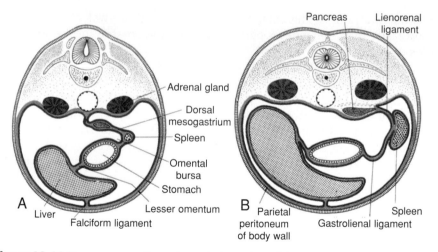

Figure 13.11 Transverse sections through the region of the stomach, liver, and spleen, showing formation of the lesser peritoneal sac, rotation of the stomach, and position of the spleen and tail of the pancreas between the two leaves of the dorsal mesogastrium. With further development, the pancreas assumes a retroperitoneal position.

13.11). Lengthening and fusion of the dorsal mesogastrium to the posterior body wall also determine the final position of the pancreas. Initially the organ grows into the dorsal mesoduodenum, but eventually its tail extends into the dorsal mesogastrium (Fig. 13.10*A*). Since this portion of the dorsal mesogastrium fuses with the dorsal body wall, the tail of the pancreas lies against

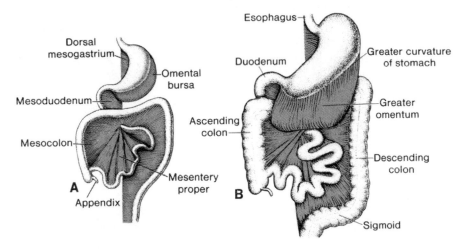

Figure 13.12 A. Derivatives of the dorsal mesentery at the end of the third month. The dorsal mesogastrium bulges out on the left side of the stomach, where it forms part of the border of the omental bursa. **B.** The greater omentum hangs down from the greater curvature of the stomach in front of the transverse colon.

this region (Fig. 13.11). Once the posterior leaf of the dorsal mesogastrium and the peritoneum of the posterior body wall degenerate along the line of fusion, the tail of the pancreas is covered by peritoneum on its anterior surface only and therefore lies in a **retroperitoneal** position. (Organs, such as the pancreas, that are originally covered by peritoneum, but later fuse with the posterior body wall to become retroperitoneal, are said to be **secondarily retroperitoneal.**)

As a result of rotation of the stomach about its anteroposterior axis, the dorsal mesogastrium bulges down (Fig. 13.12). It continues to grow down and forms a double-layered sac extending over the transverse colon and small intestinal loops like an apron (Fig. 13.13*A*). This double-leafed apron is the **greater omentum;** later its layers fuse to form a single sheet hanging from the greater curvature of the stomach (Fig. 13.13*B*). The posterior layer of the greater omentum also fuses with the mesentery of the transverse colon (Fig. 13.13*B*).

The **lesser omentum** and **falciform ligament** form from the ventral mesogastrium, which itself is derived from mesoderm of the septum transversum. When liver cords grow into the septum, it thins to form (*a*) the peritoneum of the liver, (*b*) the **falciform ligament,** extending from the liver to the ventral body wall, and (*c*) the **lesser omentum,** extending from the stomach and upper duodenum to the liver (Figs. 13.14 and 13.15). The free margin of the falciform ligament contains the umbilical vein (Fig. 13.10*A*), which is obliterated after birth to form the **round ligament of the liver** (**ligamentum teres hepatis**). The free margin of the lesser omentum connecting the duodenum and liver (**hepatoduodenal ligament**) contains the bile duct, portal vein, and hepatic artery (**portal triad**). This free margin also forms the roof of the **epiploic foramen of**

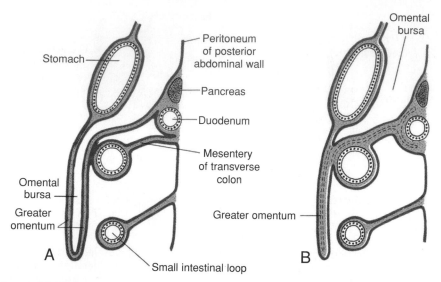

Figure 13.13 A. Sagittal section showing the relation of the greater omentum, stomach, transverse colon, and small intestinal loops at 4 months. The pancreas and duodenum have already acquired a retroperitoneal position. **B.** Similar section as in **A,** in the newborn. The leaves of the greater omentum have fused with each other and with the transverse mesocolon. The transverse mesocolon covers the duodenum, which fuses with the posterior body wall to assume a retroperitoneal position.

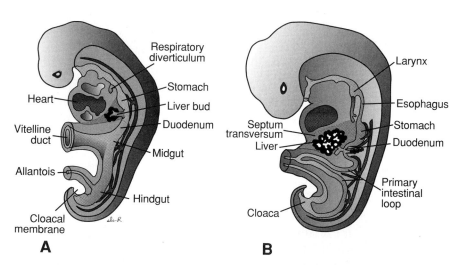

Figure 13.14 A. A 3-mm embryo (approximately 25 days) showing the primitive gastrointestinal tract and formation of the liver bud. The bud is formed by endoderm lining the foregut. **B.** A 5-mm embryo (approximately 32 days). Epithelial liver cords penetrate the mesenchyme of the septum transversum.

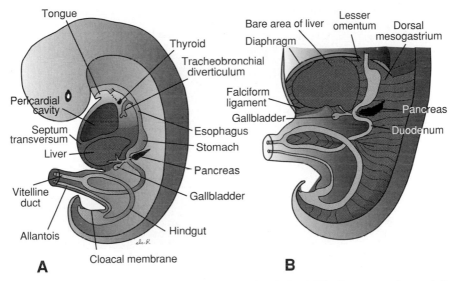

Figure 13.15 A. A 9-mm embryo (approximately 36 days). The liver expands caudally into the abdominal cavity. Note condensation of mesenchyme in the area between the liver and the pericardial cavity, foreshadowing formation of the diaphragm from part of the septum transversum. **B.** A slightly older embryo. Note the falciform ligament extending between the liver and the anterior abdominal wall and the lesser omentum extending between the liver and the foregut (stomach and duodenum). The liver is entirely surrounded by peritoneum except in its contact area with the diaphragm. This is the bare area of the liver.

Winslow, which is the opening connecting the omental bursa (lesser sac) with the rest of the peritoneal cavity (greater sac) (Fig. 13.16).

CLINICAL CORRELATES

Stomach Abnormalities

Pyloric stenosis occurs when the circular and, to a lesser degree, the longitudinal musculature of the stomach in the region of the pylorus hypertrophies. One of the most common abnormalities of the stomach in infants, pyloric stenosis is believed to develop during fetal life. There is an extreme narrowing of the pyloric lumen, and the passage of food is obstructed, resulting in severe vomiting. In a few cases the pylorus is atretic. Other malformations of the stomach, such as duplications and a prepyloric septum, are rare.

DUODENUM

The terminal part of the foregut and the cephalic part of the midgut form the duodenum. The junction of the two parts is directly distal to the origin of the liver bud (Figs. 13.14 and 13.15). As the stomach rotates, the duodenum takes

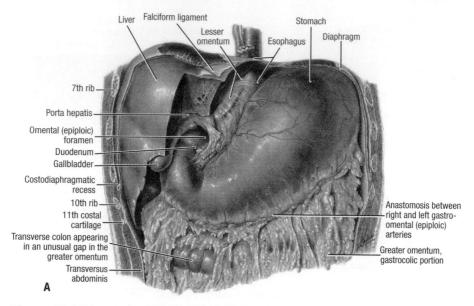

Liver   Falciform ligament   Stomach

Lesser   Diaphragm
omentum   Esophagus

7th rib

Porta hepatis

Omental (epiploic)
foramen

Duodenum

Gallbladder

Costodiaphragmatic
recess

10th rib

11th costal
cartilage

Transverse colon appearing
in an unusual gap in the
greater omentum

Transversus
abdominis

A

Anastomosis between
right and left gastro-
omental (epiploic)
arteries

Greater omentum,
gastrocolic portion

Figure 13.16 Lesser omentum extending from the liver to the lesser curvature of the stomach (hepatogastric ligament) and to the duodenum (hepatoduodenal ligament). In it's free margin anterior to the omental foramen (epiploic foramen of Winslow) are the hepatic artery, portal vein, and bile duct (portal triad).

on the form of a C-shaped loop and rotates to the right. This rotation, together with rapid growth of the head of the pancreas, swings the duodenum from its initial midline position to the left side of the abdominal cavity (Figs. 13.10*A* and 13.17). The duodenum and head of the pancreas press against the dorsal body wall, and the right surface of the dorsal mesoduodenum fuses with the adjacent peritoneum. Both layers subsequently disappear, and the duodenum and head of the pancreas become fixed in a **retroperitoneal position.** The entire pancreas thus obtains a retroperitoneal position. The dorsal mesoduodenum disappears entirely except in the region of the pylorus of the stomach, where a small portion of the duodenum (**duodenal cap**) retains its mesentery and remains intraperitoneal.

During the second month, the lumen of the duodenum is obliterated by proliferation of cells in its walls. However, the lumen is recanalized shortly thereafter (Fig. 13.18, *A* and *B*). Since the **foregut** is supplied by the **celiac artery** and the midgut is supplied by the **superior mesenteric artery,** the duodenum is supplied by branches of both arteries (Fig. 13.14).

LIVER AND GALLBLADDER

The liver primordium appears in the middle of the third week as an outgrowth of the endodermal epithelium at the distal end of the foregut (Figs. 13.14 and

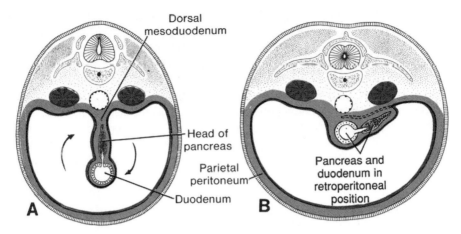

Figure 13.17 Transverse sections through the region of the duodenum at various stages of development. At first the duodenum and head of the pancreas are located in the median plane (**A**), but later they swing to the right and acquire a retroperitoneal position (**B**).

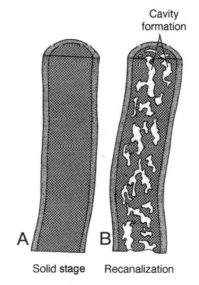

Figure 13.18 Upper portion of the duodenum showing the solid stage (**A**) and cavity formation (**B**) produced by recanalization.

13.15). This outgrowth, the **hepatic diverticulum,** or **liver bud,** consists of rapidly proliferating cells that penetrate the **septum transversum,** that is, the mesodermal plate between the pericardial cavity and the stalk of the yolk sac (Figs. 13.14 and 13.15). While hepatic cells continue to penetrate the septum, the connection between the hepatic diverticulum and the foregut (duodenum) narrows, forming the **bile duct.** A small ventral outgrowth is formed by the bile duct, and this outgrowth gives rise to the **gallbladder** and the **cystic duct** (Figs. 13.15). During further development, epithelial liver cords intermingle with the vitelline and umbilical veins, which form hepatic sinusoids. Liver cords differentiate into the **parenchyma (liver cells)** and form the lining of the biliary ducts. **Hematopoietic cells, Kupffer cells,** and **connective tissue cells** are derived from mesoderm of the septum transversum.

When liver cells have invaded the entire septum transversum, so that the organ bulges caudally into the abdominal cavity, mesoderm of the septum transversum lying between the liver and the foregut and the liver and ventral abdominal wall becomes membranous, forming the **lesser omentum** and **falciform ligament,** respectively. Together, having formed the peritoneal connection between the foregut and the ventral abdominal wall, they are known as the **ventral mesogastrium** (Fig. 13.15).

Mesoderm on the surface of the liver differentiates into visceral peritoneum except on its cranial surface (Fig. 13.15*B*). In this region, the liver remains in contact with the rest of the original septum transversum. This portion of the septum, which consists of densely packed mesoderm, will form the central tendon of the **diaphragm.** The surface of the liver that is in contact with the future diaphragm is never covered by peritoneum; it is the **bare area of the liver** (Fig. 13.15).

In the 10th week of development the weight of the liver is approximately 10% of the total body weight. Although this may be attributed partly to the large numbers of sinusoids, another important factor is its **hematopoietic function.** Large nests of proliferating cells, which produce red and white blood cells, lie between hepatic cells and walls of the vessels. This activity gradually subsides during the last 2 months of intrauterine life, and only small hematopoietic islands remain at birth. The weight of the liver is then only 5% of the total body weight.

Another important function of the liver begins at approximately the 12th week, when bile is formed by hepatic cells. Meanwhile, since the **gallbladder** and **cystic duct** have developed and the cystic duct has joined the hepatic duct to form the **bile duct** (Fig. 13.15), bile can enter the gastrointestinal tract. As a result, its contents take on a dark green color. Because of positional changes of the duodenum, the entrance of the bile duct gradually shifts from its initial anterior position to a posterior one, and consequently, the bile duct passes behind the duodenum (see Figs. 13.21 and 13.22).

Molecular Regulation of Liver Induction

All of the foregut endoderm has the potential to express liver-specific genes and to differentiate into liver tissue. However, this expression is blocked by factors produced by surrounding tissues, including ectoderm, non-cardiac mesoderm, and particularly the notochord (Fig. 13.19). The action of these inhibitors is blocked in the prospective hepatic region by **fibroblast growth factors (FGFs)**

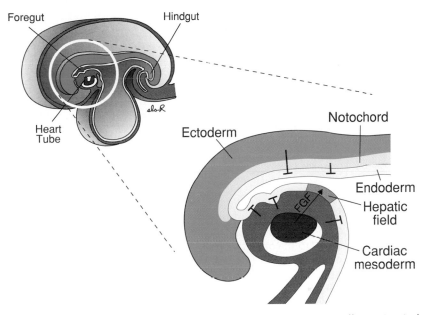

Figure 13.19 Diagrams of the cardiac and hepatic forming regions illustrating induction of liver development. All of the gut endoderm has the potential to form liver tissue, but this capacity is repressed by inhibitors secreted by neighboring mesoderm, ectoderm, and the notochord. Stimulation of hepatic development is achieved by secretion of fibroblast growth factor (FGF) by cardiac mesoderm that inhibits activity of the inhibitors, thereby specifying the hepatic field and initiating liver development. This interaction demonstrates that not all inductive processes are a result of direct signaling by an inducing molecule, but instead may occur by removal of a repressor signal.

secreted by cardiac mesoderm. Thus, the cardiac mesoderm "instructs" gut endoderm to express liver specific genes by inhibiting an inhibitory factor of these same genes. Once this "instruction" is received, cells in the liver field differentiate into both hepatocytes and biliary cell lineages, a process that is at least partially regulated by *hepatocyte nuclear transcription factors (HNF3 and 4)*.

CLINICAL CORRELATES

Liver and Gallbladder Abnormalities

Variations in liver lobulation are common but not clinically significant, **Accessory hepatic ducts** and **duplication of the gallbladder** (Fig. 13.20) are also common and usually asymptomatic. However, they become clinically important under pathological conditions. In some cases the ducts, which pass through a solid phase in their development, fail to recanalize (Fig. 13.20). This defect, **extrahepatic biliary atresia,** occurs in 1/15,000 live births. Among

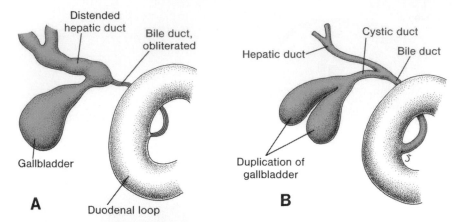

Figure 13.20 A. Obliteration of the bile duct resulting in distention of the gallbladder and hepatic ducts distal to the obliteration. **B.** Duplication of the gallbladder.

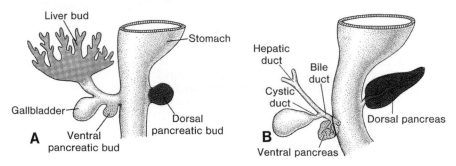

Figure 13.21 Stages in development of the pancreas. **A.** 30 days (approximately 5 mm). **B.** 35 days (approximately 7 mm). Initially the ventral pancreatic bud lies close to the liver bud, but later it moves posteriorly around the duodenum toward the dorsal pancreatic bud.

patients with extrahepatic biliary atresia, 15 to 20% have patent proximal ducts and a correctable defect, but the remainder usually die unless they receive a liver transplant. Another problem with duct formation lies within the liver itself; it is **intrahepatic biliary duct atresia** and **hypoplasia.** This rare abnormality (1/100,000 live births) may be caused by fetal infections. It may be lethal but usually runs an extended benign course.

PANCREAS

The pancreas is formed by two buds originating from the endodermal lining of the duodenum (Fig. 13.21). Whereas the **dorsal pancreatic bud** is in the dorsal mesentery, the **ventral pancreatic bud** is close to the bile duct (Fig. 13.21). When the duodenum rotates to the right and becomes C-shaped, the ventral

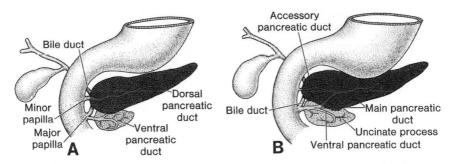

Figure 13.22 A. Pancreas during the sixth week of development. The ventral pancreatic bud is in close contact with the dorsal pancreatic bud. **B.** Fusion of the pancreatic ducts. The main pancreatic duct enters the duodenum in combination with the bile duct at the major papilla. The accessory pancreatic duct (when present) enters the duodenum at the minor papilla.

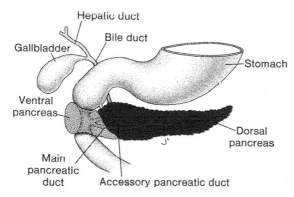

Figure 13.23 Annular pancreas. The ventral pancreas splits and forms a ring around the duodenum, occasionally resulting in duodenal stenosis.

pancreatic bud moves dorsally in a manner similar to the shifting of the entrance of the bile duct (Fig. 13.21). Finally the ventral bud comes to lie immediately below and behind the dorsal bud (Fig. 13.22). Later the parenchyma and the duct systems of the dorsal and ventral pancreatic buds fuse (Fig. 13.22*B*). The ventral bud forms the **uncinate process** and inferior part of the head of the pancreas. The remaining part of the gland is derived from the dorsal bud. The **main pancreatic duct** (of **Wirsung**) is formed by the distal part of the dorsal pancreatic duct and the entire ventral pancreatic duct (Fig. 13.22*B*). The proximal part of the dorsal pancreatic duct either is obliterated or persists as a small channel, the **accessory pancreatic duct** (of **Santorini**). The main pancreatic duct, together with the bile duct, enters the duodenum at the site of the **major papilla;** the entrance of the accessory duct (when present) is at the site of the **minor papilla.** In about 10% of cases the duct system fails to fuse, and the original double system persists.

In the third month of fetal life, **pancreatic islets** (of **Langerhans**) develop from the parenchymatous pancreatic tissue and scatter throughout the pancreas. **Insulin secretion** begins at approximately the fifth month. Glucagon- and somatostatin-secreting cells also develop from parenchymal cells. Splanchnic mesoderm surrounding the pancreatic buds forms the pancreatic connective tissue.

MOLECULAR REGULATION OF PANCREAS DEVELOPMENT

Fibroblast growth factor (**FGF**) and **activin** (a TGF-β family member) produced by the notochord repress *SHH* expression in gut endoderm destined to form pancreas. As a result, expression of the *pancreatic and duodenal homeobox 1 (PDX)* **gene,** a master gene for pancreatic development, is upregulated. Although all of the downstream effectors of pancreas development have not been determined, it appears that expression of the paired homeobox genes *PAX4* and *6* specify the endocrine cell lineage, such that cells expressing both genes become β **(insulin),** δ **(somatostatin),** and γ **(pancreatic polypeptide) cells;** whereas those expressing only *PAX6* become α **(glucagon) cells.**

CLINICAL CORRELATES

Pancreatic Abnormalities

The ventral pancreatic bud consists of two components that normally fuse and rotate around the duodenum so that they come to lie below the dorsal pancreatic bud. Occasionally, however, the right portion of the ventral bud migrates along its normal route, but the left migrates in the opposite direction. In this manner, the duodenum is surrounded by pancreatic tissue, and an **annular pancreas** is formed (Fig. 13.23). The malformation sometimes constricts the duodenum and causes complete obstruction.

Accessory pancreatic tissue may be anywhere from the distal end of the esophagus to the tip of the primary intestinal loop. Most frequently it lies in the mucosa of the stomach and in Meckel's diverticulum, where it may show all of the histological characteristics of the pancreas itself.

Midgut

In the 5-week-old embryo, the midgut is suspended from the dorsal abdominal wall by a short mesentery and communicates with the yolk sac by way of the **vitelline duct** or **yolk stalk** (Figs. 13.1 and 13.15). In the adult the midgut begins immediately distal to the entrance of the bile duct into the duodenum (Fig. 13.15) and terminates at the junction of the proximal two-thirds of the transverse colon with the distal third. Over its entire length the midgut is supplied by the **superior mesenteric artery** (Fig. 13.24).

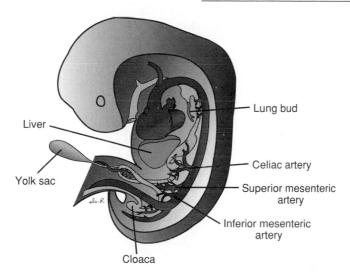

Figure 13.24 Embryo during the sixth week of development, showing blood supply to the segments of the gut and formation and rotation of the primary intestinal loop. The superior mesenteric artery forms the axis of this rotation and supplies the midgut. The celiac and inferior mesenteric arteries supply the foregut and hindgut, respectively.

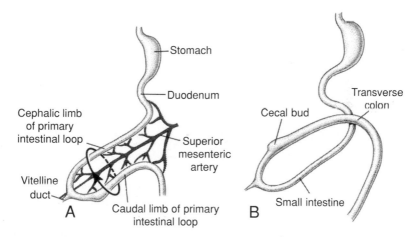

Figure 13.25 A. Primary intestinal loop before rotation (lateral view). The superior mesenteric artery forms the axis of the loop. *Arrow,* counterclockwise rotation. B. Similar view as in A, showing the primary intestinal loop after 180° counterclockwise rotation. The transverse colon passes in front of the duodenum.

Development of the midgut is characterized by rapid elongation of the gut and its mesentery, resulting in formation of the **primary intestinal loop** (Figs. 13.24 and 13.25). At its apex, the loop remains in open connection with the yolk sac by way of the narrow **vitelline duct** (Fig. 13.24). The cephalic limb of the loop develops into the distal part of the duodenum, the jejunum, and

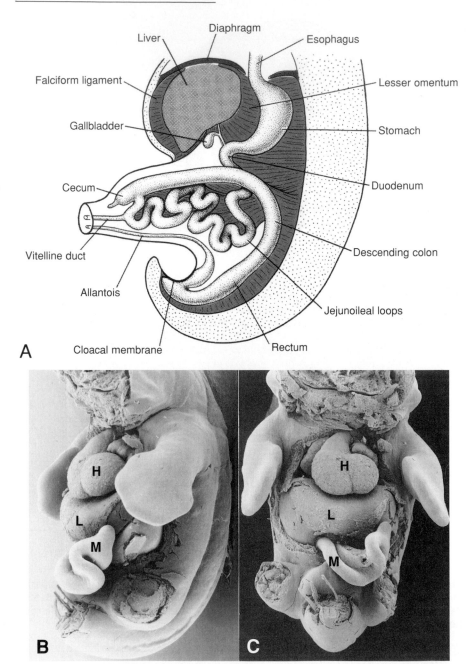

Figure 13.26 Umbilical herniation of the intestinal loops in an embryo of approximately 8 weeks (crown-rump length, 35 mm). Coiling of the small intestinal loops and formation of the cecum occur during the herniation. The first 90° of rotation occurs during herniation; the remaining 180° occurs during the return of the gut to the

(*continues on page 307*)

part of the ileum. The caudal limb becomes the lower portion of the ileum, the cecum, the appendix, the ascending colon, and the proximal two-thirds of the transverse colon.

PHYSIOLOGICAL HERNIATION

Development of the primary intestinal loop is characterized by rapid elongation, particularly of the cephalic limb. As a result of the rapid growth and expansion of the liver, the abdominal cavity temporarily becomes too small to contain all the intestinal loops, and they enter the extraembryonic cavity in the umbilical cord during the sixth week of development (**physiological umbilical herniation**) (Fig. 13.26).

ROTATION OF THE MIDGUT

Coincident with growth in length, the primary intestinal loop rotates around an axis formed by the **superior mesenteric artery** (Fig. 13.25). When viewed from the front, this rotation is counterclockwise, and it amounts to approximately 270° when it is complete (Figs. 13.24 and 13.25). Even during rotation, elongation of the small intestinal loop continues, and the jejunum and ileum form a number of coiled loops (Fig. 13.26). The large intestine likewise lengthens considerably but does not participate in the coiling phenomenon. Rotation occurs during herniation (about 90°) as well as during return of the intestinal loops into the abdominal cavity (remaining 180°) (Fig. 13.27).

RETRACTION OF HERNIATED LOOPS

During the 10th week, herniated intestinal loops begin to return to the abdominal cavity. Although the factors responsible for this return are not precisely known, it is thought that regression of the mesonephric kidney, reduced growth of the liver, and expansion of the abdominal cavity play important roles.

The proximal portion of the jejunum, the first part to reenter the abdominal cavity, comes to lie on the left side (Fig. 13.27*A*). The later returning loops gradually settle more and more to the right. The **cecal bud,** which appears at about the sixth week as a small conical dilation of the caudal limb of the

abdominal cavity in the third month. **B.** Scanning electron micrograph of a lateral view of a mouse embryo at approximately the same stage as in **A,** with the body wall and amnion removed. The heart (*H*) occupies most of the thoracic region and the liver (*L*) most of the abdomen. Herniated midgut (*M*) is just beginning to coil and protrudes from the abdomen. **C.** Frontal view of the embryo in **B.** Note the extreme size of the liver, which is serving a hematopoietic function at this time, and the initial rotation of the herniated midgut. The diaphragm between the heart and liver has been removed.

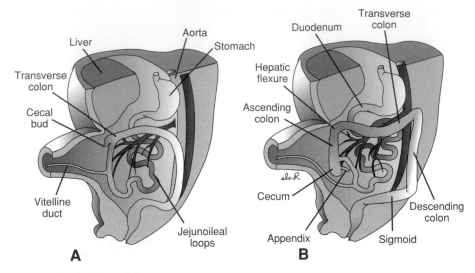

Figure 13.27 A. Anterior view of the intestinal loops after 270° counterclockwise rotation. Note the coiling of the small intestinal loops and the position of the cecal bud in the right upper quadrant of the abdomen. **B.** Similar view as in **A,** with the intestinal loops in their final position. Displacement of the cecum and appendix caudally places them in the right lower quadrant of the abdomen.

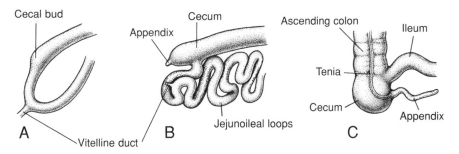

Figure 13.28 Successive stages in development of the cecum and appendix. **A.** 7 weeks. **B.** 8 weeks. **C.** Newborn.

primary intestinal loop, is the last part of the gut to reenter the abdominal cavity. Temporarily it lies in the right upper quadrant directly below the right lobe of the liver (Fig. 13.27*A*). From here it descends into the right iliac fossa, placing the **ascending colon** and **hepatic flexure** on the right side of the abdominal cavity (Fig. 13.27*B*). During this process the distal end of the cecal bud forms a narrow diverticulum, the **appendix** (Fig. 13.28).

Since the appendix develops during descent of the colon, its final position frequently is posterior to the cecum or colon. These positions of the appendix are called **retrocecal** or **retrocolic,** respectively (Fig. 13.29).

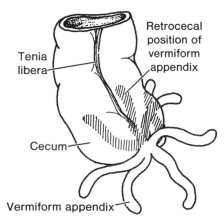

Figure 13.29 Various positions of the appendix. In about 50% of cases the appendix is retrocecal or retrocolic.

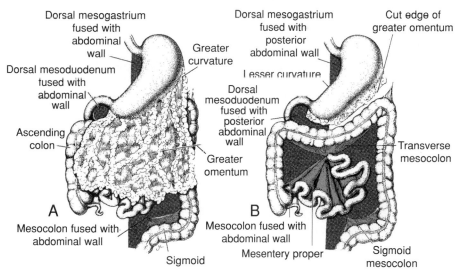

Figure 13.30 Frontal view of the intestinal loops with (**A**) and after removal of (**B**) the greater omentum. *Gray areas,* parts of the dorsal mesentery that fuse with the posterior abdominal wall. Note the line of attachment of the mesentery proper.

MESENTERIES OF THE INTESTINAL LOOPS

The mesentery of the primary intestinal loop, the **mesentery proper,** undergoes profound changes with rotation and coiling of the bowel. When the caudal limb of the loop moves to the right side of the abdominal cavity, the dorsal mesentery twists around the origin of the **superior mesenteric artery** (Fig. 13.24). Later, when the ascending and descending portions of the colon obtain their definitive positions, their mesenteries press against the peritoneum of the posterior abdominal wall (Fig. 13.30). After fusion of these layers, the ascending and descending colons are permanently anchored in a retroperitoneal position. The appendix, lower end of the cecum, and sigmoid colon, however, retain their free mesenteries (Fig. 13.30B).

The fate of the transverse mesocolon is different. It fuses with the posterior wall of the greater omentum (Fig. 13.13) but maintains its mobility. Its line of attachment finally extends from the hepatic flexure of the ascending colon to the splenic flexure of the descending colon (Fig. 13.30B).

The mesentery of the jejunoileal loops is at first continuous with that of the ascending colon (Fig. 13.12A). When the mesentery of the ascending mesocolon fuses with the posterior abdominal wall, the mesentery of the jejunoileal loops obtains a new line of attachment that extends from the area where the duodenum becomes intraperitoneal to the ileocecal junction (Fig. 13.30B).

CLINICAL CORRELATES

Abnormalities of the Mesenteries

Normally the ascending colon, except for its most caudal part (approximately 1 inch), fuses to the posterior abdominal wall and is covered by peritoneum on its anterior surface and sides. Persistence of a portion of the mesocolon gives rise to a **mobile cecum.** In the most extreme form, the mesentery of the ascending colon fails to fuse with the posterior body wall. Such a long mesentery allows abnormal movements of the gut or even **volvulus** of the cecum and colon. Similarly, incomplete fusion of the mesentery with the posterior body wall may give rise to retrocolic pockets behind the ascending mesocolon. A **retrocolic hernia** is entrapment of portions of the small intestine behind the mesocolon.

Body Wall Defects

Omphalocele (Fig. 13.31, A and B) involves herniation of abdominal viscera through an enlarged umbilical ring. The viscera, which may include liver, small and large intestines, stomach, spleen, or gallbladder, are covered by amnion. The origin of the defect is a failure of the bowel to return to the body cavity from its physiological herniation during the 6th to 10th weeks. Omphalocele occurs in 2.5/10,000 births and is associated with a high rate of mortality (25%) and severe malformations, such as cardiac anomalies (50%) and neural tube defects (40%). Approximately half of live-born infants with omphalocele have chromosomal abnormalities.

Gastroschisis (Fig. 13.31C) is a herniation of abdominal contents through the body wall directly into the amniotic cavity. It occurs lateral to the umbilicus usually on the right, through a region weakened by regression of the right umbilical vein, which normally disappears. Viscera are not covered by peritoneum or amnion, and the bowel may be damaged by exposure to amniotic fluid. Gastroschisis occurs in 1/10,000 births but is increasing in frequency, especially among young women; this increase may be related to cocaine use.

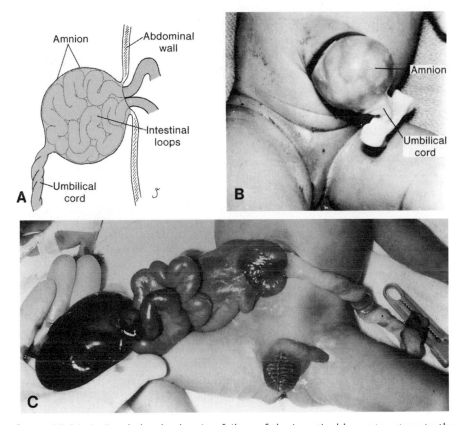

Figure 13.31 A. Omphalocele showing failure of the intestinal loops to return to the body cavity after physiological herniation. The herniated loops are covered by amnion. **B.** Omphalocele in a newborn. **C.** Newborn with gastroschisis. Loops of bowel return to the body cavity but herniate again through the body wall, usually to the right of the umbilicus in the region of the regressing right umbilical vein. Unlike omphalocele, the defect is not covered by amnion.

Unlike omphalocele, gastroschisis is not associated with chromosome abnormalities or other severe defects, so the survival rate is excellent. Volvulus (rotation of the bowel) resulting in a compromised blood supply may, however, kill large regions of the intestine and lead to fetal death.

Vitelline Duct Abnormalities

In 2 to 4% of people, a small portion of the **vitelline duct** persists, forming an outpocketing of the ileum, **Meckel's diverticulum** or **ileal diverticulum** (Fig. 13.32*A*). In the adult, this diverticulum, approximately 40 to 60 cm from the ileocecal valve on the antimesenteric border of the ileum, does not usually

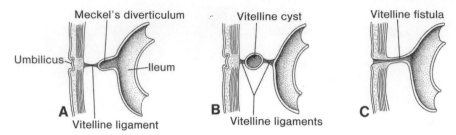

Figure 13.32 Remnants of the vitelline duct. **A.** Meckel's, or ileal, diverticulum combined with fibrous cord (vitelline ligament). **B.** Vitelline cyst attached to the umbilicus and wall of the ileum by vitelline ligaments. **C.** Vitelline fistula connecting the lumen of the ileum with the umbilicus.

cause any symptoms. However, when it contains heterotopic pancreatic tissue or gastric mucosa, it may cause ulceration, bleeding, or even perforation. Sometimes both ends of the vitelline duct transform into fibrous cords, and the middle portion forms a large cyst, an **enterocystoma,** or **vitelline cyst** (Fig. 13.32*B*). Since the fibrous cords traverse the peritoneal cavity, intestinal loops may twist around the fibrous strands and become obstructed, causing strangulation or volvulus. In another variation the vitelline duct remains patent over its entire length, forming a direct communication between the umbilicus and the intestinal tract. This abnormality is known as an **umbilical fistula,** or **vitelline fistula** (Fig. 13.32*C*). A fecal discharge may then be found at the umbilicus.

Gut Rotation Defects

Abnormal rotation of the intestinal loop may result in twisting of the intestine (**volvulus**) and a compromise of the blood supply. Normally the primary intestinal loop rotates 270° counterclockwise. Occasionally, however, rotation amounts to 90° only. When this occurs, the colon and cecum are the first portions of the gut to return from the umbilical cord, and they settle on the left side of the abdominal cavity (Fig. 13.33*A*). The later returning loops then move more and more to the right, resulting in **left-sided colon.**

Reversed rotation of the intestinal loop occurs when the primary loop rotates 90° clockwise. In this abnormality the transverse colon passes behind the duodenum (Fig. 13.33*B*) and lies behind the superior mesenteric artery.

Duplications of intestinal loops and cysts may occur anywhere along the length of the gut tube. They are most frequently found in the region of the ileum, where they may vary from a long segment to a small diverticulum. Symptoms usually occur early in life, and 33 % are associated with other defects, such as intestinal atresias, imperforate anus, gastroschisis, and omphalocele. Their origin is unknown, although they may result from abnormal proliferations of gut parenchyma.

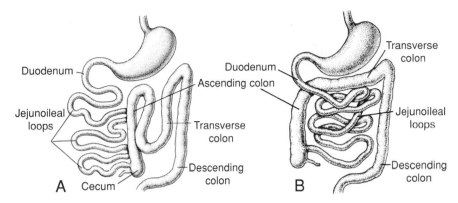

Duodenum

Jejunoileal loops

A Cecum

Duodenum

Ascending colon

Transverse colon

Descending colon

Transverse colon

Jejunoileal loops

Descending colon

B

Figure 13.33 A. Abnormal rotation of the primary intestinal loop. The colon is on the left side of the abdomen, and the small intestinal loops are on the right. The ileum enters the cecum from the right. **B.** The primary intestinal loop is rotated 90° clockwise (reversed rotation). The transverse colon passes behind the duodenum.

Gut Atresias and Stenoses

Atresias and stenoses may occur anywhere along the intestine. Most occur in the duodenum, fewest occur in the colon, and equal numbers occur in the jejunum and ileum (1/1500 births). Atresias in the upper duodenum are probably due to a lack of recanalization (Fig. 13.18). From the distal portion of the duodenum caudally, however, stenoses and atresias are most likely caused by **vascular "accidents."** These accidents may be caused by malrotation, volvulus, gastroschisis, omphalocele, and other factors. As a result, blood supply to a region of the bowel is compromised and a segment dies, resulting in narrowing or complete loss of that region. In 50% of cases a region of the bowel is lost, and in 20% a fibrous cord remains (Fig. 13.34, *A* and *B*). In another 20% there is narrowing, with a thin diaphragm separating the larger and smaller pieces of bowel (Fig. 13.34*C*). Stenoses and multiple atresias account for the remaining 10% of these defects, with a frequency of 5% each (Fig. 13.34*D*). **Apple peel atresia** accounts for 10% of atresias. The atresia is in the proximal jejunum, and the intestine is short, with the portion distal to the lesion coiled around a mesenteric remnant (Fig. 13.35). Babies with this defect have low birth weight and other abnormalities.

Hindgut

The hindgut gives rise to the distal third of the transverse colon, the descending colon, the sigmoid, the rectum, and the upper part of the anal canal. The endoderm of the hindgut also forms the internal lining of the bladder and urethra (see Chapter 14).

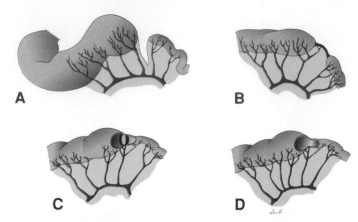

Figure 13.34 The most commonly occurring bowel atresias and stenoses. **A,** the most common, occurs in 50% of cases; **B** and **C** occur in 20% each of cases, and **D** occurs in 5% of cases. Most are caused by vascular accidents; those in the upper duodenum may be caused by a lack of recanalization. Atresias (**A,B,** and **C**) occur in 95% of cases, and stenoses (**D**), in only 5%.

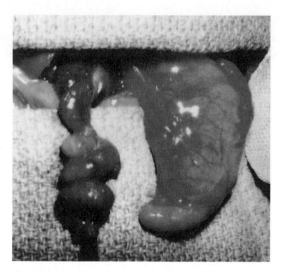

Figure 13.35 Apple peel atresia, which occurs in the jejunum and accounts for 10% of bowel atresias. The affected portion of the bowel is coiled around a remnant of mesentery.

The terminal portion of the hindgut enters into the posterior region of the cloaca, the primitive **anorectal canal;** the allantois enters into the anterior portion, the primitive **urogenital sinus** (Fig 13.36*A*). The cloaca itself is an endoderm-lined cavity covered at its ventral boundary by surface ectoderm. This boundary between the endoderm and the ectoderm forms the **cloacal membrane** (Fig. 13.36). A layer of mesoderm, the **urorectal septum,** separates

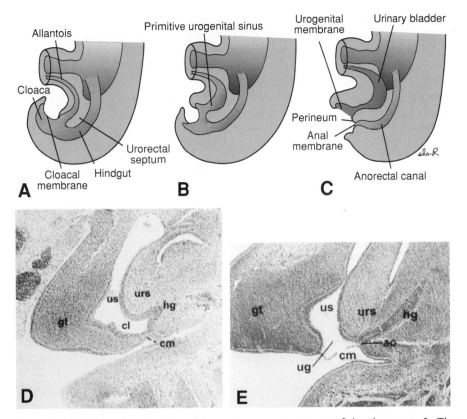

Figure 13.36 Cloacal region in embryos at successive stages of development. **A.** The hindgut enters the posterior portion of the cloaca, the future anorectal canal; the allantois enters the anterior portion, the future urogenital sinus. The urorectal septum is formed by merging of the mesoderm covering the allantois and the yolk sac (Fig. 13.1 *D*). The cloacal membrane, which forms the ventral boundary of the cloaca, is composed of ectoderm and endoderm. **B.** As caudal folding of the embryo continues, the urorectal septum moves closer to the cloacal membrane, although it never contacts this structure. **C.** Lengthening of the genital tubercle pulls the urogenital portion of the cloaca anteriorly; breakdown of the cloacal membrane creates an opening for the hindgut and one for the urogenital sinus. The tip of the urorectal septum forms the perineal body. **D.** Histological section through the cloacal region of a 6-week human embryo similar to that depicted in **B.** The cloaca (*cl*) has a smaller posterior region at the opening of the hindgut (*hg*) and a larger anterior region, the urogenital sinus (*us*). The urorectal septum (*urs*) partially divides the two regions, and the cloacal membrane (*cm*) forms a boundary at the caudal limit of the cloacal cavity; *gt,* genital tubercle. **E.** Histological section though the cloacal region of a 7-week human embryo similar to **C.** The urorectal septum (*urs*) lies close to the cloacal membrane (*cm*), which is just beginning to break down, leaving the anal opening (*ao*) for the hindgut (*hg*) and a separate opening for the urogenital sinus (*us*). The tip of the urorectal septum will form the perineal body. Growth of the genital tubercle (*gt*) will change the shape of the urogenital sinus, which will eventually be closed by fusion of the urethral folds in the male. In the female the opening will remain as the vestibule to the vagina and urethra (see Chapter 14).

the region between the allantois and hindgut. This septum is derived from the merging of mesoderm covering the yolk sac and surrounding the allantois (Figs. 13.1 and 13.36). As the embryo grows and caudal folding continues, the tip of the urorectal septum comes to lie close to the cloacal membrane, although the two structures never make contact (Figs. 13.36, *B* and *D*). At the end of the seventh week, the cloacal membrane ruptures, creating the anal opening for the hindgut and a ventral opening for the urogenital sinus. Between the two, the tip of the urorectal septum forms the perineal body (13.36*C* and *E*). At this time, proliferation of ectoderm closes the caudalmost region of the anal canal. During the ninth week, this region recanalizes. Thus, the caudal part of the anal canal originates in the ectoderm, and it is supplied by the **inferior rectal arteries,** branches of the **internal pudendal arteries.** The cranial part of the anal canal originates in the endoderm and is supplied by **the superior rectal artery,** a continuation of the **inferior mesenteric artery,** the artery of the hindgut. The junction between the endodermal and ectodermal regions of the anal canal is delineated by the **pectinate line,** just below the anal columns. At this line, the epithelium changes from columnar to stratified squamous epithelium.

CLINICAL CORRELATES

Hindgut Abnormalities

Rectoanal atresias, and **fistulas,** which occur in 1/5000 live births, are caused by abnormalities in formation of the cloaca. Thus, if the posterior portion of the cloaca is too small and consequently the posterior cloacal membrane is short, the opening of the hindgut shifts anteriorly. If the defect in the cloaca is small, the shift is small, causing a low opening of the hindgut into the vagina or urethra (Figs. 13.37, *A* and *B*). If the posterior region of the cloaca is very small, the location of the hindgut opening shifts more anteriorly to a higher location (Fig. 13.37*C*). Thus, rectoanal atresias and fistulas are due to ectopic positioning of the anal opening and not to defects in the urorectal septum. Low lesions are twice as common as high ones, with the intermediate variety being the least common. Approximately 50% of children with rectoanal atresias have other birth defects.

With **imperforate anus,** there is no anal opening. This defect occurs because of a lack of recanalization of the lower portion of the anal canal (Fig. 13.37*D*).

Congenital megacolon is due to an absence of parasympathetic ganglia in the bowel wall (**aganglionic megacolon** or **Hirschsprung disease**). These ganglia are derived from neural crest cells that migrate from the neural folds to the wall of the bowel. Mutations in the *RET* gene, a tyrosine kinase receptor involved in crest cell migration (see Chapter 19), can result in congenital megacolon. In most cases the rectum is involved, and in 80% the defect extends to the midpoint of the sigmoid. In only 10 to 20% are the transverse and right-side colonic segments involved, and in 3% the entire colon is affected.

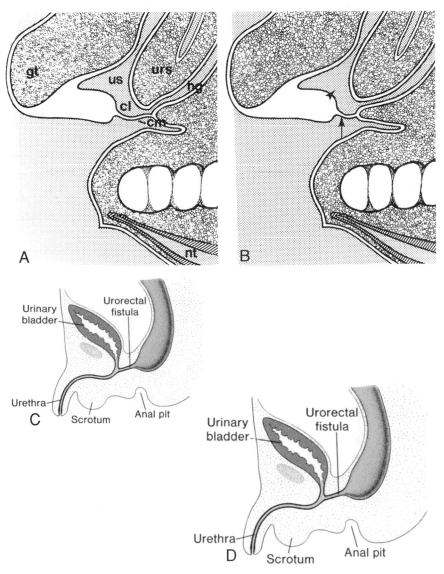

Figure 13.37 A. Normal development of the cloacal region at 7 weeks. The anterior portion of the cloaca (*cl*) forms the urogenital sinus (*us*); the posterior portion extends to the opening of the hindgut (*hg*). The cloacal membrane (*cm*) closes the cloaca and extends posteriorly beneath the end of the hindgut. *Urs,* urorectal septum; *nt,* neural tube; *gt,* genital tubercle. **B.** The cloacal region of a 7-week embryo showing a decrease in the size of the posterior portion of the cloaca and shortening of the cloacal membrane (*arrow*). Such a defect causes ectopic placement of the anal opening into the urogenital sinus (*arrowhead*) and a low urorectal fistula. **C.** High urorectal fistula resulting from a large decrease in size of the posterior portion of the cloaca and cloacal membrane that shifts the opening of the hindgut further anteriorly. **D.** Imperforate anus. The anal canal fails to recanalize, leaving a diaphragm between the upper and lower portions of the anal canal.

Summary

The epithelium of the digestive system and the parenchyma of its derivatives originate in the endoderm; connective tissue, muscular components, and peritoneal components originate in the mesoderm. Differentiation of the gut and its derivatives depends upon reciprocal interactions between the gut endoderm (epithelium) and its surrounding mesoderm. *HOX* genes in the mesoderm are induced by *sonic hedgehog* (*SHH*) secreted by gut endoderm and regulate the craniocaudal organization of the gut and its derivatives. The gut system extends from the buccopharyngeal membrane to the cloacal membrane (Fig. 13.1) and is divided into the pharyngeal gut, foregut, midgut, and hindgut. The pharyngeal gut gives rise to the pharynx and related glands (see Chapter 15).

The **foregut** gives rise to the esophagus, the trachea and lung buds, the stomach, and the duodenum proximal to the entrance of the bile duct. In addition, the liver, pancreas, and biliary apparatus develop as outgrowths of the endodermal epithelium of the upper part of the duodenum (Fig. 13.15). Since the upper part of the foregut is divided by a septum (the tracheoesophageal septum) into the esophagus posteriorly and the trachea and lung buds anteriorly, deviation of the septum may result in abnormal openings between the trachea and esophagus. The epithelial liver cords and biliary system growing out into the septum transversum (Fig. 13.15) differentiate into parenchyma. Hematopoietic cells (present in the liver in greater numbers before birth than afterward), the Kupffer cells, and connective tissue cells originate in mesoderm. The pancreas develops from a ventral bud and a dorsal bud that later fuse to form the definitive pancreas (Figs. 13.21 and 13.22). Sometimes, the two parts surround the duodenum (annular pancreas), causing constriction of the gut (Fig. 13.23).

The **midgut** forms the primary intestinal loop (Fig. 13.24), gives rise to the duodenum distal to the entrance of the bile duct, and continues to the junction of the proximal two-thirds of the transverse colon with the distal third. At its apex the primary loop remains temporarily in open connection with the yolk sac through the vitelline duct. During the sixth week, the loop grows so rapidly that it protrudes into the umbilical cord (physiological herniation) (Fig. 13.26). During the 10th week, it returns into the abdominal cavity. While these processes are occurring, the midgut loop rotates 270° counterclockwise (Fig. 13.25). Remnants of the vitelline duct, failure of the midgut to return to the abdominal cavity, malrotation, stenosis, and duplications of parts of the gut are common abnormalities.

The **hindgut** gives rise to the region from the distal third of the transverse colon to the upper part of the anal canal; the distal part of the anal canal originates from ectoderm. The hindgut enters the posterior region of the cloaca (future anorectal canal), and the allantois enters the anterior region (future urogenital sinus). Breakdown of the cloacal membrane covering this area provides communication to the exterior for the anus and urogenital sinus. Abnormalities in the size of the posterior region of the cloaca shift the entrance of the anus anteriorly, causing rectovaginal and rectourethral fistulas and atresias (Figs. 13.36 and 13.37).

Problems to Solve

1. *Prenatal ultrasound showed polyhydramnios at 36 weeks, and at birth the infant had excessive fluids in its mouth and difficulty breathing. What birth defect might cause these conditions?*

2. *Prenatal ultrasound at 20 weeks revealed a midline mass that appeared to contain intestines and was membrane bound. What diagnosis would you make, and what would be the prognosis for this infant?*

3. *At birth a baby girl has meconium in her vagina and no anal opening. What type of birth defect does she have, and what was its embryological origin?*

SUGGESTED READING

Apelqvist A, Ahlgreen U, Edlund H: Sonic hedgehog directs specialized mesoderm differentiation in the intestines and pancreas. *Curr Biol* 7:801, 1997.

Brassett C, Ellis H: Transposition of the viscera. *Clin Anat* 4:139, 1991.

Duncan SA: Transcriptional regulation of liver development. *Dev Dynam* 219:131, 2000.

Galloway J: A handle on handedness. *Nature (Lond)* 346:223, 1990.

Gualdi R, Bossard P, Zheng M, Hamada Y, Coleman JR, Zaret KS: Hepatic specification of the gut endoderm in vitro: cell signaling and transcriptional control. *Genes Dev* 10:1670, 1996.

Kluth D, Hillen M, Lambrecht W: The principles of normal and abnormal hindgut development. *J Pediatr Surg* 30:1143, 1995.

Nievelstein RAJ, Van der Werff JFA, Verbeek FJ, Vermeij-Keers C: Normal and abnormal development of the anorectum in human embryos. *Teratology* 57:70, 1998.

Severn CB: A morphological study of the development of the human liver: 1. Development of the hepatic diverticulum. *Am J Anat* 131:133, 1971.

Severn CB: A morphological study of the development of the human liver: 2. Establishment of liver parenchyma, extrahepatic ducts, and associated venous channels. *Am J Anat* 133:85, 1972.

Sosa-Pineda B, Chowdhury K, Torres M, Oliver G, Gruss P: The Pax 4 gene is essential for differentiation of insulin producing β cells in the mammalian pancreas. *Nature* 386:399, 1997.

St. Onge L, Sosa-Pineda B, Chowdhury K, Mansouri A, Gruss P: Pax 6 is required for differentiation of glucagon-producing α cells in mouse pancreas, *Nature* 387:406, 1997.

Stevenson RE, Hall JG, Goodman RM (eds): *Human Malformations and Related Anomalies*. New York, Oxford University Press, 1993.

Torfs C, Curry C, Roeper P: Gastroschisis. *J Pediatr* 116:1, 1990.

Vellguth S, van Gaudecker B, Muller-Hermelink HK: The development of the human spleen. *Cell Tissue Res* 242:579, 1985.

Yokoh Y: Differentiation of the dorsal mesentery in man. *Acta Anat* 76:56, 1970.

Urogenital System

Functionally the urogenital system can be divided into two entirely different components: the **urinary system** and the **genital system.** Embryologically and anatomically, however, they are intimately interwoven. Both develop from a common mesodermal ridge (**intermediate mesoderm**) along the posterior wall of the abdominal cavity, and initially the excretory ducts of both systems enter a common cavity, the cloaca.

Urinary System

KIDNEY SYSTEMS

Three slightly overlapping kidney systems are formed in a cranial to caudal sequence during intrauterine life in humans: the **pronephros, mesonephros,** and **metanephros.** The first of these systems is rudimentary and nonfunctional; the second may function for a short time during the early fetal period; the third forms the permanent kidney.

[handwritten margin notes: Pronephros - nonfxnal / mesonephros - short time during early fetal / metanephros - permanent kidney]

Pronephros

At the beginning of the fourth week, the pronephros is represented by 7 to 10 solid cell groups in the cervical region (Figs. 14.1 and 14.2). These groups form vestigial excretory units, nephrotomes, that regress before more caudal ones are formed. By the end of the fourth week, all indications of the pronephric system have disappeared.

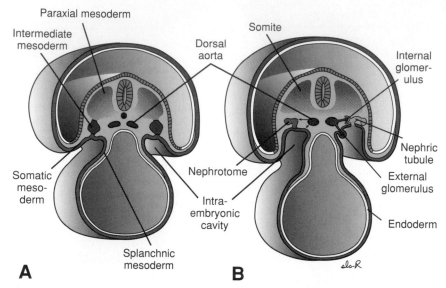

Figure 14.1 Transverse sections through embryos at various stages of development showing formation of nephric tubules. **A.** 21 days. **B.** 25 days. Note formation of external and internal glomeruli and the open connection between the intraembryonic cavity and the nephric tubule.

Mesonephros

The mesonephros and mesonephric ducts are derived from intermediate meso-derm from upper thoracic to upper lumbar (L3) segments (Fig. 14.2). Early in the fourth week of development, during regression of the pronephric sys-tem, the first excretory tubules of the mesonephros appear. They lengthen rapidly, form an S-shaped loop, and acquire a tuft of capillaries that will form a glomerulus at their medial extremity (Fig. 14.3*A*). Around the glomerulus the tubules form **Bowman's capsule,** and together these struc-tures constitute a **renal corpuscle.** Laterally the tubule enters the longitudi-nal collecting duct known as the **mesonephric** or **wolffian duct** (Figs. 14.2 and 14.3).

In the middle of the second month the mesonephros forms a large ovoid organ on each side of the midline (Fig. 14.3). Since the developing gonad is on its medial side, the ridge formed by both organs is known as the **uro-genital ridge** (Fig. 14.3). While caudal tubules are still differentiating, cra-nial tubules and glomeruli show degenerative changes, and by the end of the second month the majority have disappeared. In the male a few of the caudal tubules and the mesonephric duct persist and participate in forma-tion of the genital system, but they disappear in the female (see Genital System; p. 337).

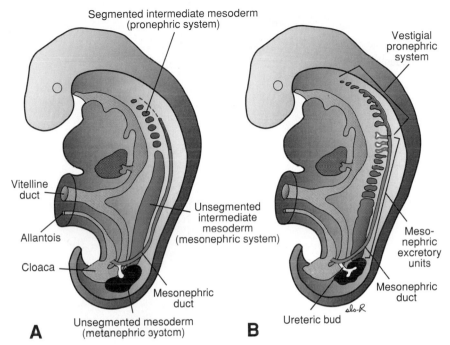

Figure 14.2 A. Relationship of the intermediate mesoderm of the pronephric, mesonephric, and metanephric systems. In cervical and upper thoracic regions intermediate mesoderm is segmented; in lower thoracic, lumbar, and sacral regions it forms a solid, unsegmented mass of tissue, the nephrogenic cord. Note the longitudinal collecting duct, formed initially by the pronephros but later by the mesonephros. **B.** Excretory tubules of the pronephric and mesonephric systems in a 5-week-old embryo.

Metanephros: The Definitive Kidney

The third urinary organ, the **metanephros,** or **permanent kidney,** appears in the fifth week. Its excretory units develop from **metanephric mesoderm** (Fig. 14.4) in the same manner as in the mesonephric system. The development of the duct system differs from that of the other kidney systems.

Collecting System. Collecting ducts of the permanent kidney develop from the **ureteric bud,** an outgrowth of the mesonephric duct close to its entrance to the cloaca (Fig. 14.4). The bud penetrates the metanephric tissue, which is molded over its distal end as a cap (Fig. 14.4). Subsequently the bud dilates, forming the primitive **renal pelvis,** and splits into cranial and caudal portions, the future **major calyces** (Fig. 14.5, *A* and *B*).

Each calyx forms two new buds while penetrating the metanephric tissue. These buds continue to subdivide until 12 or more generations of tubules have formed (Fig. 14.5). Meanwhile, at the periphery more tubules form until the

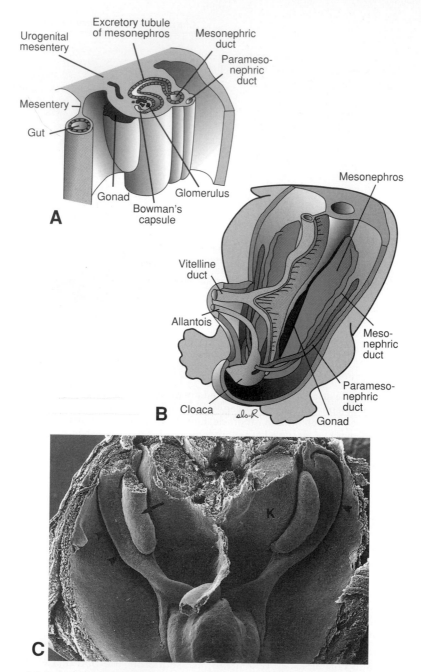

Figure 14.3 A. Transverse section through the urogenital ridge in the lower thoracic region of a 5-week embryo showing formation of an excretory tubule of the mesonephric system. Note the appearance of Bowman's capsule and the gonadal ridge. The mesonephros and gonad are attached to the posterior abdominal wall by a broad urogenital mesentery. **B.** Relation of the gonad and the mesonephros. Note the size of the mesonephros. The mesonephric duct (wolffian duct) runs along the lateral side of the mesonephros. **C.** Scanning electron micrograph of a mouse embryo showing the genital ridge (*arrow*) and mesonephric duct (*arrowheads*). *K*, kidneys.

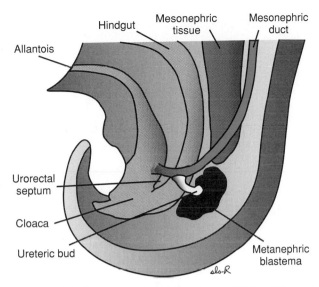

Figure 14.4 Relation of the hindgut and cloaca at the end of the fifth week. The ureteric bud penetrates the metanephric mesoderm (blastema).

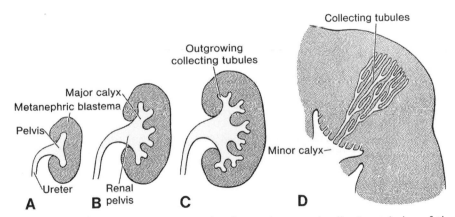

Figure 14.5 Development of the renal pelvis, calyces, and collecting tubules of the metanephros. **A.** 6 weeks. **B.** At the end of the sixth week. **C.** 7 weeks, **D.** Newborn. Note the pyramid form of the collecting tubules entering the minor calyx.

end of the fifth month. The tubules of the second order enlarge and absorb those of the third and fourth generations, forming the **minor calyces** of the renal pelvis. During further development, collecting tubules of the fifth and successive generations elongate considerably and converge on the minor calyx, forming the **renal pyramid** (Fig. 14.5D). **The ureteric bud gives rise to the ureter, the renal pelvis, the major and minor calyces, and approximately 1 million to 3 million collecting tubules.**

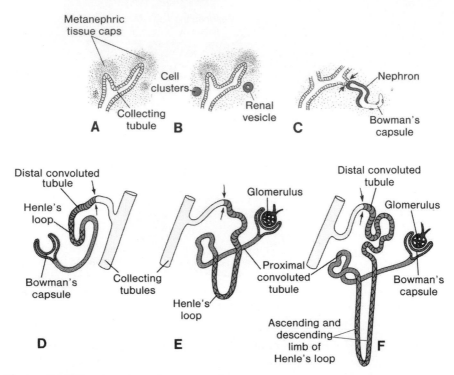

Figure 14.6 Development of a metanephric excretory unit. *Arrows,* the place where the excretory unit (*blue*) establishes an open communication with the collecting system (*yellow*), allowing flow of urine from the glomerulus into the collecting ducts.

Excretory System. Each newly formed collecting tubule is covered at its distal end by a **metanephric tissue cap** (Fig. 14.6*A*). Under the inductive influence of the tubule, cells of the tissue cap form small vesicles, the **renal vesicles,** which in turn give rise to small S-shaped tubules (Fig. 14.6, *B* and *C*). Capillaries grow into the pocket at one end of the S and differentiate into **glomeruli.** These tubules, together with their glomeruli, form **nephrons,** or **excretory units.** The proximal end of each nephron forms **Bowman's capsule,** which is deeply indented by a glomerulus (Fig. 14.6, *C* and *D*). The distal end forms an open connection with one of the collecting tubules, establishing a passageway from Bowman's capsule to the collecting unit. Continuous lengthening of the excretory tubule results in formation of the **proximal convoluted tubule, loop of Henle,** and **distal convoluted tubule** (Fig. 14.6, *E* and *F*). Hence, the kidney develops from two sources: (*a*) metanephric mesoderm, which provides excretory units; and (*b*) the ureteric bud, which gives rise to the collecting system.

Nephrons are formed until birth, at which time there are approximately 1 million in each kidney. Urine production begins early in gestation, soon after differentiation of the glomerular capillaries, which start to form by the

10th week. At birth the kidneys have a lobulated appearance, but the lobulation disappears during infancy as a result of further growth of the nephrons, although there is no increase in their number.

MOLECULAR REGULATION OF KIDNEY DEVELOPMENT

As with most organs, differentiation of the kidney involves epithelial mesenchymal interactions. In this example, epithelium of the ureteric bud from the mesonephros interacts with mesenchyme of the metanephric blastema (Fig. 14.7). The mesenchyme expresses **WT1**, a transcription factor that makes this tissue competent to respond to induction by the ureteric bud. *WT1* also regulates production of **glial-derived neurotrophic factor (GDNF)** and **hepatocyte growth factor (HGF, or scatter factor)** by the mesenchyme, and these proteins stimulate growth of the ureteric buds (Fig. 14.7A). The **tyrosine kinase receptors RET,** for GDNF, and **MET,** for HGF, are synthesized by the epithelium of the ureteric buds, establishing signaling pathways between the two tissues. In turn, the buds induce the mesenchyme via **fibroblast growth factor-2 (FGF-2)** and **bone morphogenetic protein-**7 (**BMP-**7) (Fig. 14.7A). Both of these growth factors block apoptosis and stimulate proliferation in the metanephric mesenchyme while maintaining production of *WT1*. Conversion of the

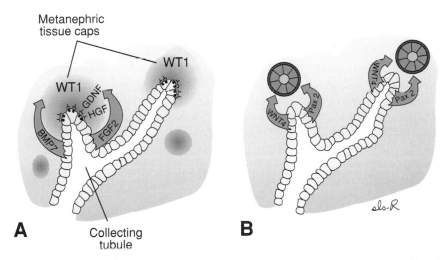

Figure 14.7 Genes involved in differentiation of the kidney. **A.** *WT1*, expressed by the mesenchyme, enables this tissue to respond to induction by the ureteric bud. GDNF and HGF, also produced by the mesenchyme, interact through their receptors, RET and MET, respectively, in the ureteric bud epithelium, to stimulate growth of the bud and maintain the interactions. The growth factors FGF2 and BMP7 stimulate proliferation of the mesenchyme and maintain *WT1* expression. **B.** PAX2 and WNT4, produced by the ureteric bud, cause the mesenchyme to epithelialize in preparation for excretory tubule differentiation. Laminin and type IV collagen form a basement membrane for the epithelial cells.

mesenchyme to an epithelium for nephron formation is also mediated by the ureteric buds, in part through modification of the extracellular matrix. Thus **fibronectin, collagen I,** and **collagen III** are replaced with **laminin** and **type IV collagen,** characteristic of an epithelial basal lamina (Fig. 14.7*B*). In addition, the cell adhesion molecules **syndecan** and **E-cadherin,** which are essential for condensation of the mesenchyme into an epithelium, are synthesized. Regulatory genes for conversion of the mesenchyme into an epithelium appear to involve *PAX2* and *WNT4* (Fig. 14.7*B*).

CLINICAL CORRELATES

Renal Tumors and Defects

Wilms' tumor is a cancer of the kidneys that usually affects children by 5 years of age but may also occur in the fetus. Wilms' tumor is due to mutations in the *WT1* gene on 11p13, and it may be associated with other abnormalities and syndromes. For example, **WAGR syndrome** is characterized by aniridia, hemihypertrophy, and Wilms' tumor. Similarly, **Denys-Drash syndrome** consists of renal failure, pseudohermaphrodism, and Wilms' tumor.

Renal dysplasias and agenesis are a spectrum of severe malformations that represent the primary diseases requiring dialysis and transplantation in the first years of life. **Multicystic dysplastic kidney** is one example of this group of abnormalities in which numerous ducts are surrounded by undifferentiated cells. Nephrons fail to develop and the ureteric bud fails to branch, so that the collecting ducts never form. In some cases these defects cause involution of the kidneys and **renal agenesis.** Renal agenesis may also occur if the ureteric bud fails to contact and/or induce the metanephric mesoderm. Bilateral renal agenesis, which occurs in 1/10,000 births, results in renal failure. The baby presents with **Potter sequence,** characterized by anuria, oligohydramnios (decreased volume of amniotic fluid), and hypoplastic lungs secondary to the oligohydramnios. In 85% of cases other severe defects, including absence or abnormalities of the vagina and uterus, vas deferens, and seminal vesicles, accompany this condition. Common associated defects in other systems include cardiac anomalies, tracheal and duodenal atresias, cleft lip and palate, and brain abnormalities.

In **congenital polycystic kidney** (Fig. 14.8) numerous cysts form. It may be inherited as an autosomal recessive or autosomal dominant disorder or may be caused by other factors. **Autosomal recessive polycystic kidney disease,** which occurs in 1/5,000 births, is a progressive disorder in which cysts form from collecting ducts. The kidneys become very large, and renal failure occurs in infancy or childhood. In **autosomal dominant polycystic kidney disease,** cysts form from all segments of the nephron and usually do not cause renal failure until adulthood. The autosomal dominant disease is more common (1/500 to 1/1,000 births) but less progressive than the autosomal recessive disease.

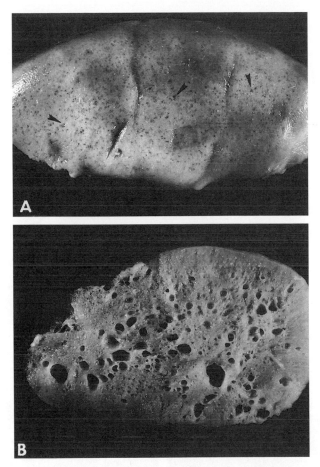

Figure 14.8 A. Surface view of a fetal kidney with multiple cysts (*arrowheads*) characteristic of polycystic kidney disease. **B.** Section of the kidney in **A,** showing multiple cysts.

Duplication of the ureter results from early splitting of the ureteric bud (Fig. 14.9). Splitting may be partial or complete, and metanephric tissue may be divided into two parts, each with its own renal pelvis and ureter. More frequently, however, the two parts have a number of lobes in common as a result of intermingling of collecting tubules. In rare cases one ureter opens into the bladder, and the other is ectopic, entering the vagina, urethra, or vestibule (Fig. 14.9*C*). This abnormality results from development of two ureteric buds. One of the buds usually has a normal position, while the abnormal bud moves down together with the mesonephric duct. Thus it has a low, abnormal entrance in the bladder, urethra, vagina, or epididymal region.

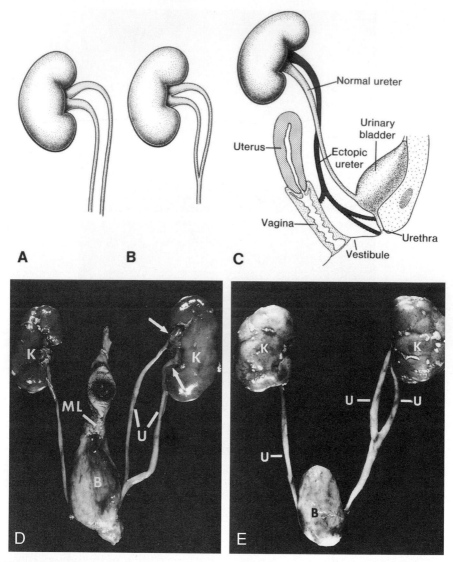

Figure 14.9 A and **B.** A complete and a partial double ureter. **C.** Possible sites of ectopic ureteral openings in the vagina, urethra, and vestibule. **D** and **E.** Photomicrographs of complete and partial duplications of the ureters (*U*). *Arrows,* duplicated hilum; *B,* bladder; *K,* kidneys; *ML,* median umbilical ligament.

POSITION OF THE KIDNEY

The kidney, initially in the pelvic region, later shifts to a more cranial position in the abdomen. This **ascent of the kidney** is caused by diminution of body curvature and by growth of the body in the lumbar and sacral regions

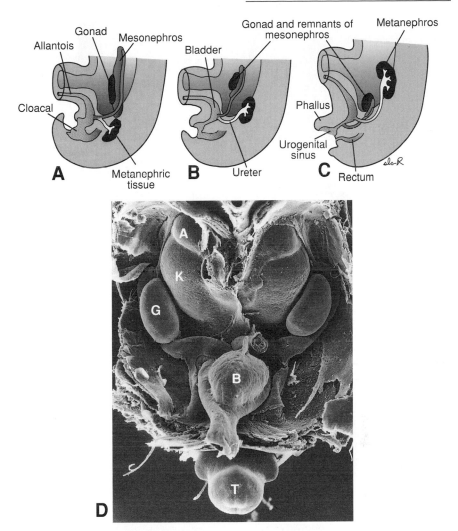

Figure 14.10 A to **C.** Ascent of the kidneys. Note the change in position between the mesonephric and metanephric systems. The mesonephric system degenerates almost entirely, and only a few remnants persist in close contact with the gonad. In both male and female embryos, the gonads descend from their original level to a much lower position. **D.** Scanning electron micrograph of a mouse embryo showing the kidneys in the pelvis. *B*, bladder; *K*, kidney *A*, adrenal gland; *G*, gonad; *T*, tail.

(Fig. 14.10). In the pelvis the metanephros receives its arterial supply from a pelvic branch of the aorta. During its ascent to the abdominal level, it is vascularized by arteries that originate from the aorta at continuously higher levels. The lower vessels usually degenerate, but some may remain.

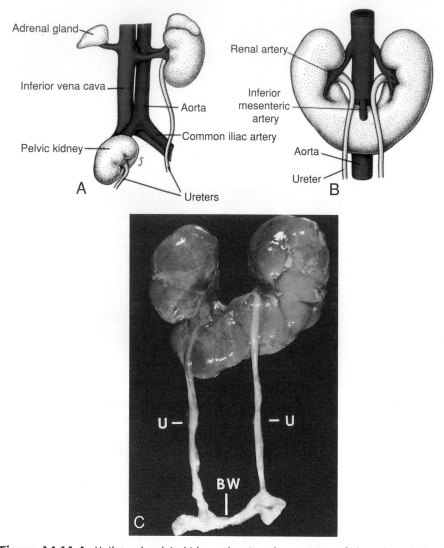

Figure 14.11 A. Unilateral pelvic kidney showing the position of the adrenal gland on the affected side. **B** and **C.** Drawing and photomicrograph, respectively, of horseshoe kidneys showing the position of the inferior mesenteric artery. *BW,* bladder wall; *U*, ureters.

CLINICAL CORRELATES

Abnormal Location of the Kidneys

During their ascent the kidneys pass through the arterial fork formed by the umbilical arteries, but occasionally one of them fails to do so. Remaining in the pelvis close to the common iliac artery, it is known as a **pelvic kidney** (Fig. 14.11 *A*). Sometimes the kidneys are pushed so close together during their passage through the arterial fork that the lower poles fuse, forming a **horseshoe kidney** (Fig. 14.11, *B* and *C*). The horseshoe kidney is usually at the level of the lower lumbar vertebrae, since its ascent is prevented by the root of the inferior mesenteric artery (Fig. 14.11*B*). The ureters arise from the anterior surface of the kidney and pass ventral to the isthmus in a caudal direction. Horseshoe kidney is found in 1/600 people.

Accessory renal arteries are common; they derive from the persistence of embryonic vessels that formed during ascent of the kidneys. These arteries usually arise from the aorta and enter the superior or inferior poles of the kidneys.

FUNCTION OF THE KIDNEY

The definitive kidney formed from the metanephros becomes functional near the 12th week. Urine is passed into the amniotic cavity and mixes with the amniotic fluid. The fluid is swallowed by the fetus and recycles through the kidneys. During fetal life, the kidneys are not responsible for excretion of waste products, since the placenta serves this function.

BLADDER AND URETHRA

During the fourth to seventh weeks of development the **cloaca** divides into the **urogenital sinus** anteriorly and the **anal canal** posteriorly (Fig. 14.12)

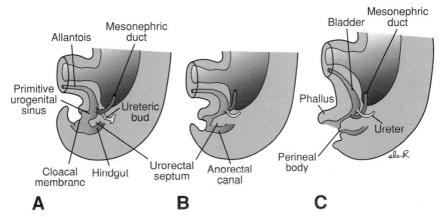

A   **B**   **C**

Figure 14.12 Divisions of the cloaca into the urogenital sinus and anorectal canal. The mesonephric duct is gradually absorbed into the wall of the urogenital sinus, and the ureters enter separately. **A.** At the end of the fifth week. **B.** 7 weeks. **C.** 8 weeks.

(see Chapter 13). The **urorectal septum** is a layer of mesoderm between the primitive anal canal and the urogenital sinus. The tip of the septum will form the **perineal body** (Fig. 14.12*C*). Three portions of the urogenital sinus can be distinguished: The upper and largest part is the **urinary bladder** (Fig. 14.13*A*). Initially the bladder is continuous with the allantois, but when the lumen of the allantois is obliterated, a thick fibrous cord, the **urachus,** remains and connects the apex of the bladder with the umbilicus (Fig. 14.13*B*). In the adult, it is known as the **median umbilical ligament.** The next part is a rather narrow canal, the **pelvic part of the urogenital sinus,** which in the male gives rise to the **prostatic** and **membranous** parts of the **urethra.** The last part is the **phallic part** of the urogenital sinus. It is flattened from side to side, and as the genital tubercle grows, this part of the sinus will be pulled ventrally (Fig. 14.13*A*). (Development of the phallic part of the urogenital sinus differs greatly between the two sexes: see Genital System; p. 337.)

During differentiation of the cloaca, the caudal portions of the mesonephric ducts are absorbed into the wall of the urinary bladder (Fig. 14.14). Consequently, the ureters, initially outgrowths from the mesonephric ducts, enter the bladder separately (Fig. 14.14*B*). As a result of ascent of the kidneys, the orifices of the ureters move farther cranially; those of the mesonephric ducts move close together to enter the prostatic urethra and in the male become the **ejaculatory ducts** (Fig. 14.14, *C* and *D*). Since both the mesonephric ducts and ureters originate in the mesoderm, the mucosa of the bladder formed by incorporation of the ducts (the **trigone** of the bladder) is also mesodermal. With time the mesodermal lining of the trigone is replaced by endodermal epithelium, so that finally the inside of the bladder is completely lined with endodermal epithelium.

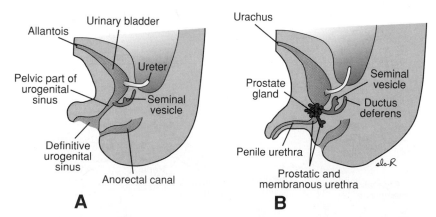

Figure 14.13 A. Development of the urogenital sinus into the urinary bladder and definitive urogenital sinus. **B.** In the male the definitive urogenital sinus develops into the penile urethra. The prostate gland is formed by buds from the urethra, and seminal vesicles are formed by budding from the ductus deferens.

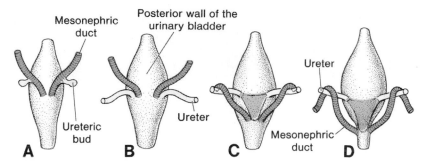

Figure 14.14 Dorsal views of the bladder showing the relation of the ureters and mesonephric ducts during development. Initially the ureters are formed by an outgrowth of the mesonephric duct **(A),** but with time they assume a separate entrance into the urinary bladder **(B-D).** Note the trigone of the bladder formed by incorporation of the mesonephric ducts **(C** and **D).**

URETHRA

The epithelium of the urethra in both sexes originates in the endoderm; the surrounding connective and smooth muscle tissue is derived from splanchnic mesoderm. At the end of the third month, epithelium of the prostatic urethra begins to proliferate and forms a number of outgrowths that penetrate the surrounding mesenchyme. In the male, these buds form the **prostate gland** (Fig. 14.13*B*). In the female, the cranial part of the urethra gives rise to the **urethral** and **paraurethral glands.**

CLINICAL CORRELATES

Bladder Defects

When the lumen of the intraembryonic portion of the allantois persists, a **urachal fistula** may cause urine to drain from the umbilicus (Fig. 14.15*A*). If only a local area of the allantois persists, secretory activity of its lining results in a cystic dilation, a **urachal cyst** (Fig. 14.15*B*). When the lumen in the upper part persists, it forms a **urachal sinus.** This sinus is usually continuous with the urinary bladder (Fig. 14.15*C*).

Exstrophy of the bladder (Fig. 14.16*A*) is a ventral body wall defect in which the bladder mucosa is exposed. Epispadias is a constant feature (see Fig. 14.33), and the open urinary tract extends along the dorsal aspect of the penis through the bladder to the umbilicus. Exstrophy of the bladder may be caused by a lack of mesodermal migration into the region between the umbilicus and genital tubercle, followed by rupture of the thin layer of ectoderm. This anomaly is rare, occurring in 2/100,000 live births.

Exstrophy of the cloaca (Fig. 14.16*B*) is a more severe ventral body wall defect in which migration of mesoderm to the midline is inhibited and the tail

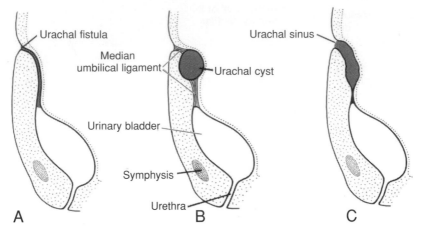

Figure 14.15 A. Urachal fistula. **B.** Urachal cyst. **C.** Urachal sinus. The sinus may or may not be in open communication with the urinary bladder.

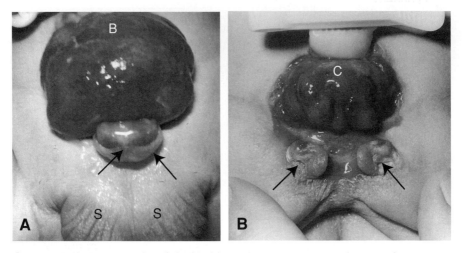

Figure 14.16 A. Exstrophy of the bladder (*B.*) *Arrows*, penis with epispadias; *S*, scrotum. **B.** Cloacal exstrophy in a newborn. *C*, cloaca; *arrows*, unfused genital swellings.

(caudal) fold fails to progress. As a result, there is an extended thin layer of ectoderm that ruptures. The defect includes exstrophy of the bladder, spinal defects with or without meningomyelocele, imperforate anus, and usually omphalocele. Occurrence is rare (1/30,000), and causation has not been defined, although the defect is associated with early amniotic rupture.

Genital System

Sex differentiation is a complex process that involves many genes, including some that are autosomal. The key to sexual dimorphism is the Y chromosome, which contains the **SRY (sex-determining region on Y) gene** on its short arm (Yp11). The protein product of this gene is a transcription factor initiating a cascade of downstream genes that determine the fate of rudimentary sexual organs. The SRY protein is the **testis-determining factor;** under its influence male development occurs; in its absence female development is established.

GONADS

Although the sex of the embryo is determined genetically at the time of fertilization, the gonads do not acquire male or female morphological characteristics until the seventh week of development.

Gonads appear initially as a pair of longitudinal ridges, the **genital** or **gonadal ridges** (Fig. 14.17). They are formed by proliferation of the epithelium and a condensation of underlying mesenchyme. **Germ cells** do not appear in the genital ridges until the sixth week of development.

Primordial germ cells first appear at an early stage of development among endoderm cells in the wall of the yolk sac close to the allantois (Fig. 14.18*A*). They migrate by ameboid movement along the dorsal mesentery of the hindgut (Fig. 14.18, *B* and *C*), arriving at the primitive gonads at the beginning of the fifth week and invading the genital ridges in the sixth week. If they fail to reach the ridges, the gonads do not develop. Hence the primordial germ cells have an inductive influence on development of the gonad into ovary or testis.

Shortly before and during arrival of primordial germ cells, the epithelium of the genital ridge proliferates, and epithelial cells penetrate the underlying mesenchyme. Here they form a number of irregularly shaped cords, the **primitive sex cords** (Fig. 14.19). In both male and female embryos, these cords are connected to surface epithelium, and it is impossible to differentiate between the male and female gonad. Hence, the gonad is known as the **indifferent gonad.**

Testis

If the embryo is genetically male, the primordial germ cells carry an XY sex chromosome complex. Under influence of the *SRY* gene on the Y chromosome, which encodes the testis-determining factor, the primitive sex cords continue to proliferate and penetrate deep into the medulla to form the **testis** or **medullary cords** (Figs. 14.20*A*, Fig. 14.21). Toward the hilum of the gland the cords break up into a network of tiny cell strands that later give rise to tubules of the **rete testis** (Fig. 14.20, *A* and *B*). During further development, a dense layer

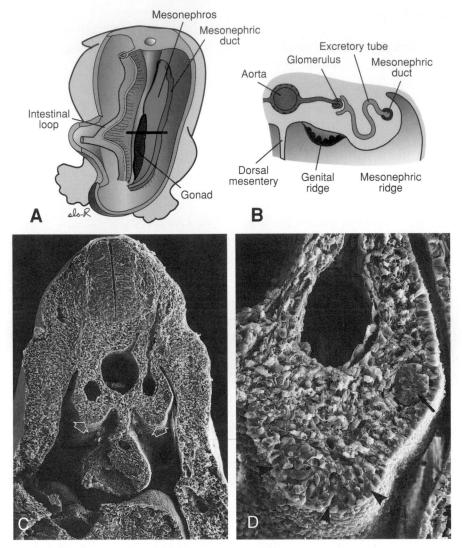

Figure 14.17 A. Relation of the genital ridge and the mesonephros showing location of the mesonephric duct. **B.** Transverse section through the mesonephros and genital ridge at the level indicated in **A. C.** Scanning electron micrograph of a mouse embryo showing the genital ridge (*arrows*). **D.** High magnification of the genital ridge showing the mesonephric duct (*arrow*) and the developing gonad (*arrowheads*).

of fibrous connective tissue, the **tunica albuginea,** separates the testis cords from the surface epithelium (Fig. 14.20).

In the fourth month, the testis cords become horseshoe shaped, and their extremities are continuous with those of the rete testis (Fig. 14.20*B*). Testis cords are now composed of primitive germ cells and **sustentacular cells of Sertoli** derived from the surface epithelium of the gland.

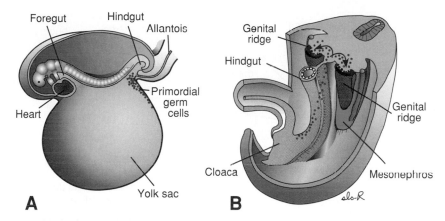

Figure 14.18 A. A 3-week-old embryo showing the primordial germ cells in the wall of the yolk sac close to the attachment of the allantois. **B.** Migrational path of the primordial germ cells along the wall of the hindgut and the dorsal mesentery into the genital ridge.

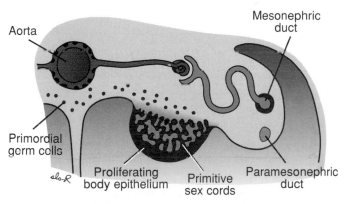

Figure 14.19 Transverse section through the lumbar region of a 6-week embryo showing the indifferent gonad with the primitive sex cords. Some of the primordial germ cells are surrounded by cells of the primitive sex cords.

Interstitial cells of Leydig, derived from the original mesenchyme of the gonadal ridge, lie between the testis cords. They begin development shortly after onset of differentiation of these cords. By the eighth week of gestation, Leydig cells begin production of **testosterone,** and the testis is able to influence sexual differentiation of the genital ducts and external genitalia.

Testis cords remain solid until puberty, when they acquire a lumen, thus forming the **seminiferous tubules.** Once the seminiferous tubules are canalized, they join the rete testis tubules, which in turn enter the **ductuli efferentes.** These efferent ductules are the remaining parts of the excretory tubules of the mesonephric system. They link the rete testis and the mesonephric or wolffian duct, which becomes the **ductus deferens** (Fig. 14.20*B*).

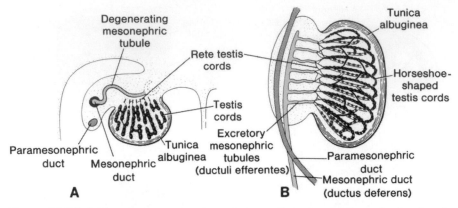

Figure 14.20 A. Transverse section through the testis in the eighth week, showing the tunica albuginea, testis cords, rete testis, and primordial germ cells. The glomerulus and Bowman's capsule of the mesonephric excretory tubule are degenerating. **B.** Testis and genital duct in the fourth month. The horseshoe-shaped testis cords are continuous with the rete testis cords. Note the ductuli efferentes (excretory mesonephric tubules), which enter the mesonephric duct.

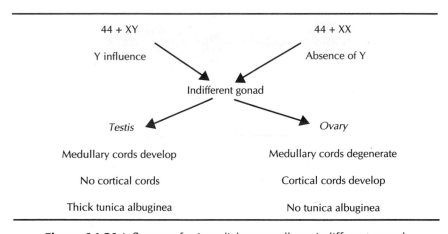

Figure 14.21 Influence of primordial germ cells on indifferent gonad.

Ovary

In female embryos with an XX sex chromosome complement and no Y chromosome, primitive sex cords dissociate into irregular cell clusters (Figs. 14.21 and 14.22*A*). These clusters, containing groups of primitive germ cells, occupy the medullary part of the ovary. Later they disappear and are replaced by a vascular stroma that forms the **ovarian medulla** (Fig. 14.22).

The surface epithelium of the female gonad, unlike that of the male, continues to proliferate. In the seventh week, it gives rise to a second generation of cords, **cortical cords,** which penetrate the underlying mesenchyme but remain

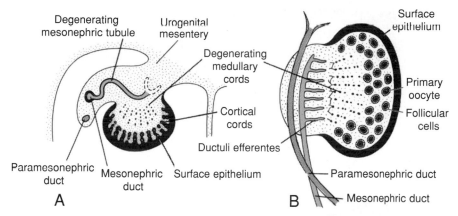

Figure 14.22 A. Transverse section of the ovary at the seventh week, showing degeneration of the primitive (medullary) sex cords and formation of the cortical cords. **B.** Ovary and genital ducts in the fifth month. Note degeneration of the medullary cords. The excretory mesonephric tubules (efferent ductules) do not communicate with the rete. The cortical zone of the ovary contains groups of oogonia surrounded by follicular cells.

close to the surface (Fig. 14.22*A*). In the fourth month, these cords split into isolated cell clusters, with each surrounding one or more primitive germ cells (Fig. 14.22*B*). Germ cells subsequently develop into oogonia, and the surrounding epithelial cells, descendants of the surface epithelium, form **follicular cells** (see Chapter 1).

It may thus be stated that the genetic sex of an embryo is determined at the time of fertilization, depending on whether the spermatocyte carries an X or a Y chromosome. In embryos with an XX sex chromosome configuration, medullary cords of the gonad regress, and a secondary generation of cortical cords develops (Figs. 14.21 and 14.22). In embryos with an XY sex chromosome complex, medullary cords develop into testis cords, and secondary cortical cords fail to develop (Figs. 14.20 and 14.21).

GENITAL DUCTS

Indifferent Stage

Initially both male and female embryos have two pairs of genital ducts: **mesonephric (wolffian) ducts** and **paramesonephric (müllerian) ducts.** The paramesonephric duct arises as a longitudinal invagination of the epithelium on the anterolateral surface of the urogenital ridge (Fig. 14.23). Cranially the duct opens into the abdominal cavity with a funnel-like structure. Caudally it first runs lateral to the mesonephric duct, then crosses it ventrally to grow caudomedially (Fig.14.23). In the midline it comes in close contact with the paramesonephric duct from the opposite side. The two ducts are initially

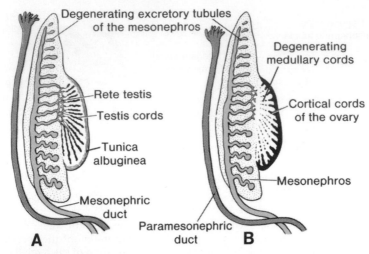

Figure 14.23 Genital ducts in the sixth week in the male **(A)** and female **(B)**. The mesonephric and paramesonephric ducts are present in both. Note the excretory tubules of the mesonephros and their relation to the developing gonad in both sexes.

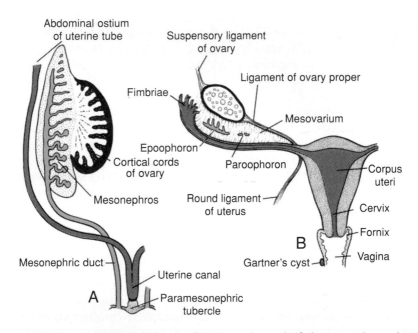

Figure 14.24 A. Genital ducts in the female at the end of the second month. Note the paramesonephric (müllerian) tubercle and formation of the uterine canal. **B.** Genital ducts after descent of the ovary. The only parts remaining from the mesonephric system are the epoophoron, paroophoron, and Gartner's cyst. Note the suspensory ligament of the ovary, ligament of the ovary proper, and round ligament of the uterus.

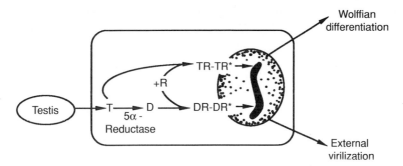

Figure 14.25 Androgen action at the cellular level. Receptor complexes with testosterone (*T*) and dihydrotestosterone (*D*) interact with DNA to control differentiation of the wolffian duct and external genitalia, respectively. *R*, androgen receptor; *R**, transformed androgen receptor hormone complex.

separated by a septum but later fuse to form the **uterine canal** (Fig. 14.24*A*). The caudal tip of the combined ducts projects into the posterior wall of the urogenital sinus, where it causes a small swelling, the paramesonephric or müllerian tubercle (Fig. 14.24*A*). The mesonephric ducts open into the urogenital sinus on either side of the müllerian tubercle.

Molecular Regulation of Genital Duct Development

SRY is the master gene for testes development and appears to act directly on the gonadal ridge and indirectly on the mesonephric ducts. Thus, it induces the testes to secrete a chemotactic factor that causes tubules from the mesonephric duct to penetrate the gonadal ridge and stimulate further testicular development. In fact, without penetration by these tubules, differentiation of the testes fails. *SRY* also upregulates ***steroidogenesis factor 1 (SF1),*** which acts through another transcription factor, ***SOX9,*** to induce differentiation of Sertoli and Leydig cells. Sertoli cells then produce **müllerian inhibiting substance (MIS,** also called **antimüllerian hormone, AMH)** that causes regression of the paramesonephric (müllerian) ducts. Leydig cells produce **testosterone,** which enters cells of target tissues where it may remain or be converted to **dihydrotestosterone** by a 5α reductase enzyme. Testosterone and dihydrotestosterone bind to a specific high-affinity intracellular receptor protein, and ultimately this hormone receptor complex binds to DNA to regulate transcription of tissue-specific genes and their protein products (Fig. 14.25). Testosterone receptor complexes mediate virilization of the mesonephric ducts, whereas dihydrotestosterone receptor complexes modulate differentiation of the male external genitalia (Fig. 14.26).

Sexual differentiation in females was once thought to be a default mechanism that occurred in the absence of a Y chromosome, but it now appears that there are specific genes that induce ovarian development. For example, ***DAX1***, a member of the nuclear hormone receptor family, is located on the short arm of the X chromosome and acts by downregulating ***SF1*** activity, thereby

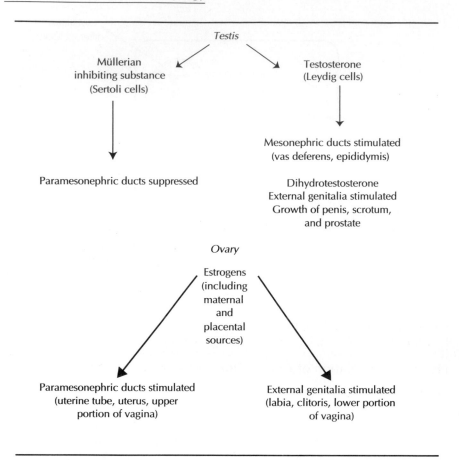

Figure 14.26 Influence of the sex glands on further sex differentiation.

preventing differentiation of Sertoli and Leydig cells. The secreted growth factor **WNT4** also contributes to ovarian differentiation, and its early expression in the gonadal ridge is maintained in females but downregulated in males. In the absence of MIS production by Sertoli cells, the paramesonephric (müllerian) ducts are stimulated by estrogens to form the uterine tubes, uterus, cervix, and upper vagina. Estrogens also act on the external genitalia at the indifferent stage to form the labia majora, labia minora, clitoris, and lower vagina (Fig. 14.26).

Genital Ducts in the Male

As the mesonephros regresses, a few excretory tubules, the **epigenital tubules,** establish contact with cords of the rete testis and finally form the **efferent ductules** of the testis (Fig. 14.27). Excretory tubules along the caudal pole

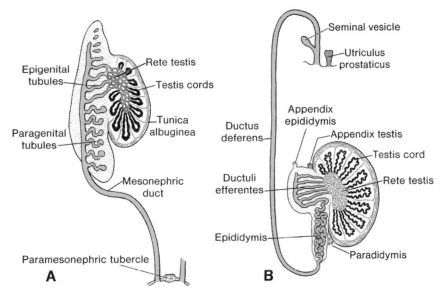

Figure 14.27 A. Genital ducts in the male in the fourth month. Cranial and caudal (paragenital tubule) segments of the mesonephric system regress. **B.** Genital ducts after descent of the testis. Note the horseshoe-shaped testis cords, rete testis, and efferent ductules entering the ductus deferens. The paradidymis is formed by remnants of the paragenital mesonephric tubules. The paramesonephric duct has degenerated except for the appendix testis. The prostatic utricle is an outpocketing from the urethra.

of the testis, the **paragenital tubules,** do not join the cords of the rete testis (Fig. 14.27*B*). Their vestiges are collectively known as the **paradidymis.**

Except for the most cranial portion, the **appendix epididymis,** the mesonephric ducts persist and form the main genital ducts (Fig. 14.27). Immediately below the entrance of the efferent ductules, the mesonephric ducts elongate and become highly convoluted, forming the **(ductus) epididymis.** From the tail of the epididymis to the outbudding of the **seminal vesicle,** the mesonephric ducts obtain a thick muscular coat and form the **ductus deferens.** The region of the ducts beyond the seminal vesicles is the **ejaculatory duct.** The paramesonephric ducts in the male degenerate except for a small portion at their cranial ends, the **appendix testis** (Fig. 14.27*B*).

Genital Ducts in the Female

The paramesonephric ducts develop into the main genital ducts of the female. Initially, three parts can be recognized in each duct: (*a*) a cranial vertical portion that opens into the abdominal cavity, (*b*) a horizontal part that crosses the mesonephric duct, and (*c*) a caudal vertical part that fuses with its partner from the opposite side (Fig. 14.24*A*). With descent of the ovary, the first two parts develop into the **uterine tube** (Fig. 14.24*B*) and the caudal parts fuse

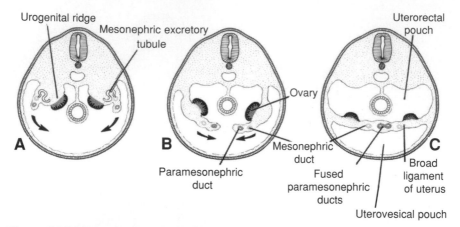

Figure 14.28 Transverse sections through the urogenital ridge at progressively lower levels. **A.** and **B.** The paramesonephric ducts approach each other in the midline and fuse. **C.** As a result of fusion, a transverse fold, the broad ligament of the uterus, forms in the pelvis. The gonads come to lie at the posterior aspect of the transverse fold.

to form the **uterine canal.** When the second part of the paramesonephric ducts moves mediocaudally, the urogenital ridges gradually come to lie in a transverse plane (Fig. 14.28, *A* and *B*). After the ducts fuse in the midline, a broad transverse pelvic fold is established (Fig. 14.28*C*). This fold, which extends from the lateral sides of the fused paramesonephric ducts toward the wall of the pelvis, is **the broad ligament of the uterus.** The uterine tube lies in its upper border, and the ovary lies on its posterior surface (Fig. 14.28*C*). The uterus and broad ligaments divide the pelvic cavity into the **uterorectal pouch** and the **uterovesical pouch.** The fused paramesonephric ducts give rise to the **corpus** and **cervix** of the uterus. They are surrounded by a layer of mesenchyme that forms the muscular coat of the uterus, the **myometrium,** and its peritoneal covering, the **perimetrium.**

VAGINA

Shortly after the solid tip of the paramesonephric ducts reaches the urogenital sinus (Figs. 14.29*A* and 14.30*A*), two solid evaginations grow out from the pelvic part of the sinus (Figs. 14.29*B* and 14.30*B*). These evaginations, the **sinovaginal bulbs,** proliferate and form a solid **vaginal plate.** Proliferation continues at the cranial end of the plate, increasing the distance between the uterus and the urogenital sinus. By the fifth month, the vaginal outgrowth is entirely canalized. The winglike expansions of the vagina around the end of the uterus, the **vaginal fornices,** are of paramesonephric origin (Fig. 14.30*C*). Thus, the vagina has a dual origin, with the upper portion derived from the uterine canal and the lower portion derived from the urogenital sinus.

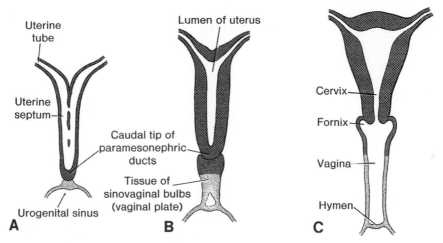

Figure 14.29 Formation of the uterus and vagina. **A.** 9 weeks. Note the disappearance of the uterine septum. **B.** At the end of the third month. Note the tissue of the sinovaginal bulbs. **C.** Newborn. The fornices and the upper portion of the vagina are formed by vacuolization of the paramesonephric tissue, and the lower protion of the vagina is formed by vacuolization of the sinovaginal bulbs.

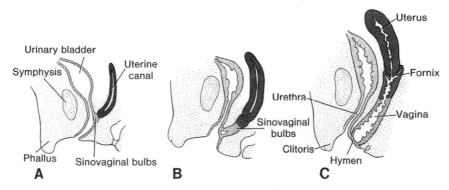

Figure 14.30 Sagittal sections showing formation of the uterus and vagina at various stages of development. **A.** Nine weeks. **B.** End of third month. **C.** Newborn.

The lumen of the vagina remains separated from that of the urogenital sinus by a thin tissue plate, the **hymen** (Figs. 14.29*C* and 14.30*C*), which consists of the epithelial lining of the sinus and a thin layer of vaginal cells. It usually develops a small opening during perinatal life.

The female may retain some remnants of the cranial and caudal excretory tubules in the mesovarium, where they form the **epoophoron** and **paroophoron,** respectively (Fig. 14.24*B*). The mesonephric duct disappears except for a small cranial portion found in the epoophoron and occasionally a small caudal portion that may be found in the wall of the uterus or vagina. Later in life it may form **Gartner's cyst** (Fig. 14.24*B*).

CLINICAL CORRELATES

Uterine and Vaginal Defects

Duplications of the uterus result from lack of fusion of the paramesonephric ducts in a local area or throughout their normal line of fusion. In its extreme form the uterus is entirely double (**uterus didelphys**) (Fig. 14.31*A*); in the least severe form, it is only slightly indented in the middle (**uterus arcuatus**) (Fig. 14.31*B*). One of the relatively common anomalies is the **uterus bicornis,** in which the uterus has two horns entering a common vagina (Fig. 14.31*C*). This condition is normal in many mammals below the primates.

In patients with complete or partial atresia of one of the paramesonephric ducts, the rudimentary part lies as an appendage to the well-developed side. Since its lumen usually does not communicate with the vagina, complications are common (uterus bicornis unicollis with one rudimentary horn) (Fig. 14.31*D*). If the atresia involves both sides, an atresia of the cervix may result (Fig. 14.31*E*). If the sinovaginal bulbs fail to fuse or do not develop at all, a double vagina or atresia of the vagina, respectively, results (Fig. 14.31, *A* and *F*). In the latter case, a small vaginal pouch originating from the paramesonephric ducts usually surrounds the opening of the cervix.

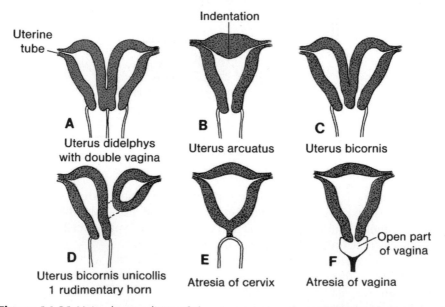

Figure 14.31 Main abnormalities of the uterus and vagina, caused by persistence of the uterine septum or obliteration of the lumen of the uterine canal.

EXTERNAL GENITALIA

Indifferent Stage

In the third week of development, mesenchyme cells originating in the region of the primitive streak migrate around the cloacal membrane to form a pair of slightly elevated **cloacal folds** (Fig. 14.32*A*). Cranial to the cloacal membrane the folds unite to form the **genital tubercle.** Caudally the folds are subdivided into **urethral folds** anteriorly and **anal folds** posteriorly (Fig. 14.32*B*).

In the meantime, another pair of elevations, the **genital swellings,** becomes visible on each side of the urethral folds. These swellings later form the **scrotal swellings** in the male (Fig. 14.33*A*) and the **labia majora** in the female (see Fig. 14.36*B*). At the end of the sixth week, however, it is impossible to distinguish between the two sexes (Fig. 14.34*C*).

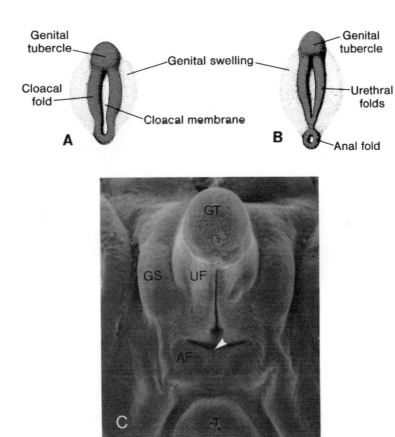

Figure 14.32 A and **B.** Indifferent stages of the external genitalia. **A.** Approximately 4 weeks. **B.** Approximately 6 weeks. **C.** Scanning electron micrograph of the external genitalia of a human embryo at approximately the seventh week. *AF*, anal fold; *arrowhead*, anal opening; *GS*, genital swelling; *GT*, genital tubercle; *T*, tail; *UF*, urethral fold.

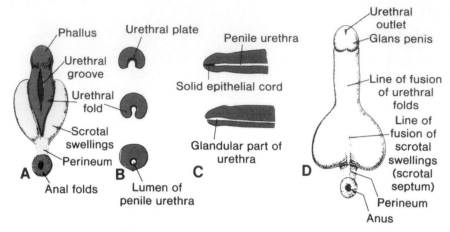

Figure 14.33 A. Development of external genitalia in the male at 10 weeks. Note the deep urethral groove flanked by the urethral folds. B. Transverse sections through the phallus during formation of the penile urethra. The urogenital groove is bridged by the urethral folds. C. Development of the glandular portion of the penile urethra. D. Newborn.

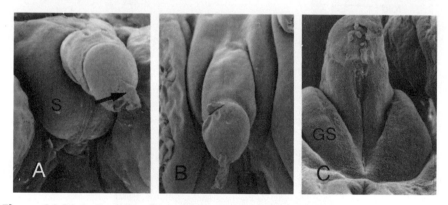

Figure 14.34 A. Genitalia of a male fetus at 14 weeks, showing fusion of the scrotal swellings (*S*). *Arrow*, epithelial tag. B and C. Dorsal and ventral views, respectively, of the genitalia of a female fetus at 11 weeks. The genital tubercle at this stage is longer than in the male (A), and the genital swellings (*GS*) remain unfused.

External Genitalia in the Male

Development of the external genitalia in the male is under the influence of androgens secreted by the fetal testes and is characterized by rapid elongation of the genital tubercle, which is now called the **phallus** (Figs. 14.33*A* and 14.34*A*). During this elongation, the phallus pulls the urethral folds forward so that they form the lateral walls of the **urethral groove**. This groove extends along the caudal aspect of the elongated phallus but does not reach the most distal part, the glans. The epithelial lining of the groove, which originates in the endoderm, forms the **urethral plate** (Fig. 14.33*B*).

At the end of the third month the two urethral folds close over the urethral plate, forming the **penile urethra** (Figs. 14.33*B* and 14.34*A*). This canal does not extend to the tip of the phallus. This most distal portion of the urethra is formed during the fourth month, when ectodermal cells from the tip of the glans penetrate inward and form a short epithelial cord. This cord later obtains a lumen, thus forming the **external urethral meatus** (Fig. 14.33*C*).

The genital swellings, known in the male as the scrotal swellings, arise in the inguinal region. With further development they move caudally, and each swelling then makes up half of the scrotum. The two are separated by the **scrotal septum** (Figs. 14.33*D* and 14.34*A*).

CLINICAL CORRELATES

Defects in the Male Genitalia

In **hypospadias** fusion of the urethral folds is incomplete, and abnormal openings of the urethra occur along the inferior aspect of the penis, usually near the glans, along the shaft, or near the base of the penis (Fig.14.35). In rare cases the urethral meatus extends along the scrotal raphe. When fusion of the urethral folds fails entirely, a wide sagittal slit is found along the entire length of the penis and the scrotum. The two scrotal swellings then closely resemble the labia majora. The incidence of hypospadias is 3–5/1000 births, and this rate represents a doubling over the past 15 to 20 years. Reasons for the increase are not known, but one hypothesis suggests it could be a result of a rise in environmental estrogens (endocrine disruptors, see Chapter 7).

Epispadias is a rare abnormality (1/30,000 births) in which the urethral meatus is found on the dorsum of the penis. Instead of developing at the cranial margin of the cloacal membrane, the genital tubercle seems to form in the region of the urorectal septum. Hence a portion of the cloacal membrane is found cranial to the genital tubercle, and when this membrane ruptures, the outlet of the urogenital sinus comes to lie on the cranial aspect of the penis (Fig. 14.35*C*). Although epispadias may occur as an isolated defect, it is most often associated with exstrophy of the bladder.

In **exstrophy of the bladder,** of which epispadias is a constant feature, the bladder mucosa is exposed to the outside (Figs. 14.16*A* and 14.35*C*). Normally the abdominal wall in front of the bladder is formed by primitive streak mesoderm, which migrates around the cloacal membrane. When this migration does not occur, rupture of the cloacal membrane extends cranially, creating exstrophy of the bladder.

Micropenis occurs when there is insufficient androgen stimulation for growth of the external genitalia. Micropenis is usually caused by primary hypogonadism or hypothalamic or pituitary dysfunction. By definition, the penis is 2.5 standard deviations below the mean in length as measured along

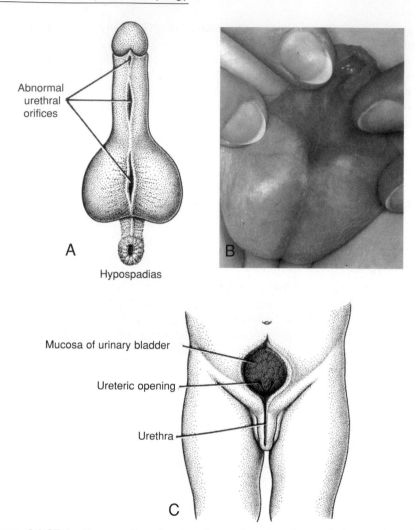

Abnormal urethral orifices

A

Hypospadias

B

Mucosa of urinary bladder

Ureteric opening

Urethra

C

Figure 14.35 A. Hypospadias showing the various locations of abnormal urethral orifices. **B.** Patient with hypospadias. The urethra is open on the ventral surface of the penis. **C.** Epispadias combined with exstrophy of the bladder. Bladder mucosa is exposed.

the dorsal surface from the pubis to the tip with the penis stretched to resistance. **Bifid penis** or **double penis** may occur if the genital tubercle splits.

External Genitalia in the Female

Estrogens stimulate development of the external genitalia of the female. The genital tubercle elongates only slightly and forms the **clitoris** (Figs. 14.34*B* and

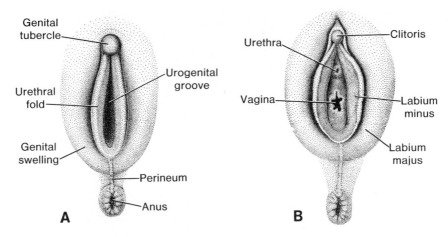

Figure 14.36 Development of the external genitalia in the female at 5 months **(A)** and in the newborn **(B).**

14.36*A*); urethral folds do not fuse, as in the male, but develop into the **labia minora.** Genital swellings enlarge and form the **labia majora.** The urogenital groove is open and forms the **vestibule** (Figs. 14.34*C* and 14.36*B*). Although the genital tubercle does not elongate extensively in the female, it is larger than in the male during the early stages of development (Fig. 14.34,*A* and *B*). In fact, using tubercle length as a criterion (as monitored by ultrasound) has resulted in mistakes in identification of the sexes during the third and fourth months of gestation.

CLINICAL CORRELATES

Defects in Sex Differentiation

Klinefelter syndrome, with a karyotype of 47,XXY (or other variants, e.g., XXXY), is the most common major abnormality of sexual differentiation, occurring with a frequency of 1/500 males. Patients are characterized by infertility, gynecomastia, varying degrees of impaired sexual maturation, and in some cases underandrogenization. Nondisjunction of the XX homologues is the most common causative factor.

In **gonadal dysgenesis** oocytes are absent and the ovaries appear as streak gonads. Individuals are phenotypically female but may have a variety of chromosomal complements, including XY. **XY female gonadal dysgenesis (Swyer syndrome)** results from point mutations or deletions of the *SRY* gene. Individuals appear to be normal females but do not menstruate and do not develop secondary sexual characteristics at puberty. Patients with **Turner syndrome** also have gonadal dysgenesis. They have a 45,X karyotype and short stature, high arched palate, webbed neck, shieldlike chest, cardiac and renal anomalies,

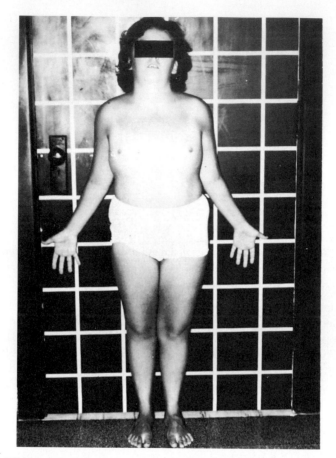

Figure 14.37 Patient with Turner syndrome, which is characterized by a 45,X chromosome complement. Note the absence of sexual maturation. Other typical features are webbed neck, broad chest with widely spaced nipples, and short stature.

and inverted nipples (Fig. 14.37). Absence of oocytes in 45,X cases is due to increased oocyte loss and not to germ cell abnormalities.

Since sexual development of males and females begins in an identical fashion, it is not surprising that abnormalities in differentiation and sex determination occur. In some cases these abnormalities result in individuals with characteristics of both sexes, known as **hermaphrodites.** True hermaphrodites have both testicular and ovarian tissue, usually combined as ovotestes. In 70 % of cases the karyotype is 46,XX, and there is usually a uterus. External genitalia are ambiguous or predominantly female, and most of these individuals are raised as females.

In **pseudohermaphrodites,** the genotypic sex is masked by a phenotypic appearance that closely resembles the other sex. When the pseudohermaphrodite has a testis, the patient is called a male pseudo-hermaphrodite; when an ovary is present, the patient is called a female pseu-dohermaphrodite.

Female pseudohermaphroditism is most commonly caused by **congenital adrenal hyperplasia (adrenogenital syndrome).** Biochemical abnormalities in the adrenal glands result in decreased steroid hormone production and an increase in adrenocorticotropic hormone (ACTH). In most cases, 21-hydroxylation is inhibited, such that 17-hydroxyprogesterone (17-OHP) is not converted to 11-deoxycortisol. ACTH levels increase in response to defective cortisol production, which leads to ever-increasing amounts of 17-OHP. In turn, there is excessive production of androgens. Patients have a 46,XX chromosome complement, chromatin-positive nuclei, and ovaries, but excessive production of androgens masculinizes the external genitalia. This masculinization may vary from enlargement of the clitoris to almost male genitalia (Fig. 14.38). Frequently there is clitoral hypertrophy and partial fusion of

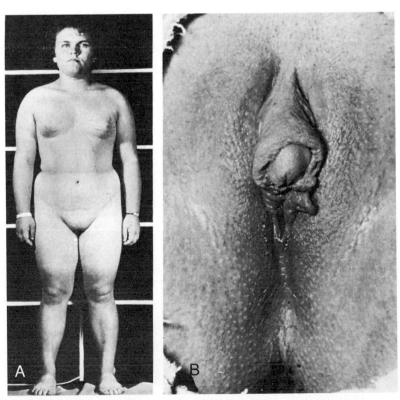

Figure 14.38 A. Patient with female pseudohermaphrodism caused by congenital adrenal hyperplasia (adrenogenital syndrome). **B.** External genitalia show fusion of the labia majora and enlargement of the clitoris.

the labia majora, giving the appearance of a scrotum, and a small persistent urogenital sinus.

Male pseudohermaphrodites have a 46,XY chromosome complement, and their cells are usually chromatin-negative. Reduced production of androgenic hormones and MIS are responsible for this condition. Internal and external sex characteristics vary considerably, depending on the degree of development of external genitalia and the presence of paramesonephric derivatives.

Androgen insensitivity syndrome (formerly testicular feminization) occurs in patients who have a 46,XY chromosome complement but have the external appearance of normal females (Fig. 14.39). This disorder results from a lack of androgen receptors or failure of tissues to respond to receptor-dihydrotestosterone complexes. Consequently, androgens produced by the testes are ineffective in inducing differentiation of male genitalia. Since these patients have testes and MIS is present, the paramesonephric system is suppressed, and uterine tubes and uterus are absent. The vagina is short and blind. The testes are frequently found in the inguinal or labial regions, but spermatogenesis does not occur. Furthermore, there is an increased risk of tumor formation in these structures, and 33% of these individuals develop malignancies prior to age 50. This syndrome is an X-linked recessive disorder that occurs in 1/20,000 live births.

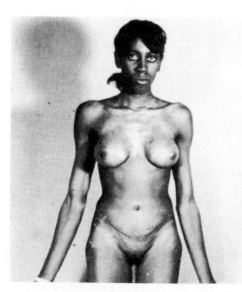

Figure 14.39 Patient with androgen insensitivity syndrome, which is characterized by a 46,XY chromosome complement.

DESCENT OF THE TESTES

Toward the end of the second month, the **urogenital mesentery** attaches the testis and mesonephros to the posterior abdominal wall (Fig. 14.3*A*). With degeneration of the mesonephros the attachment serves as a mesentery for the gonad (Fig. 14.28*B*). Caudally it becomes ligamentous and is known as the **caudal genital ligament** (Fig. 14.40*A*). Also extending from the caudal pole of the testis is a mesenchymal condensation rich in extracellular matrices, the **gubernaculum** (Fig. 14.40). Prior to descent of the testis, this band of mesenchyme terminates in the inguinal region between the differentiating internal and external abdominal oblique muscles. Later, as the testis begins to descend toward the inguinal ring, an extra-abdominal portion of the gubernaculum forms and grows from the inguinal region toward the scrotal swellings. When the testis passes through the inguinal canal, this extra-abdominal portion contacts the scrotal floor (the gubernaculum forms in females also, but in normal cases it remains rudimentary).

Factors controlling descent of the testis are not entirely clear. It appears, however, that outgrowth of the extra-abdominal portion of the gubernaculum produces intra-abdominal migration, that an increase in intra-abdominal pressure due to organ growth produces passage through the inguinal canal, and that regression of the extra-abdominal portion of the gubernaculum completes movement of the testis into the scrotum. Normally, the testes reach the inguinal region by approximately 12 weeks gestation, migrate through the inguinal canal by 28 weeks, and reach the scrotum by 33 weeks (Fig. 14.40). The process is influenced by hormones, including androgens and MIS. During descent, blood supply to the testis from the aorta is retained, and testicular vessels extend from their original lumbar position to the testis in the scrotum.

Independently from descent of the testis, the peritoneum of the abdominal cavity forms an evagination on each side of the midline into the ventral abdominal wall. This evagination, the **processus vaginalis,** follows the course of the gubernaculum testis into the scrotal swellings (Fig. 14.40*B*). Hence the processus vaginalis, accompanied by the muscular and fascial layers of the body wall, evaginates into the scrotal swelling, forming the **inguinal canal** (Fig. 14.41).

The testis descends through the inguinal ring and over the rim of the pubic bone and is present in the scrotum at birth. The testis is then covered by a reflected fold of the processus vaginalis (Fig. 14.40*D*). The peritoneal layer covering the testis is the **visceral layer of the tunica vaginalis;** the remainder of the peritoneal sac forms the **parietal layer of the tunica vaginalis** (Fig. 14.40*D*). The narrow canal connecting the lumen of the vaginal process with the peritoneal cavity is obliterated at birth or shortly thereafter.

In addition to being covered by peritoneal layers derived from the processus vaginalis, the testis becomes ensheathed in layers derived from the anterior abdominal wall through which it passes. Thus, the **transversalis fascia** forms the **internal spermatic fascia,** the **internal abdominal oblique muscle** gives rise to the **cremasteric fascia and muscle,** and the **external abdominal oblique**

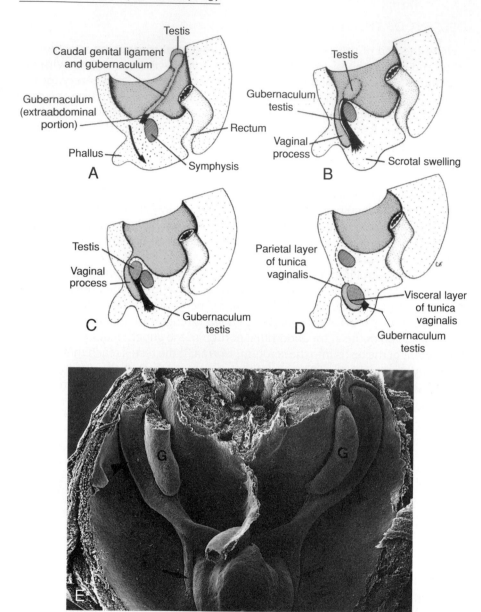

Figure 14.40 Descent of the testis. **A.** During the second month. **B.** In the middle of the third month. Peritoneum lining the coelomic cavity evaginates into the scrotal swelling, where it forms the vaginal process (tunica vaginalis). **C.** In the seventh month. **D.** Shortly after birth. **E.** Scanning electron micrograph of a mouse embryo showing the primitive gonad (*G*), mesonephric duct (*arrowheads*), and gubernaculum (*arrows*).

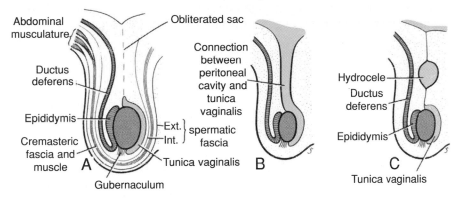

Figure 14.41 A. Testis, epididymis, ductus deferens, and various layers of the abdominal wall that surround the testis in the scrotum. **B.** Vaginal process in open communication with the peritoneal cavity. In such a case, portions of the intestinal loops often descend toward and occasionally into the scrotum, causing an inguinal hernia. **C.** Hydrocele.

muscle forms the **external spermatic fascia** (Fig. 14.41*A*). The transversus abdominis muscle does not contribute a layer, since it arches over this region and does not cover the path of migration.

CLINICAL CORRELATES

Hernias and Cryptorchism

The connection between the abdominal cavity and the processus vaginalis in the scrotal sac normally closes in the first year after birth (Fig. 14.40*D*). If this passageway remains open, intestinal loops may descend into the scrotum, causing a **congenital inguinal hernia.** (Fig, 14.41*B*). Sometimes obliteration of this passageway is irregular, leaving small cysts along its course. Later these cysts may secrete fluid, forming a **hydrocele of the testis and/or spermatic cord** (Fig. 14,41*C*).

In 97% of male newborns, testes are present in the scrotum before birth. In most of the remainder, descent will be completed during the first 3 months postnatally. However, in less than 1% of infants, one or both testes fail to descend. The condition is called **cryptorchidism** and may be caused by decreased androgen (testosterone) production. The undescended testes fail to produce mature spermatozoa and the condition is associated with a 3% to 5% incidence of renal anomalies.

DESCENT OF THE OVARIES

Descent of the gonads is considerably less in the female than in the male, and the ovaries finally settle just below the rim of the true pelvis. The cranial genital ligament forms the **suspensory ligament** of the ovary, whereas the caudal genital ligament forms the **ligament of the ovary proper** and the **round ligament of the uterus** (Fig. 14.24). the latter extends into the labia majora.

Summary

The urinary and genital systems both develop from mesodermal tissue. Three urinary systems develop in a temporal sequence from cranial to caudal segments:

The **pronephros,** which forms in the cervical region, is vestigial.

The **mesonephros,** which forms in the thoracic and lumbar regions, is large and is characterized by excretory units (nephrons) and its own collecting duct, the mesonephric or wolffian duct. In the human it may function briefly, but most of the system disappears. Ducts and tubules from the mesonephros form the conduit for sperm from the testes to the urethra. In the female, these ducts regress.

The **metanephros,** or permanent kidney, develops from two sources. It forms its own excretory tubules or nephrons like the other systems, but its collecting system originates from the **ureteric bud,** an outgrowth of the mesonephric duct. This bud gives rise to the ureter, renal pelvis, calyces, and the entire collecting system (Fig. 14.5). Connection between the collecting and excretory tubule systems is essential for normal development (Fig. 14.6). *WT1,* expressed by the mesenchyme, makes this tissue competent to respond to induction by the ureteric bud. Interactions between the bud and mesenchyme occur through production of GDNF and HGF by the mesenchyme with their tyrosine kinase receptors RET and MET, respectively, produced by the ureteric epithelium. PAX2 and WNT4, produced by the ureteric bud, cause epithelialization of the metanephric mesenchyme in preparation for excretory tubule differentiation (Fig. 14.7). Early division of the ureteric bud may lead to bifid or supernumerary kidneys with ectopic ureters (Fig. 14.9). Abnormal positions of the kidney, such as pelvic and horseshoe kidney, are also well known (Fig. 14.11).

The genital system consists of (*a*) gonads or primitive sex glands, (*b*) genital ducts, and (*c*) external genitalia. All three components go through an **indifferent stage** in which they may develop into either a male or a female. The *SRY* gene on the Y chromosome produces testes-determining factor and regulates male sexual development. Genes downstream from SRY include ***steroidogenesis factor (SF1)*** and ***SOX9*** that stimulate differentiation of Sertoli and Leydig cells in the testes. Expression of the SRY gene causes (*a*) development of the medullary (testis) cords, (*b*) formation of the tunica albuginea, and (*c*) failure of the cortical (ovarian) cords to develop. In the absence of the *SRY* gene, the combination of *DAX1* expression, to downregulate SF1, and continued expression of *WNT4* in the gonadal ridge, causes formation of ovaries with (*a*) typical cortical cords, (*b*) disappearance of the medullary (testis) cords, and (*c*) failure of the tunica albuginea to develop (Fig. 14.21). When primordial germ cells fail to reach the indifferent gonad, the gonad remains indifferent or is absent.

The indifferent duct system and external genitalia develop under the influence of hormones. **Testoterone** produced by Leydig cells in the testes stimulates development of the mesonephric ducts (vas deferens epididymis), while **MIS** produced by Sertoli cells in the testes causes regression of the paramesonephric ducts (female duct system). **Dihydrotestosterone** stimulates development of the external genitalia, penis, scrotum, and prostate (Fig. 14.26). **Estrogens** influence development of the paramesonephric female system, including the uterine tube, uterus, cervix, and upper portion of the vagina. They also stimulate differentiation of the external genitalia, including the clitoris, labia, and lower portion of the vagina (Fig. 14.26). Errors in production of or sensitivity to hormones of the testes lead to a predominance of female characteristics under influence of the maternal and placental estrogens.

Problems to Solve

1. *During development of the urinary system, three systems form. What are they, and what parts of each, if any, remain in the newborn?*

2. *At birth an apparently male baby has no testicles in the scrotum. Later it is determined that both are in the abdominal cavity. What is the term given to this condition, and can you explain the embryological origin of this defect?*

3. *It is said that male and female external genitalia have homologies. What are they, and what are their embryological origins?*

4. *After several years of trying to become pregnant, a young woman seeks consultation. Examination reveals a bicornate uterus. How could such an abnormality occur?*

SUGGESTED READING

Behringer RR, Finegold MJ, Cate RL: Müllerian-inhibiting substance function during mammalian sexual development. *Cell* 79:415, 1994.

Griffin JE, Wilson JD: Disorders of sexual differentiation. *In* Walsh PC, et al (eds): *Campbell's Urology.* Philadelphia, WB Saunders, 1986.

Haqq CM, et al.: Molecular basis of mammalian sexual determination: activation of mullerian inhibiting substance gene expression by SRY. *Science* 266:1494, 1994.

McElreavey K, et al.: A regulatory cascade hypothesis for mammalian sex determination: SRY represses a negative regulator of male development. *Proc Natl Acad Sci* 90:3368, 1993.

Mesrobian HGJ, Rushton HG, Bulas D: Unilateral renal agenesis may result from in utero regression of multicystic dysplasia. *J Urol* 150:793, 1993.

Mittwoch U: Sex determination and sex reversal: genotype, phenotype, dogma and semantics. *Hum Genet* 89:467, 1992.

O'Rahilly R: The development of the vagina in the human. *In* Blandau RJ, Bergsma D (eds): *Morphogenesis and Malformation of the Genital Systems.* New York, Alan R Liss, 1977:123.

Paulozzi LJ, Erickson JD, Jackson RJ: Hypospadias trends in two US surveillance systems, Pediatrics 100:831, 1997.

Persuad TVN: Embryology of the female genital tract and gonads. *In* Copeland LJ, Jarrell J, McGregor J (eds): *Textbook of Gynecology.* Philadelphia, WB Saunders, 1992.

Saxen L, Sariola H, Lehtonen E: Sequential cell and tissue interactions governing organogenesis of the kidney. *Anat Embryol* 175:1, 1986.

Stevenson RE, Hall JG, Goodman RM (eds): *Human Malformations and Related Anomalies,* vol 2. New York, Oxford University, 1993.

Swain A, Narvaez V, Burgoyne P, Camerino G, Lovell-Badge R: Dax I antagonizes Sry action in mammalian sex determination, *Nature* 391:761, 1998.

Tilmann C, Capel B: Mesonephric cell migration induces testes cord formation and Sertoli cell differentiation in the mammalian gonad. *Dev* 126:2883, 1999.

Vaino S, Heikkila M, Kispert A, Chin N, McMahon AP: Female development in mammals is regulated by WNT4 signaling. *Nature* 397:405, 1999.

Van der Werff JFA, Nievelstein RAJ, Brands E, Linjsterburg AJM, Vermeij-Keers C: Normal development of the male anterior urethra. *Teratology* 61:172, 2000.

Vilain E, Jaubert F, Fellows M, McElreavey K: Pathology of 46,XY pure gonadal dysgenesis: absence of testes differentiation associated with mutations in the testes determining factor. *Differentiation* 52:151, 1993.

Wensing CJG, Colenbrander B: Normal and abnormal testicular descent. *Oxf Rev Reprod Biol* 130, 1986.

Woolf AS: Clinical impact and biological basis of renal malformations. *Seminars Nephrol* 15:361, 1995.

chapter 15

Head and Neck

Mesenchyme for formation of the head region is derived from **paraxial** and **lateral plate mesoderm, neural crest,** and thickened regions of ectoderm known as **ectodermal placodes.** Paraxial mesoderm (**somites** and **somitomeres**) forms the floor of the brain case and a small portion of the occipital region (Fig. 15.1) (see Chapter 8), all voluntary muscles of the craniofacial region (see Chapter 9), the dermis and connective tissues in the dorsal region of the head, and the meninges caudal to the prosencephalon. Lateral plate mesoderm forms the laryngeal cartilages (arytenoid and cricoid) and connective tissue in this region. Neural crest cells originate in the neuroectoderm of forebrain, midbrain, and hindbrain regions and migrate ventrally into the pharyngeal arches and rostrally around the forebrain and optic cup into the facial region (Fig. 15.2). In these locations they form midfacial and pharyngeal arch skeletal structures (Fig. 15.1) and all other tissues in these regions, including cartilage, bone, dentin, tendon, dermis, pia and arachnoid, sensory neurons, and glandular stroma. Cells from **ectodermal placodes,** together with neural crest, form neurons of the fifth, seventh, ninth, and tenth cranial sensory ganglia.

The most typical feature in development of the head and neck is formed by the **pharyngeal** or **branchial arches.** These arches appear in the fourth and fifth weeks of development and contribute to the characteristic external appearance of the embryo (Table 15.1 and Fig. 15.3). Initially, they consist of bars of mesenchymal tissue separated by deep clefts known as **pharyngeal (branchial) clefts** (Figs. 15.3C;

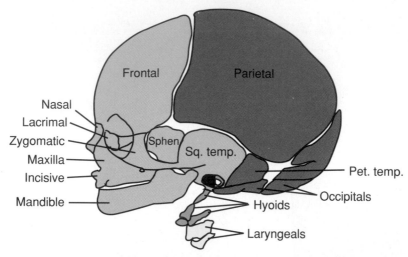

Figure 15.1 Skeletal structures of the head and face. Mesenchyme for these structures is derived from neural crest (*blue*), lateral plate mesoderm (*yellow*), and paraxial mesoderm (somites and somitomeres) (*red*).

see also Fig. 15.6). Simultaneously, with development of the arches and clefts, a number of outpocketings, the **pharyngeal pouches,** appear along the lateral walls of the pharyngeal gut, the most cranial part of the foregut (Fig. 15.4; see also Fig. 15.6). The pouches penetrate the surrounding mesenchyme, but do not establish an open communication with the external clefts (see Fig. 15.6). Hence, although development of pharyngeal arches, clefts, and pouches resembles formation of gills in fishes and amphibia, in the human embryo real gills (branchia) are never formed. Therefore, the term **pharyngeal** (arches, clefts, and pouches) has been adopted for the human embryo.

Pharyngeal arches not only contribute to formation of the neck, but also play an important role in formation of the face. At the end of the fourth week, the center of the face is formed by the stomodeum, surrounded by the first pair of pharyngeal arches (Fig. 15.5). When the embryo is 42 weeks old, five mesenchymal prominences can be recognized: the **mandibular prominences** (first pharyngeal arch), caudal to the stomodeum; the **maxillary prominences** (dorsal portion of the first pharyngeal arch), lateral to the stomodeum; and the **frontonasal prominence,** a slightly rounded elevation cranial to the stomodeum. Development of the face is later complemented by formation of the **nasal prominences** (Fig. 15.5). In all cases, differentiation of structures derived from arches pouches, clefts, and prominences is dependent upon epithelial-mesenchymal interactions. In some instances, signals for these interactions are similar to those involved in limb development, including fibroblast growth factor (FGFs) for outgrowth and *sonic hedgehog (SHH)* and WNTs for patterning.

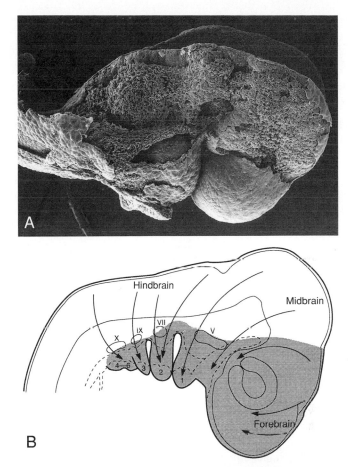

Figure 15.2 A. Scanning electron micrograph showing cranial neural crest cells migrating into the facial region beneath the ectoderm that has been removed. **B.** Migration pathways of neural crest cells from forebrain, midbrain, and hindbrain regions into their final locations (*shaded areas*) in the pharyngeal arches and face. Regions of ectodermal thickenings (placodes), which will assist crest cells in formation of the fifth (V), seventh (VII), ninth (IX), and tenth (X) cranial sensory ganglia, are also illustrated.

Pharyngeal Arches

Each pharyngeal arch consists of a core of mesenchymal tissue covered on the outside by surface ectoderm and on the inside by epithelium of endodermal origin (Fig. 15.6). In addition to mesenchyme derived from paraxial and lateral plate mesoderm, the core of each arch receives substantial numbers of **neural crest cells,** which migrate into the arches to contribute to **skeletal components** of the face. The original mesoderm of the arches gives rise to the musculature of the face and neck. Thus, each pharyngeal arch is characterized by its own

TABLE 15.1 Derivatives of the Pharyngeal Arches and
Their Innervation

Pharyngeal Arch	Nerve	Muscles	Skeleton
1 mandibular (maxillary and mandibular processes)	V. Trigeminal: maxillary and mandibular divisions	Mastication (temporal; masseter; medial, lateral pterygoids); mylohyoid; anterior belly of digastric; tensor palatine, tensor tympani	Premaxilla, maxilla, zygomatic bone, part of temporal bone, Meckel's cartilage, mandible malleus, incus, anterior ligament of malleus, sphenomandibular ligament
2 hyoid	VII. Facial	Facial expression (buccinator; auricularis; frontalis; platysma; orbicularis oris; orbicularis oculi); posterior belly of digastric; stylohyoid; stapedius	Stapes; styloid process; stylohyoid ligament; lesser horn and upper portion of body of hyoid bone
3	IX. Glossopharyngeal	Stylopharyngeus	Greater horn and lower portion of body of hyoid bone
4–6	X. Vagus · Superior laryngeal branch (nerve to fourth arch) · Recurrent laryngeal branch (nerve to sixth arch)	Cricothyroid; levator palatine; constrictors of pharynx Intrinsic muscles of larynx	Laryngeal cartilages (thyroid, cricoid, arytenoid, corniculate, cuneiform)

muscular components. The muscular components of each arch have their own
cranial nerve, and wherever the muscle cells migrate, they carry their **nerve
component** with them (Figs. 15.6 and 15.7). In addition, each arch has its
own **arterial component** (Figs. 15.4 and 15.6). (Derivatives of the pharyngeal
arches and their nerve supply are summarized in Table 15.1).

FIRST PHARYNGEAL ARCH

The **first pharyngeal arch** consists of a dorsal portion, the **maxillary process,**
which extends forward beneath the region of the eye, and a ventral portion,

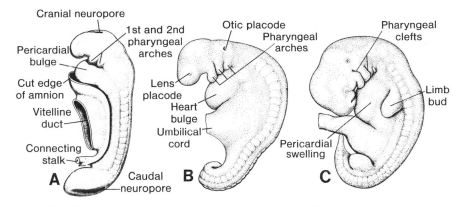

Figure 15.3 Development of the pharyngeal arches. **A.** 25 days. **B.** 28 days. **C.** 5 weeks.

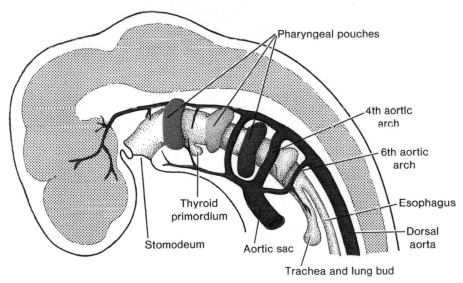

Figure 15.4 Pharyngeal pouches as outpocketings of the foregut and the primordium of the thyroid gland and aortic arches.

the **mandibular process,** which contains **Meckel's cartilage** (Figs. 15.5 and 15.8*A*). During further development, Meckel's cartilage disappears except for two small portions at its dorsal end that persist and form the **incus** and **malleus** (Figs. 15.8*B* and 15.9). Mesenchyme of the maxillary process gives rise to the **premaxilla, maxilla, zygomatic bone,** and part of the **temporal bone** through membranous ossification (Fig. 15.8*B*). The **mandible** is also formed by membranous ossification of mesenchymal tissue surrounding Meckel's cartilage. In addition, the first arch contributes to formation of the bones of the middle ear (see Chapter 16).

Musculature of the first pharyngeal arch includes the **muscles of mastication** (temporalis, masseter, and pterygoids), **anterior belly of the digastric,**

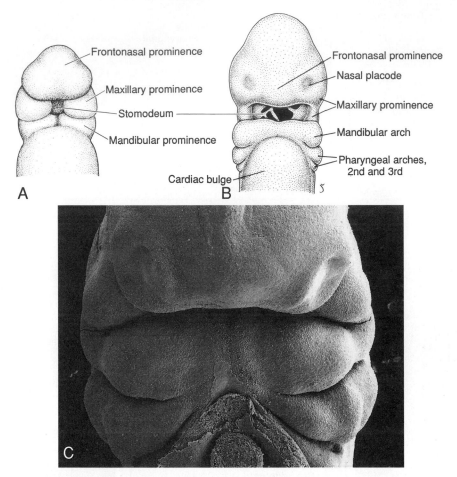

Figure 15.5 A. Frontal view of an embryo of approximately 24 days. The stomodeum, temporarily closed by the buccopharyngeal membrane, is surrounded by five mesenchymal prominences. **B.** Frontal view of a slightly older embryo showing rupture of the buccopharyngeal membrane and formation of the nasal placodes on the frontonasal prominence. **C.** Scanning electron micrograph of a human embryo similar to that shown in **B.**

mylohyoid, tensor tympani, and **tensor palatini.** The **nerve** supply to the muscles of the first arch is provided by the **mandibular branch of the trigeminal nerve** (Fig. 15.7). Since mesenchyme from the first arch also contributes to the dermis of the face, sensory supply to the skin of the face is provided by **ophthalmic, maxillary,** and **mandibular branches of the trigeminal nerve.**

Muscles of the arches do not always attach to the bony or cartilaginous components of their own arch but sometimes migrate into surrounding regions. Nevertheless, the origin of these muscles can always be traced, since their nerve supply is derived from the arch of origin.

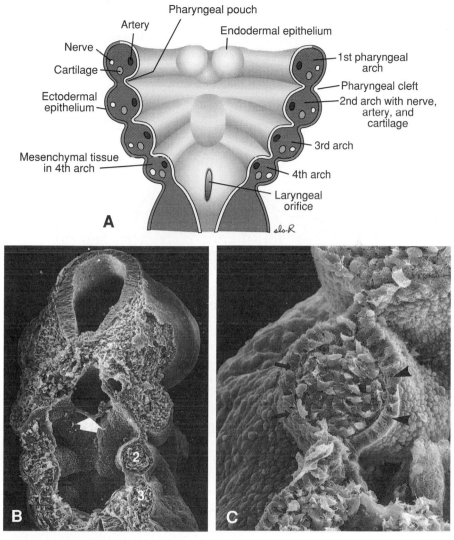

Figure 15.6 A. Pharyngeal arches. Each arch contains a cartilaginous component, a cranial nerve, an artery, and a muscular component. **B.** Scanning electron micrograph of the pharyngeal region of a mouse embryo, showing the pharyngeal arches, pouches, and clefts. The first three arches (I, II, and III) are visible. A remnant of the buccopharyngeal membrane (*arrow*) is present at the entrance to the oral cavity. **C.** Higher magnification of the pharyngeal arches of a mouse embryo. Pharyngeal arches consist of a core of mesoderm lined by endoderm internally (*arrowheads*) and ectoderm externally (*arrows*). Pouches and clefts occur between the arches, where endoderm and ectoderm appose each other.

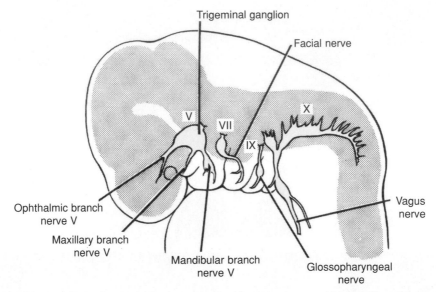

Figure 15.7 Each pharyngeal arch is supplied by its own cranial nerve. The trigeminal nerve supplying the first pharyngeal arch has three branches: the ophthalmic, maxillary, and mandibular. The nerve of the second arch is the facial nerve; that of the third, the glossopharyngeal nerve. The musculature of the fourth arch is supplied by the superior laryngeal branch of the vagus nerve, and that of the sixth arch, by the recurrent branch of the vagus nerve.

SECOND PHARYNGEAL ARCH

The cartilage of the **second** or **hyoid arch (Reichert's cartilage)** (Fig. 15.8*B*) gives rise to the **stapes, styloid process of the temporal bone, stylohyoid ligament,** and ventrally, the **lesser horn** and **upper part of the body of the hyoid bone** (Fig. 15.9). Muscles of the hyoid arch are the **stapedius, stylohyoid, posterior belly of the digastric, auricular,** and **muscles of facial expression.** The **facial nerve,** the nerve of the second arch, supplies all of these muscles.

THIRD PHARYNGEAL ARCH

The **cartilage** of the third pharyngeal arch produces the **lower part of the body** and **greater horn of the hyoid bone** (Fig. 15.9). The **musculature** is limited to the **stylopharyngeus muscles.** These muscles are innervated by the **glossopharyngeal nerve,** the nerve of the third arch (Fig. 15.7).

FOURTH AND SIXTH PHARYNGEAL ARCHES

Cartilaginous components of the fourth and sixth pharyngeal arches fuse to form the **thyroid, cricoid, arytenoid, corniculate,** and **cuneiform cartilages**

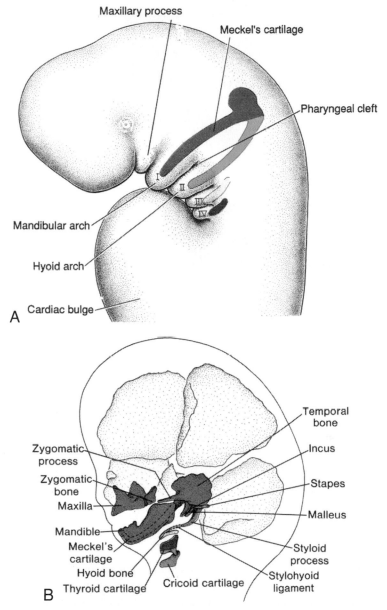

Figure 15.8 A. Lateral view of the head and neck region of a 4-week embryo demonstrating the cartilages of the pharyngeal arches participating in formation of the bones of the face and neck. **B.** Various components of the pharyngeal arches later in development. Some of the components ossify; others disappear or become ligamentous. The maxillary process and Meckel's cartilage are replaced by the maxilla and mandible, respectively, which develop by membranous ossification.

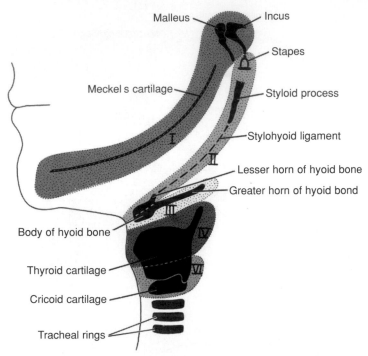

Figure 15.9 Definitive structures formed by the cartilaginous components of the various pharyngeal arches.

of the **larynx** (Fig. 15.9). **Muscles** of the fourth arch (**cricothyroid, levator palatini,** and **constrictors of the pharynx**) are innervated by the **superior laryngeal branch of the vagus,** the nerve of the fourth arch. Intrinsic muscles of the larynx are supplied by the **recurrent laryngeal branch of the vagus,** the nerve of the sixth arch.

Pharyngeal Pouches

The human embryo has five pairs of pharyngeal pouches (Figs. 15.6 and 15.10). The last one of these is atypical and often considered as part of the fourth. Since the **epithelial endodermal lining** of the pouches gives rise to a number of important organs, the fate of each pouch is discussed separately.

FIRST PHARYNGEAL POUCH

The first pharyngeal pouch forms a stalklike diverticulum, the **tubotympanic recess,** which comes in contact with the epithelial lining of the first pharyngeal cleft, the future **external auditory meatus** (Fig. 15.10). The distal portion of

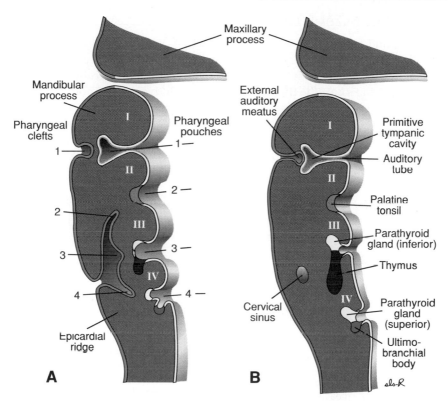

Figure 15.10 A. Development of the pharyngeal clefts and pouches. The second arch grows over the third and fourth arches, burying the second, third, and fourth pharyngeal clefts. **B.** Remnants of the second, third, and fourth pharyngeal clefts form the cervical sinus, which is normally obliterated. Note the structures formed by the various pharyngeal pouches.

the diverticulum widens into a saclike structure, the **primitive tympanic** or **middle ear cavity,** and the proximal part remains narrow, forming the **auditory (eustachian) tube.** The lining of the tympanic cavity later aids in formation of the **tympanic membrane** or **eardrum** (see Chapter 16).

SECOND PHARYNGEAL POUCH

The epithelial lining of the second pharyngeal pouch proliferates and forms buds that penetrate into the surrounding mesenchyme. The buds are secondarily invaded by mesodermal tissue, forming the primordium of the **palatine tonsil** (Fig. 15.10). During the third and fifth months, the tonsil is infiltrated by lymphatic tissue. Part of the pouch remains and is found in the adult as the **tonsillar fossa.**

THIRD PHARYNGEAL POUCH

The third and fourth pouches are characterized at their distal extremity by a dorsal and a ventral wing (Fig. 15.10). In the fifth week, epithelium of the dorsal wing of the third pouch differentiates into the **inferior parathyroid gland,** while the ventral wing forms the **thymus** (Fig. 15.10). Both gland primordia lose their connection with the pharyngeal wall, and the thymus then migrates in a caudal and a medial direction, pulling the **inferior parathyroid** with it (Fig. 15.11). Although the main portion of the thymus moves rapidly to its final position in the anterior part of the thorax, where it fuses with its counterpart from the opposite side, its tail portion sometimes persists either embedded in the thyroid gland or as isolated thymic nests.

Growth and development of the thymus continue until puberty. In the young child, the thymus occupies considerable space in the thorax and lies behind the sternum and anterior to the pericardium and great vessels. In older

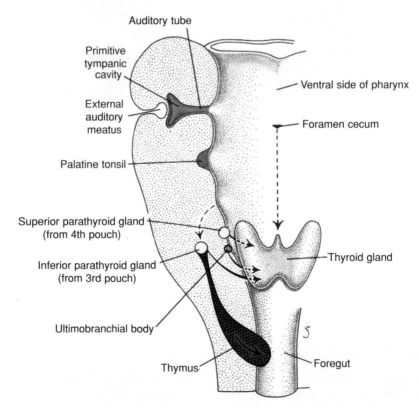

Figure 15.11 Migration of the thymus, parathyroid glands, and ultimobranchial body. The thyroid gland originates in the midline at the level of the foramen cecum and descends to the level of the first tracheal rings.

persons, it is difficult to recognize, since it is atrophied and replaced by fatty tissue.

The parathyroid tissue of the third pouch finally comes to rest on the dorsal surface of the thyroid gland and forms the **inferior parathyroid gland** (Fig. 15.11).

FOURTH PHARYNGEAL POUCH

Epithelium of the dorsal wing of the fourth pharyngeal pouch forms the **superior parathyroid gland.** When the parathyroid gland loses contact with the wall of the pharynx, it attaches itself to the dorsal surface of the caudally migrating thyroid as the **superior parathyroid gland** (Fig. 15.11).

FIFTH PHARYNGEAL POUCH

The fifth pharyngeal pouch, the last to develop, is usually considered to be a part of the fourth pouch. It gives rise to the **ultimobranchial body,** which is later incorporated into the thyroid gland. Cells of the ultimobranchial body give rise to the **parafollicular,** or **C, cells** of the thyroid gland. These cells secrete **calcitonin,** a hormone involved in regulation of the calcium level in the blood.

Pharyngeal Clefts

The 5-week embryo is characterized by the presence of four pharyngeal clefts (Fig. 15.6), of which only one contributes to the definitive structure of the embryo. The dorsal part of the first cleft penetrates the underlying mesenchyme and gives rise to the **external auditory meatus** (Figs. 15.10 and 15.11). The epithelial lining at the bottom of the meatus participates in formation of the **eardrum** (see Chapter 16).

Active proliferation of mesenchymal tissue in the second arch causes it to overlap the third and fourth arches. Finally, it merges with the **epicardial ridge** in the lower part of the neck (Fig. 15.10), and the second, third, and fourth clefts lose contact with the outside (Fig. 15.10*B*). The clefts form a cavity lined with ectodermal epithelium, the **cervical sinus,** but with further development this sinus disappears.

Molecular Regulation of Facial Development

As indicated, much of the face is derived from neural crest cells that migrate into the pharyngeal arches from the edges of the cranial neural folds. In the hindbrain, crest cells originate from segmented regions known as **rhombomeres.** There are eight of these segments in the hindbrain (R1 to R8), and crest cells

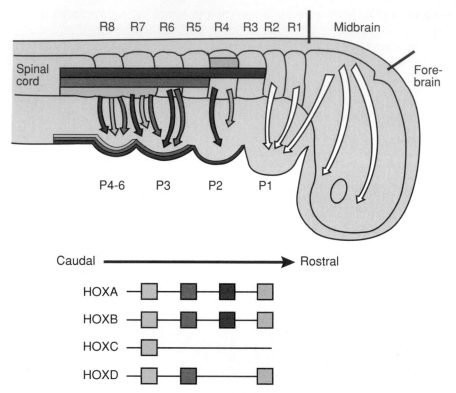

Figure 15.12 Patterns of neural crest cell migration into the pharyngeal arches and of *HOX* gene expression in rhombomeres of the hindbrain. *HOX* genes are expressed in overlapping patterns with those at the 3′ end having the most rostral boundaries. This pattern specifies derivatives of each of the rhombomeres, including crest cells and their pathways of migration. The first arch is also populated by crest cells from the midbrain. These cells express *OTX2*, a transcription factor containing a homeodomain.

from specific segments populate specific arches (Fig. 15.12). Crest cells from R1 and R2 migrate to the first arch, cells from R4 go to the second arch, those from R6 and 7 to the third arch, and those from R8 to the fourth and sixth arches. In addition, the first arch receives crest cells originating in the midbrain. Few if any crest cells form from R3 and R5. Most of the cells from these rhombomeres undergo cell death by **apoptosis,** while a few migrate with crest cells originating from adjacent segments.

Patterning of pharyngeal arches (with exception of the first arch) is regulated by ***HOX*** genes carried by migrating neural crest (Fig. 15.12). Expression of *HOX* genes in the hindbrain occurs in specific overlapping patterns, such that the most 3′ genes in a cluster have the most rostral boundaries (Fig. 15.12). Since 3′ genes are the first to be expressed, a temporal relationship for *HOX* gene expression is also established in a rostrocaudal sequence. In addition, paralogous genes, for example *HOXA3, HOXB3,* and *HOXD3* (see Chapter 5 and Fig. 5.22), share identical expression domains. These expression patterns determine the organization of cranial ganglia and nerves and pathways of neural crest migration. Initially, crest cells express the *HOX* genes from their segment of origin, but maintenance of this specific expression is dependent upon interaction of these cells with mesoderm in the pharyngeal arches. For example, crest cells from the second arch express *HOXA2,* and if these cells interact with second arch mesoderm, then this expression continues (Fig. 15.13). However, if second arch crest is placed into the first arch, this expression is downregulated. Thus, although an overlapping *HOX* code is essential to specifying the identity of the arches and their derivatives, crest cells alone do not establish or maintain the expression pattern. How the code is translated to control

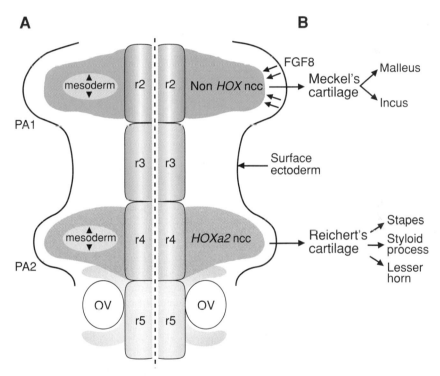

Figure 15.13 Schematic drawing showing the relationship between the first two pharyngeal arches (PA1 and PA2), segmentation of a region of the hindbrain at rhombomeres 2-5 (r2-r5), and pathways of neural crest cell migration (colors). A *HOX* code is established in the hindbrain that specifies the arches and pathways of neural crest cell migration (with exception of PA1). Maintenance of the code in the arches is dependent upon an interaction between crest cells and arch-specific mesoderm. Patterning of the arches into their derivatives requires epithelial-mesenchymal interactions and includes molecular signals from the surface ectoderm, i.e., fibroblast growth factors (FGFs) acting upon underlying mesenchyme cells. OV, otic vesicle.

differentiation of the arches is not known, but a host of upstream and downstream genes must be involved. **Sonic hedgehog** may be one of the upstream regulators, since it is expressed in the arches and has been shown to regulate *HOX* gene expression. **Retinoids (retinoic acid)** can also regulate *HOX* gene expression in a concentration-dependent manner with genes at the 3' end being more responsive than those in more 5' regions. Regulation occurs through **retinoic acid response elements (RAREs)**, which are binding sites for retinoic acid in the promoter regions of the *HOX* genes. Deficiencies and excesses of retinoids disrupt migration and the axial identity of hindbrain crest cells, resulting in severe craniofacial defects.

In addition to the *HOX* genes, *OTX2* may participate in morphogenesis of the first arch. This gene, a transcription factor important for brain development, contains a homeodomain and is expressed in forebrain and midbrain regions (see Chapter 19). Neural crest cells that migrate from the midbrain to the first arch carry *OTX2* with them into this region. Presumably, *OTX2* and possibly *HOX* genes in the first arch interact to pattern this structure.

CLINICAL CORRELATES

Birth Defects Involving the Pharyngeal Region
Ectopic Thymic and Parathyroid Tissue

Since glandular tissue derived from the pouches undergoes migration, it is not unusual for accessory glands or remnants of tissue to persist along the pathway. This is true particularly for thymic tissue, which may remain in the neck, and for the parathyroid glands. The inferior parathyroids are more variable in position than the superior ones and are sometimes found at the bifurcation of the common carotid artery.

Branchial Fistulas

Branchial fistulas occur when the second pharyngeal arch fails to grow caudally over the third and fourth arches, leaving remnants of the second, third, and fourth clefts in contact with the surface by a narrow canal (Fig. 15.14*A*). Such a fistula, found on the lateral aspect of the neck directly anterior to the **sternocleidomastoid muscle,** usually provides drainage for a **lateral cervical cyst** (Fig. 15.14*B*). These cysts, remnants of the cervical sinus, are most often just below the angle of the jaw (Fig. 15.15), although they may be found anywhere along the anterior border of the sternocleidomastoid muscle. Frequently a lateral cervical cyst is not visible at birth but becomes evident as it enlarges during childhood.

Internal branchial fistulas are rare; they occur when the cervical sinus is connected to the lumen of the pharynx by a small canal, which usually opens in the tonsillar region (Fig. 15.14*C*). Such a fistula results from a rupture of the membrane between the second pharyngeal cleft and pouch at some time during development.

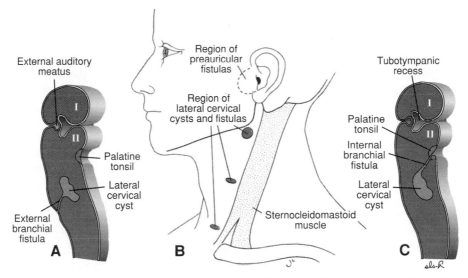

Figure 15.14 A. Lateral cervical cyst opening at the side of the neck by way of a fistula. **B.** Lateral cervical cysts and fistulas in front of the sternocleidomastoid muscle. Note also the region of preauricular fistulas. **C.** A lateral cervical cyst opening into the pharynx at the level of the palatine tonsil.

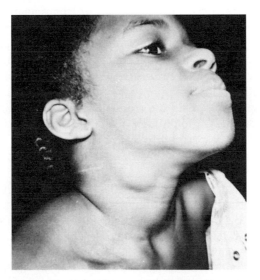

Figure 15.15 Patient with a lateral cervical cyst. These cysts are always on the lateral side of the neck in front of the sternocleidomastoid muscle. They commonly lie under the angle of the mandible and do not enlarge until later in life.

Neural Crest Cells and Craniofacial Defects

Neural crest cells (Fig. 15.2) are essential for formation of much of the craniofacial region. Consequently, disruption of crest cell development results in severe craniofacial malformations. Since crest cells also contribute to the **conotruncal endocardial cushions,** which septate the outflow tract of the heart into pulmonary and aortic channels, many infants with craniofacial defects also have cardiac abnormalities, including persistent truncus arteriosus, tetralogy of Fallot, and transposition of the great vessels. Unfortunately, crest cells appear to be a particularly vulnerable cell population and are easily killed by compounds such as alcohol and retinoic acid. One reason for this vulnerability may be that they are deficient in superoxide dismutase (SOD) and catalase enzymes that are responsible for scavenging free radicals that damage cells. Examples of craniofacial defects involving crest cells include the following:

Treacher Collins syndrome (**mandibulofacial dysostosis**) is characterized by malar hypoplasia due to underdevelopment of the zygomatic bones, mandibular hypoplasia, down-slanting palpebral fissures, lower eyelid colobomas, and malformed external ears (Fig. 15.16*A*). Treacher Collins is inherited as an autosomal dominant trait, with 60% arising as new mutations. However, phenocopies can be produced in laboratory animals following exposure to teratogenic doses of retinoic acid, suggesting that some cases in humans may be caused by teratogens.

Robin sequence may occur independently or in association with other syndromes and malformations. Like Treacher Collins syndrome, Robin sequence alters first-arch structures, with development of the mandible most severely affected. Infants usually have a triad of micrognathia, cleft palate, and glossoptosis (posteriorly placed tongue) (Fig. 15.16*B*). Robin sequence may be due to genetic and/or environmental factors. It may also occur as a deformation, as for example when the chin is compressed against the chest in cases of oligohydramnios. The primary defect includes poor growth of the mandible and, as a result, a posteriorly placed tongue that fails to drop from between the palatal shelves, preventing their fusion. Robin sequence occurs in approximately 1/8500 births.

DiGeorge anomaly occurs in approximately 1 in 2000 to 3000 births and represents the most severe example of a collection of disorders that also includes **velocardiofacial syndrome** (**VCFS**) and **conotruncal anomalies face syndrome** (Fig. 15.16*C*). All of these disorders are part of a spectrum called **CATCH22** because they include **c**ardiac defects, **a**bnormal facies, **t**hymic hypoplasia, **c**left palate, and **h**ypocalcemia and are a result of a deletion on the long arm of chromosome 22 (22q11). Patients with complete DiGeorge anomaly have immunological deficiencies, hypocalcemia, and a poor prognosis. Origin of the defects is caused by abnormal development of neural crest cells that contribute to formation of all of the affected structures. In addition

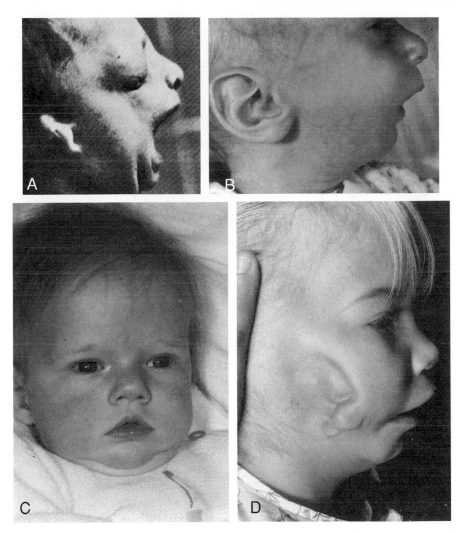

Figure 15.16 Patients with craniofacial defects thought to arise from insults to neural crest cells. **A.** Treacher Collins syndrome (mandibulofacial dysostosis). Note underdevelopment of the zygomatic bones, small mandible, and malformed ears. **B.** Robin sequence. Note the very small mandible (micrognathia). **C.** DiGeorge anomaly. In addition to craniofacial defects, such as hypertelorism and microstomia, these individuals have partial or complete absence of the thymus. **D.** Hemifacial microsomia (oculoauriculovertebral spectrum, or Goldenhar syndrome).

to genetic causes, exposure to retinoids (vitamin A), alcohol, and maternal diabetes can produce the defects.

Hemifacial microsomia (oculoauriculovertebral spectrum, Goldenhar syndrome) includes a number of craniofacial abnormalities that usually involve the maxillary, temporal, and zygomatic bones, which are small and flat. Ear (anotia, microtia), eye (tumors and dermoids in the eyeball), and vertebral (fused and hemivertebrae, spina bifida) defects are commonly observed in these patients (Fig. 15.16D). Asymmetry is present in 65% of the cases, which occur in 1/5600 births. Other malformations, which occur in 50% of cases, include cardiac abnormalities, such as tetralogy of Fallot and ventricular septal defects. Causes of hemifacial microsomia are unknown.

Tongue

The tongue appears in embryos of approximately 4 weeks in the form of two **lateral lingual swellings** and one **medial swelling,** the **tuberculum impar** (Fig. 15.17, *A* and *C*). These three swellings originate from the first pharyngeal arch. A second median swelling, the **copula,** or **hypobranchial eminence,** is formed by mesoderm of the second, third, and part of the fourth arch. Finally, a third median swelling, formed by the posterior part of the fourth arch, marks development of the epiglottis. Immediately behind this swelling is the **laryngeal orifice,** which is flanked by the **arytenoid swellings** (Fig. 15.17,*A* and *C*).

As the lateral lingual swellings increase in size, they overgrow the tuberculum impar and merge, forming the anterior two-thirds, or body, of the tongue (Fig. 15.17, *B* and *D*). Since the mucosa covering the body of the tongue originates from the first pharyngeal arch, **sensory innervation** to this area is by the **mandibular branch of the trigeminal nerve.** The body of the tongue is separated from the posterior third by a V-shaped groove, the **terminal sulcus** (Fig. 15.17, *B* and *D*).

The posterior part, or root, of the tongue originates from the second, third, and part of the fourth pharyngeal arch. The fact that **sensory innervation** to this part of the tongue is supplied by the **glossopharyngeal nerve** indicates that tissue of the third arch overgrows that of the second.

The epiglottis and the extreme posterior part of the tongue are innervated by the **superior laryngeal nerve,** reflecting their development from the fourth arch. Some of the tongue muscles probably differentiate in situ, but most are derived from myoblasts originating in **occipital somites.** Thus, tongue musculature is innervated by the **hypoglossal nerve.**

The general sensory innervation of the tongue is easy to understand. The body is supplied by the trigeminal nerve, the nerve of the first arch; that of the root is supplied by the glossopharyngeal and vagus nerves, the nerves of the third and fourth arches, respectively. **Special sensory innervation (taste)**

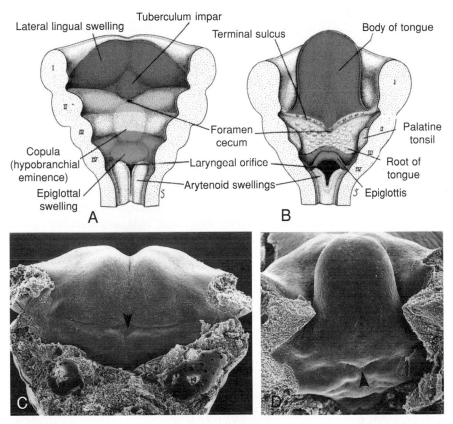

Figure 15.17 Ventral portion of the pharyngeal arches seen from above showing development of the tongue. *I* to *I V*, the cut pharyngeal arches. **A.** 5 weeks (approximately 6 mm). **B.** 5 months. Note the foramen cecum, site of origin of the thyroid primordium. **C** and **D.** Scanning electron micrographs of similar stages of tongue development in human embryos. A depression marks the location of the foramen cecum (*arrowheads*).

to the anterior two thirds of the tongue is provided by the **chorda tympani branch of the facial nerve,** while the posterior third is supplied by the glossopharyngeal nerve.

CLINICAL CORRELATES

Tongue-Tie

In **ankyloglossia (tongue-tie)** the tongue is not freed from the floor of the mouth. Normally, extensive cell degeneration occurs, and the frenulum is the only tissue that anchors the tongue to the floor of the mouth. In the most common form of ankyloglossia, the frenulum extends to the tip of the tongue.

Thyroid Gland

The thyroid gland appears as an epithelial proliferation in the floor of the pharynx between the tuberculum impar and the copula at a point later indicated by the **foramen cecum** (Figs. 15.17 and 15.18*A*). Subsequently the thyroid descends in front of the pharyngeal gut as a bilobed diverticulum (Fig. 15.18). During this migration, the thyroid remains connected to the tongue by a narrow canal, the **thyroglossal duct.** This duct later disappears.

   With further development, the thyroid gland descends in front of the hyoid bone and the laryngeal cartilages. It reaches its final position in front of the trachea in the seventh week (Fig. 15.18*B*). By then it has acquired a small median isthmus and two lateral lobes. The thyroid begins to function at approximately the end of the third month, at which time the first follicles containing colloid become visible. **Follicular cells** produce the colloid that serves as a source of **thyroxine** and **triiodothyronine. Parafollicular,** or **C, cells** derived from the **ultimobranchial body** (Fig. 15.10) serve as a source of calcitonin.

CLINICAL CORRELATES

Thyroglossal Duct and Thyroid Abnormalities

A **thyroglossal cyst** may lie at any point along the migratory pathway of the thyroid gland but is always near or in the **midline** of the neck. As indicated by its name, it is a cystic remnant of the thyroglossal duct, Although approximately 50% of these cysts are close to or just inferior to the body of the hyoid bone (Figs. 15.19 and 15.20), they may also be found at the base of the tongue or close to the thyroid cartilage. Sometimes a thyroglossal cyst is connected to the outside by a fistulous canal, a **thyroglossal fistula.** Such a fistula usually arises secondarily after rupture of a cyst but may be present at birth.

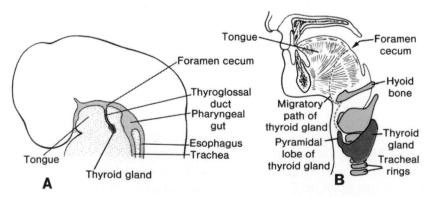

Figure 15.18 A. The thyroid primordium arises as an epithelial diverticulum in the midline of the pharynx immediately caudal to the tuberculum impar. **B.** Position of the thyroid gland in the adult. *Broken line,* the path of migration.

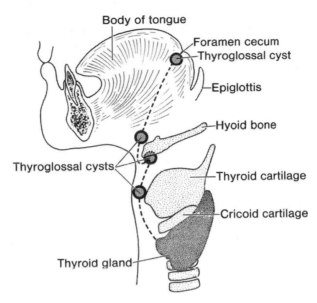

Figure 15.19 Thyroglossal cysts. These cysts, most frequently found in the hyoid region, are always close to the midline.

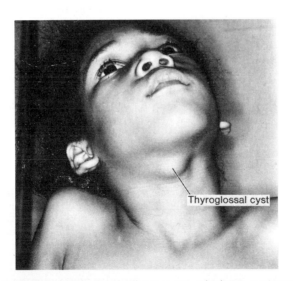

Figure 15.20 Thyroglossal cyst. These cysts, which are remnants of the thyroglossal duct, may be anywhere along the migration pathway of the thyroid gland. They are commonly found behind the arch of the hyoid bone. An important diagnostic characteristic is their midline location.

Aberrant thyroid tissue may be found anywhere along the path of descent of the thyroid gland. It is commonly found in the base of the tongue, just behind the foramen cecum, and is subject to the same diseases as the thyroid gland itself.

Face

At the end of the fourth week, **facial prominences** consisting primarily of neural crest-derived mesenchyme and formed mainly by the first pair of pharyngeal arches appear. **Maxillary prominences** can be distinguished lateral to the stomodeum, and **mandibular prominences** can be distinguished caudal to this structure (Fig. 15.21). The **frontonasal prominence,** formed by proliferation of mesenchyme ventral to the brain vesicles, constitutes the upper border of the stomodeum. On both sides of the frontonasal prominence, local thickenings of the surface ectoderm, the **nasal (olfactory) placodes,** originate under inductive influence of the ventral portion of the forebrain (Fig. 15.21).

During the fifth week, the nasal placodes invaginate to form **nasal pits.** In so doing, they create a ridge of tissue that surrounds each pit and forms the **nasal prominences.** The prominences on the outer edge of the pits are the **lateral nasal prominences;** those on the inner edge are the **medial nasal prominences** (Fig. 15.22)

During the following 2 weeks, the maxillary prominences continue to increase in size. Simultaneously, they grow medially, compressing the medial nasal prominences toward the midline. Subsequently the cleft between the medial nasal prominence and the maxillary prominence is lost, and the two fuse (Fig. 15.23). Hence, the upper lip is formed by the two medial nasal prominences and the two maxillary prominences. The lateral nasal prominences do not participate in formation of the upper lip. The lower lip and jaw form from the mandibular prominences that merge across the midline.

Initially, the maxillary and lateral nasal prominences are separated by a deep furrow, the **nasolacrimal groove** (Figs. 15.22 and 15.23). Ectoderm in the floor of this groove forms a solid epithelial cord that detaches from the overlying ectoderm. After canalization, the cord forms the **nasolacrimal duct;** its upper end widens to form the **lacrimal sac.** Following detachment of the cord, the maxillary and lateral nasal prominences merge with each other. The nasolacrimal duct then runs from the medial corner of the eye to the inferior meatus of the nasal cavity, and the maxillary prominences enlarge to form the **cheeks** and **maxillae.**

The **nose** is formed from five facial prominences (Fig. 15.23): the frontal prominence gives rise to the bridge; the merged medial nasal prominences provide the crest and tip; and the lateral nasal prominences form the sides (alae) (Table 15.2, p. 389).

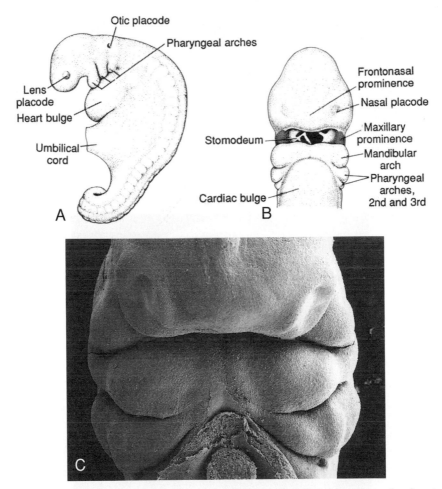

Figure 15.21 A. Lateral view of an embryo at the end of the fourth week, showing position of the pharyngeal arches. **B.** Frontal view of a 4.5-week embryo showing the mandibular and maxillary prominences. The nasal placodes are visible on either side of the frontonasal prominence. **C.** Scanning electron micrograph of a human embryo at a stage similar to that of **B.**

Intermaxillary Segment

As a result of medial growth of the maxillary prominences, the two medial nasal prominences merge not only at the surface but also at a deeper level. The structure formed by the two merged prominences is the **intermaxillary segment.** It is composed of (*a*) a **labial component,** which forms the philtrum of the upper lip; (*b*) an **upper jaw component,** which carries the four incisor teeth; and (*c*) a **palatal component,** which forms the triangular primary palate (Fig. 15.24). The intermaxillary segment is continuous

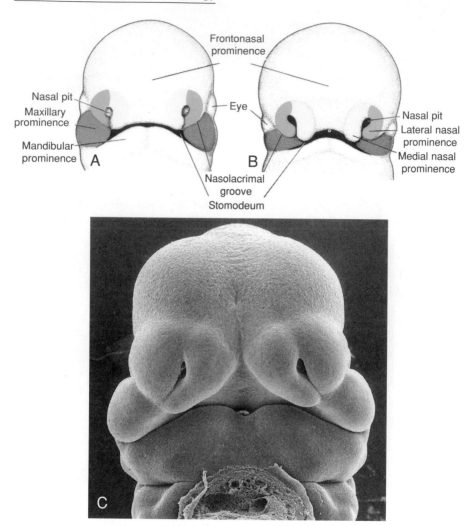

Figure 15.22 Frontal aspect of the face. **A.** 5-week embryo. **B.** 6-week embryo. The nasal prominences are gradually separated from the maxillary prominence by deep furrows. **C.** Scanning electron micrograph of a mouse embryo at a stage similar to that of **B.**

with the rostral portion of the **nasal septum,** which is formed by the frontal prominence.

Secondary Palate

Although the primary palate is derived from the intermaxillary segment (Fig. 15.24), the main part of the definitive palate is formed by two shelflike

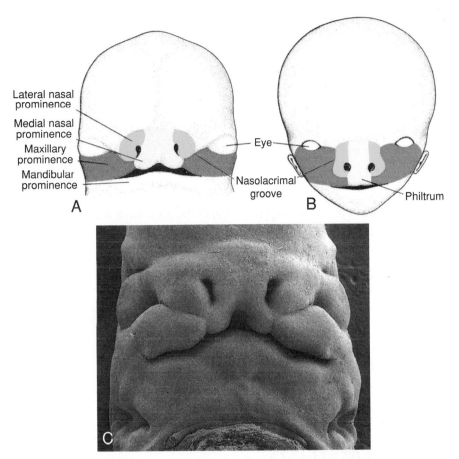

Figure 15.23 Frontal aspect of the face. **A.** 7-week embryo. Maxillary prominences have fused with the medial nasal prominences. **B.** 10-week embryo. **C.** Scanning electron micrograph of a human embryo at a stage similar to that of **A.**

TABLE 15.2 Structures Contributing to Formation of the Face

Prominence	Structures Formed
Frontonasal[a]	Forehead, bridge of nose, medial and lateral nasal prominences
Maxillary	Cheeks, lateral portion of upper lip
Medial nasal	Philtrum of upper lip, crest and tip of nose
Lateral nasal	Alae of nose
Mandibular	Lower lip

[a] The frontonasal prominence is a single unpaired structure; the other prominences are paired.

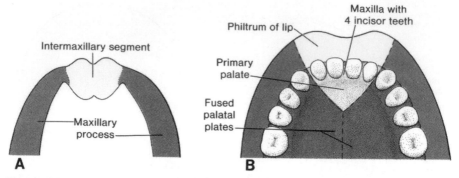

Figure 15.24 A. Intermaxillary segment and maxillary processes. **B.** The intermaxillary segment giving rise to the philtrum of the upper lip, the median part of the maxillary bone with its four incisor teeth, and the triangular primary palate.

outgrowths from the maxillary prominences. These outgrowths, the **palatine shelves,** appear in the sixth week of development and are directed obliquely downward on each side of the tongue (Fig. 15.25). In the seventh week, however, the palatine shelves ascend to attain a horizontal position above the tongue and fuse, forming the **secondary palate** (Figs. 15.26 and 15.27).

Anteriorly, the shelves fuse with the triangular primary palate, and the **incisive foramen** is the midline landmark between the primary and secondary palates (Fig. 15.27*B*). At the same time as the palatine shelves fuse, the nasal septum grows down and joins with the cephalic aspect of the newly formed palate (Fig. 15.27).

CLINICAL CORRELATES

Facial Clefts

Cleft lip and cleft palate are common defects that result in abnormal facial appearance and defective speech. The **incisive foramen** is considered the dividing landmark between the **anterior** and **posterior** cleft deformities. Those anterior to the incisive foramen include **lateral cleft lip, cleft upper jaw,** and **cleft** between the **primary and secondary palates** (Figs. 15.28, *B* and *D*, and 15.29, *A* and *B*). Such defects are due to a partial or complete lack of fusion of the maxillary prominence with the medial nasal prominence on one or both sides. Those that lie posterior to the incisive foramen include **cleft (secondary) palate** and **cleft uvula** (Figs. 15.28*E* and 15.29, *C* and *D*). Cleft palate results from a lack of fusion of the palatine shelves, which may be due to smallness of the shelves, failure of the shelves to elevate, inhibition of the fusion process itself, or failure of the tongue to drop from between the shelves because of micrognathia. The third category is formed by a combination of clefts lying anterior as well as posterior to the incisive foramen (Fig. 15.28*F*).

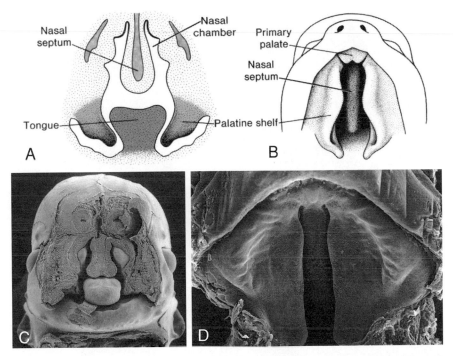

Figure 15.25 A. Frontal section through the head of a 6.5-week-old embryo. The palatine shelves are in the vertical position on each side of the tongue. **B.** Ventral view of the palatine shelves after removal of the lower jaw and the tongue. Note the clefts between the primary triangular palate and the palatine shelves, which are still vertical. **C.** Scanning electron micrograph of a mouse embryo at a stage similar to that of **A. D.** Palatal shelves at a stage slightly older than those in **B.** The shelves have elevated, but they are widely separated. The primary palate has fused with the secondary palatal shelves.

Anterior clefts vary in severity from a barely visible defect in the vermilion of the lip to extension into the nose (Fig. 15.29*A*). In severe cases the cleft extends to a deeper level, forming a cleft of the upper jaw, and the maxilla is split between the lateral incisor and the canine tooth. Frequently such a cleft extends to the incisive foramen (Fig. 15.28, *C* and *D*). Likewise, posterior clefts vary in severity from cleavage of the entire secondary palate (Fig. 15.29*D*) to cleavage of the uvula only.

Oblique facial clefts are produced by failure of the maxillary prominence to merge with its corresponding lateral nasal prominence. When this occurs, the nasolacrimal duct is usually exposed to the surface (Fig. 15.29*E*).

Median cleft lip, a rare abnormality, is caused by incomplete merging of the two medial nasal prominences in the midline. This anomaly is usually accompanied by a deep groove between the right and left sides of the nose

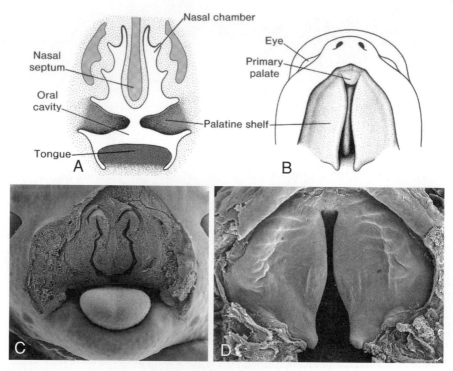

Figure 15.26 A. Frontal section through the head of a 7.5-week embryo. The tongue has moved downward, and the palatine shelves have reached a horizontal position. **B.** Ventral view of the palatine shelves after removal of the lower jaw and tongue. The shelves are horizontal. Note the nasal septum. **C.** Scanning electron micrograph of a mouse embryo at a stage similar to that of **A. D.** Palatal shelves at a stage similar to that of **B.**

(Fig. 15.29*F*). Infants with midline clefts are often **mentally retarded** and may have brain abnormalities that include varying degrees of loss of midline structures. Loss of midline tissue may be so extensive that the lateral ventricles fuse (**holoprosencephaly**). These defects are induced very early in development, at the beginning of neurulation (days 19–21) when the midline of the forebrain is being established.

Most cases of cleft lip and cleft palate are multifactorial. Cleft lip (approximately 1/1000 births) occurs more frequently in males (80%) than in females; its incidence increases slightly with maternal age, and it varies among populations. If normal parents have one child with a cleft lip, the chance that the next baby will have the same defect is 4%. If two siblings are affected, the risk for the next child increases to 9%. If one of the parents has a cleft lip and they have one child with the same defect, the probability that the next baby will be affected rises to 17%.

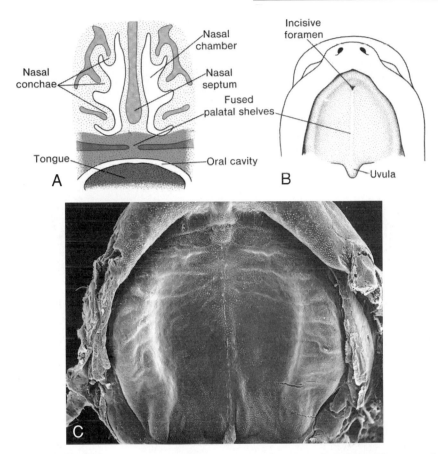

Figure 15.27 A. Frontal section through the head of a 10-week embryo. The two palatine shelves have fused with each other and with the nasal septum. **B.** Ventral view of the palate. The incisive foramen forms the midline between the primary and secondary palate. **C.** Scanning electron micrograph of the palatal shelves of a mouse embryo at a stage similar to that of **B.**

The frequency of isolated **cleft palate** is much lower than that of cleft lip (1/2500 births), occurs more often in females (67%) than in males, and is not related to maternal age. If the parents are normal and have one child with a cleft palate, the probability of the next child being affected is about 2%. If, however, there is a similarly affected child and a relative or parent both with a cleft palate, the probability increases to 7% and 15%, respectively. In females, the palatal shelves fuse approximately 1 week later than in males. This difference may explain why isolated cleft palate occurs more frequently in females than in males. **Anticonvulsant drugs,** such as **phenobarbital and diphenylhydantoin,** given during pregnancy increase the risk of cleft palate.

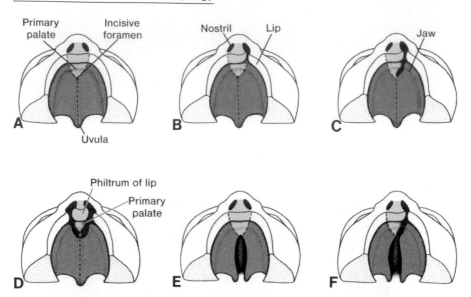

Figure 15.28 Ventral view of the palate, gum, lip, and nose. **A.** Normal. **B.** Unilateral cleft lip extending into the nose. **C.** Unilateral cleft involving the lip and jaw and extending to the incisive foramen. **D.** Bilateral cleft involving the lip and jaw. **E.** Isolated cleft palate. **F.** Cleft palate combined with unilateral anterior cleft lip.

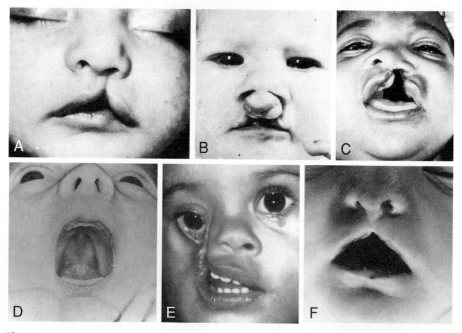

Figure 15.29 A. Incomplete cleft lip. **B.** Bilateral cleft lip. **C.** Cleft lip, cleft jaw, and cleft palate. **D.** Isolated cleft palate. **E.** Oblique facial cleft. **F.** Midline cleft lip.

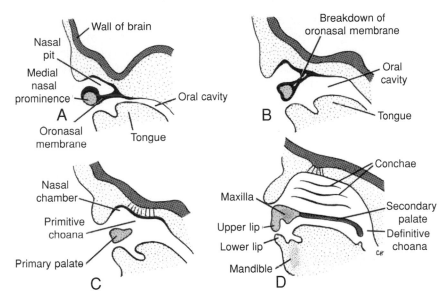

Figure 15.30 A. Sagittal section through the nasal pit and lower rim of the medial nasal prominence of a 6-week embryo. The primitive nasal cavity is separated from the oral cavity by the oronasal membrane. **B.** Similar section as in **A** showing the oronasal membrane breaking down. **C.** A 7-week embryo with a primitive nasal cavity in open connection with the oral cavity. **D.** Sagittal section through the face of a 9-week embryo showing separation of the definitive nasal and oral cavities by the primary and secondary palate. Definitive choanae are at the junction of the oral cavity and the pharynx.

Nasal Cavities

During the sixth week, the nasal pits deepen considerably, partly because of growth of the surrounding nasal prominences and partly because of their penetration into the underlying mesenchyme (Fig. 15.30A). At first the **oronasal membrane** separates the pits from the primitive oral cavity by way of the newly formed foramina, the **primitive choanae** (Fig. 15.30C). These choanae lie on each side of the midline and immediately behind the primary palate. Later, with formation of the secondary palate and further development of the primitive nasal chambers (Fig. 15.30D), the **definitive choanae** lie at the junction of the nasal cavity and the pharynx.

Paranasal air sinuses develop as diverticula of the lateral nasal wall and extend into the maxilla, ethmoid, frontal, and sphenoid bones. They reach their maximum size during puberty and contribute to the definitive shape of the face.

Teeth

The shape of the face is determined not only by expansion of the paranasal sinuses but also by growth of the mandible and maxilla to accommodate the teeth. Teeth themselves arise from an epithelial-mesenchymal interaction between overlying oral epithelium and underlying mesenchyme derived from neural crest cells. By the sixth week of development, the basal layer of the epithelial lining of the oral cavity forms a C-shaped structure, the **dental lamina,** along the length of the upper and lower jaws. This lamina subsequently gives rise to a number of **dental buds** (Fig. 15.31A), 10 in each jaw, which form the primordia of the ectodermal components of the teeth. Soon the deep surface of the buds invaginates, resulting in the **cap stage of tooth development** (Fig. 15.31B). Such a cap consists of an outer layer, the **outer dental epithelium,** an inner layer, the **inner dental epithelium,** and a central core of loosely woven tissue, the **stellate reticulum.** The **mesenchyme,** which originates in the **neural crest** in the indentation, forms the **dental papilla** (Fig. 15.31B).

As the dental cap grows and the indentation deepens, the tooth takes on the appearance of a bell (**bell stage**) (Fig. 15.31C). Mesenchyme cells of the papilla adjacent to the inner dental layer differentiate into **odontoblasts,** which later produce **dentin.** With thickening of the dentin layer, odontoblasts retreat into the dental papilla, leaving a thin cytoplasmic process (**dental process**) behind in the dentin (Fig. 15.31D). The odontoblast layer persists throughout the life of the tooth and continuously provides predentin. The remaining cells of the dental papilla form the **pulp** of the tooth (Fig. 15.31D).

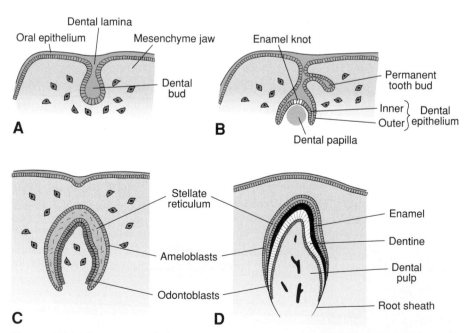

Figure 15.31 Formation of the tooth at successive stages of development. **A.** Bud stage; 8 weeks. **B.** Cap stage; 10 weeks. **C.** Bell stage; 3 months. **D.** 6 months.

In the meantime, epithelial cells of the inner dental epithelium differentiate into **ameloblasts (enamel formers)**. These cells produce long enamel prisms that are deposited over the dentin (Fig. 15.31*D*). Furthermore, a cluster of these cells in the inner dental epithelium forms the **enamel knot** that regulates early tooth development (Fig. 15.31*B*).

Enamel is first laid down at the apex of the tooth and from here spreads toward the neck. When the enamel thickens, the ameloblasts retreat into the stellate reticulum. Here they regress, temporarily leaving a thin membrane (**dental cuticle**) on the surface of the enamel. After eruption of the tooth, this membrane gradually sloughs off.

Formation of the root of the tooth begins when the dental epithelial layers penetrate into the underlying mesenchyme and form the **epithelial root sheath** (Fig. 15.31*D*). Cells of the dental papilla lay down a layer of dentin continuous with that of the crown (Fig. 15.32). As more and more dentin is deposited, the pulp chamber narrows and finally forms a canal containing blood vessels and nerves of the tooth.

Mesenchymal cells on the outside of the tooth and in contact with dentin of the root differentiate into **cementoblasts** (Fig. 15.32*A*). These cells produce a thin layer of specialized bone, the **cementum**. Outside of the cement layer, mesenchyme gives rise to the **periodontal ligament** (Fig. 15.32), which holds the tooth firmly in position and functions as a shock absorber.

With further lengthening of the root, the crown is gradually pushed through the overlying tissue layers into the oral cavity (Fig. 15.32*B*). The eruption of **deciduous** or **milk teeth** occurs 6 to 24 months after birth.

Buds for the **permanent teeth**, which lie on the lingual aspect of the milk teeth, are formed during the third month of development. These buds remain dormant until approximately the sixth year of postnatal life (Fig. 15.33). Then

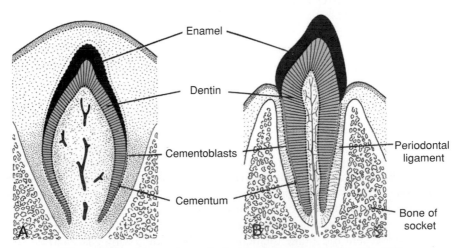

Figure 15.32 The tooth just before birth (**A**) and after eruption (**B**).

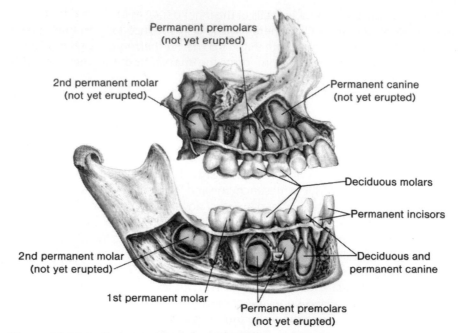

Permanent premolars
(not yet erupted)

2nd permanent molar
(not yet erupted)

Permanent canine
(not yet erupted)

Deciduous molars

Permanent incisors

2nd permanent molar
(not yet erupted)

Deciduous and
permanent canine

1st permanent molar

Permanent premolars
(not yet erupted)

Figure 15.33 Replacement of deciduous teeth by permanent teeth in a child of 8 or 9 years.

they begin to grow, pushing against the underside of the milk teeth and aiding in the shedding of them. As a permanent tooth grows, the root of the overlying deciduous tooth is resorbed by osteoclasts.

Molecular Regulation of Tooth Development

Teeth are present only in vertebrates and parallel the evolutionary appearance of the neural crest. Tooth development represents a classic example of an epithelial-mesenchymal interaction, in this case between the overlying epithelium and underlying neural crest derived mesenchyme. Regulation of tooth patterning from incisors to molars is generated by a combinatorial expression of *HOX* genes expressed in the mesenchyme. With respect to each tooth's individual development, the epithelium governs differentiation to the bud stage, at which time this regulatory function is transferred to the mesenchyme. Signals for development involve growth factors including: **WNTs, bone morphogenetic proteins (BMPs)**, and **fibroblast growth factors (FGFs)**; the secreted factor *sonic hedgehog* (*SHH*); and transcription factors, such as **MSX1 and 2** that interact in a complex pathway to produce cell differentiation and patterning for each tooth. Teeth also appear to have a signaling center that represents the "organizer" for tooth development much like the activity of the node during gastrulation (see Chapter 4). This organizer region is called the **enamel knot**, and it appears in a circumscribed region of the dental epithelium at the tips of the tooth buds. It then enlarges at the cap stage into a tightly packed group of cells but undergoes apoptosis (cell death) and disappears by the end of this stage (Fig. 15.31). While it is present, it expresses FGF-4, SHH, and BMP2 and 4. FGF-4 may regulate outgrowth of cusps much as it participates in limb outgrowth produced by the apical ectodermal ridge (AER); while BMP-4 may regulate the timing of apoptosis in knot cells.

CLINICAL CORRELATES

Tooth Abnormalities

Natal teeth have erupted by the time of birth. Usually they involve the mandibular incisors, which may be abnormally formed and have little enamel.

Teeth may be abnormal in number, shape, and size. They may be discolored by foreign substances, such as **tetracyclines,** or be deficient in enamel, a condition often caused by **vitamin D deficiency (rickets).** Many factors affect tooth development, including genetic and environmental influences.

Summary

Pharyngeal (branchial) arches, consisting of bars of mesenchymal tissue separated by pharyngeal pouches and clefts, give the head and neck their typical appearance in the fourth week (Fig. 15.3). Each arch contains its own artery (Fig. 15.4), cranial nerve (Fig. 15.7), muscle element, and cartilage bar or skeletal element (Figs. 15.8 and 15.9; Table 15.1, p. 366). Endoderm of the **pharyngeal pouches** gives rise to a number of endocrine glands and part of the middle ear. In subsequent order the pouches give rise to (*a*) the **middle ear cavity** and **auditory tube** (pouch 1), (*b*) the stroma of the **palatine tonsil** (pouch 2), (*c*) the **inferior parathyroid glands** and **thymus** (pouch 3), and (*d*) the **superior parathyroid glands** and **ultimobranchial body** (pouches 4 and 5) (Fig. 15.10).

Pharyngeal clefts give rise to only one structure, the **external auditory meatus.**

Molecular control of arch development resides in *HOX* **genes,** whose **pharyngeal arch code** is carried to the arches by neural crest cells migrating from hindbrain segments known as rhombomeres (Fig. 15.12). This code is then maintained by interactions between crest cells and arch mesoderm (Fig. 15.13).

The **thyroid gland** originates from an epithelial proliferation in the floor of the tongue and descends to its level in front of the tracheal rings in the course of development.

The paired **maxillary** and **mandibular** prominences and the **frontonasal prominence** are the first prominences of the facial region. Later, medial and lateral nasal prominences form around the nasal placodes on the frontonasal prominence. All of these structures are important, since they determine, through fusion and specialized growth, the size and integrity of the mandible, upper lip, palate, and nose (Table 15.2, p. 389). Formation of the upper lip occurs by fusion of the two maxillary prominences with the two medial nasal prominences (Figs. 15.22 and 15.23). The intermaxillary segment is formed by merging of the two medial nasal prominences in the midline. This segment is composed of (*a*) the **philtrum,** (*b*) the **upper jaw component,** which carries the four incisor teeth, and (*c*) the **palatal component,** which forms the triangular primary palate. The nose is derived from (*a*) the **frontonasal prominence,** which forms the **bridge,** (*b*) the **medial nasal prominences,** which provide the **crest and tip,** and (*c*) the **lateral nasal prominences,** which form the **alae** (Fig. 15.23). Fusion of the **palatal shelves,** which form from the **maxillary prominences,** creates the **hard (secondary)** and **soft palate.** A series of cleft deformities may result from partial or incomplete fusion of these mesenchymal tissues, which may be caused by hereditary factors and drugs (diphenylhydantoin).

The adult form of the face is influenced by development of **paranasal sinuses, nasal conchae,** and **teeth.** Teeth develop from epithelial-mesenchymal interactions between oral epithelium and neural crest derived mesenchyme. **Enamel** is made by **ameloblasts** (Figs. 15.31 and 15.32). It lies on a thick layer of **dentin** produced by **odontoblasts,** a neural crest derivative. **Cementum** is formed by **cementoblasts,** another mesenchymal derivative found in the root of the tooth. The first teeth (**deciduous teeth,** or **milk teeth**) appear 6 to 24 months after birth, and the definitive or **permanent teeth,** which supplant the milk teeth, are formed during the third month of development (Fig. 15.33).

Problems to Solve

1. *Why are neural crest cells considered such an important cell population for craniofacial development?*

2. *You are called as a consultant for a child with a very small mandible and ears that are represented by small protuberances bilaterally. The baby has had numerous episodes of pneumonia and is small for its age. What might your diagnosis be, and what might have caused these abnormalities?*

3. *A child is born with a median cleft lip. Should you be concerned about any other abnormalities?*

4. *A child presents with a midline swelling beneath the arch of the hyoid bone. What might this swelling be, and what is its basis embryologically?*

SUGGESTED READING

Francis-West P, Ladher R, Barlow A, Graveson A: Signaling interactions during facial development. *Mech Dev* 75:3, 1998.

Freidberg J: Pharyngeal cleft sinuses and cysts, and other benign neck lesions. *Pediatr Clin North Am* 36:1451, 1989.

Gorlin RJ, Cohen MM, Levin LS (eds): *Syndromes of the Head and Neck.* 4th ed. New York, Oxford University, 2002.

Hong R, The DiGeorge anomaly (Catch22, DiGeorge/Velocardiofacial syndrome). *Sem Hematol* 35:282, 1998.

Jiang X, Iseki S, Maxson RE, Sucov HM, Morriss-Kay GM: Tissue origins and interactions in the mammalian skull vault. *Dev Biol* 241:106, 2002.

Lumsden A, Sprawson N, Graham A: Segmental origin and migration of neural crest cells in the hindbrain region of the chick embryo. *Development* 113:1281, 1991.

Nichols DH: Mesenchyme formation from the trigeminal placodes of the mouse embryo. *Am J Anat* 176:1931, 1986.

Noden DM: Cell movements and control of patterned tissue assembly during craniofacial development. *J Craniofac Genet Dev Biol* 11:192, 1991.

Osumi-Yamashita N, Ninomiya Y, Doi H, Eto K: The contribution of both forebrain and midbrain crest cells to the mesenchyme in the frontonasal mass of mouse embryos. *Dev Biol* 164(2):409, 1994.

Sulik KK, Cook CS, Webster WS: Teratogens and craniofacial malformations: relationships to cell death. *Dev Suppl* 103:213, 1988.

Sulik KK, et al.: Fetal alcohol syndrome and DiGeorge anomaly: critical ethanol exposure periods for craniofacial malformation as illustrated in an animal model. *Am J Med Genet* 2(suppl):97, 1986.

Sulik KK, Schoenwokf GC: Highlights of craniofacial morphogenesis in mammalian embryos, as revealed by scanning electron microscopy. *Scanning Electron Microsc* 4:1735, 1985.

Thesleff I, Sharpe P: Signaling networks regulating dental development. *Mech Dev* 67:111, 1997.

Thorogood P: The head and face. *In* Thorogood P (ed): *Embryos, Genes, and Birth Defects.* New York, Wiley & Sons, 1997.

Trainor PA, Krumlauf R: Hox genes, neural crest cells and branchial arch patterning. *Curr Op Cell Biol* 13:698, 2001.

Webster WS, Lipson AH, Sulik KK: Interference with gastrulation during the third week of pregnancy as a cause of some facial abnormalities and CNS defects. *Am J Med Genet* 31:505, 1988.

Wilkie AOM, Morriss-Kay GM: Genetics of craniofacial development and malformation. *Nature Rev Genet* 2:458, 2001.

chapter 16

Ear

In the adult, the ear forms one anatomical unit serving both hearing and equilibrium. In the embryo, however, it develops from three distinctly different parts: (*a*) the **external ear,** the sound-collecting organ; (*b*) the **middle ear,** a sound conductor from the external to the internal ear; and (*c*) the **internal ear,** which converts sound waves into nerve impulses and registers changes in equilibrium.

Internal Ear

The first indication of the developing ear can be found in embryos of approximately 22 days as a thickening of the surface ectoderm on each side of the rhombencephalon (Fig. 16.1). These thickenings, the **otic placodes,** invaginate rapidly and form the **otic** or **auditory vesicles (otocysts)** (Fig. 16.2). During later development, each vesicle divides into (*a*) a ventral component that gives rise to the **saccule** and **cochlear duct** and (*b*) a dorsal component that forms the **utricle, semicircular canals,** and **endolymphatic duct** (Figs. 16.3–16.6). Together these epithelial structures form the **membranous labyrinth.**

SACCULE, COCHLEA, AND ORGAN OF CORTI

In the sixth week of development, the saccule forms a tubular outpocketing at its lower pole (Fig. 16.3, *C–E* and *G*). This outgrowth, the **cochlear duct,** penetrates the surrounding mesenchyme in a spiral

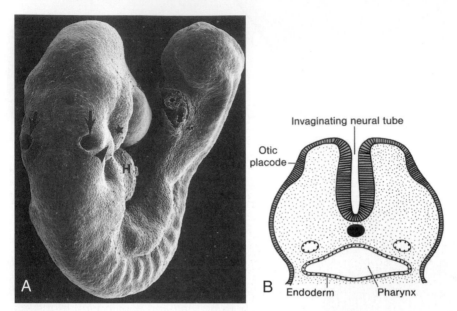

Figure 16.1 A. Scanning electron micrograph of a mouse embryo equivalent to approximately 28 days of human development. The otic placodes, as shown in **B**, are invaginating to form the otic pits (*arrows*). *Arrowhead*, second arch; *H*, heart; *star*, mandibular prominence. **B**. Region of the rhombencephalon showing the otic placodes in a 22-day embryo.

fashion until at the end of the eighth week it has completed 2.5 turns (Fig. 16.3, *D* and *E*). Its connection with the remaining portion of the saccule is then confined to a narrow pathway, the **ductus reuniens** (Fig. 16.3*E*; see also Fig. 16.8).

Mesenchyme surrounding the cochlear duct soon differentiates into cartilage (Fig. 16.4*A*). In the 10th week, this cartilaginous shell undergoes vacuolization, and two perilymphatic spaces, the **scala vestibuli** and **scala tympani,** are formed (Fig. 16.4, *B* and *C*). The cochlear duct is then separated from the scala vestibuli by the **vestibular membrane** and from the scala tympani by the **basilar membrane** (Fig. 16.4*C*). The lateral wall of the cochlear duct remains attached to the surrounding cartilage by the **spiral ligament,** whereas its median angle is connected to and partly supported by a long cartilaginous process, the **modiolus,** the future axis of the bony cochlea (Fig. 16.4*B*).

Initially, epithelial cells of the cochlear duct are alike (Fig. 16.4*A*). With further development, however, they form two ridges: the **inner ridge,** the future **spiral limbus,** and the **outer ridge** (Fig. 16.4*B*). The outer ridge forms one row of inner and three or four rows of outer **hair cells,** the sensory cells of the auditory system (Fig. 16.5). They are covered by the **tectorial membrane,** a fibrillar gelatinous substance attached to the spiral limbus that rests with its tip on the hair cells (Fig. 16.5). The sensory cells and tectorial membrane together constitute the **organ of Corti.** Impulses received by this organ are transmitted

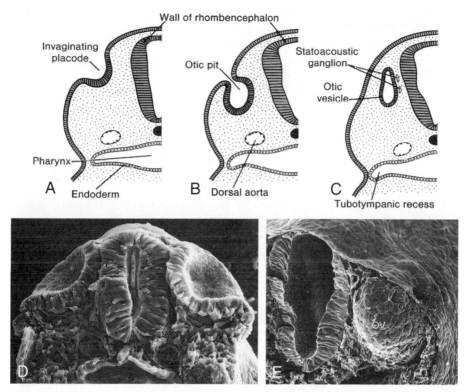

Figure 16.2 A to **C.** Transverse sections through the region of the rhombencephalon showing formation of the otic vesicles. **A.** 24 days. **B.** 27 days. **C.** 4.5 weeks. Note the statoacoustic ganglia. **D** and **E.** Scanning electron micrographs of mouse embryos equivalent to stages depicted in **A** and **B** showing development of the otic vesicles (*OV*).

to the spiral ganglion and then to the nervous system by the **auditory fibers of cranial nerve VIII** (Figs. 16.4 and 16.5).

UTRICLE AND SEMICIRCULAR CANALS

During the sixth week of development, **semicircular canals** appear as flattened outpocketings of the utricular part of the otic vesicle (Fig. 16.6, *A* and *B*). Central portions of the walls of these outpocketings eventually appose each other (Fig. 16.6, *C* and *D*) and disappear, giving rise to three semicircular canals (Fig. 16.6; see also Fig. 16.8). Whereas one end of each canal dilates to form the **crus ampullare,** the other, the **crus nonampullare,** does not widen (Fig. 16.6). Since two of the latter type fuse, however, only five crura enter the utricle, three with an ampulla and two without.

Cells in the ampullae form a crest, the **crista ampullaris,** containing sensory cells for maintenance of equilibrium. Similar sensory areas, the **maculae**

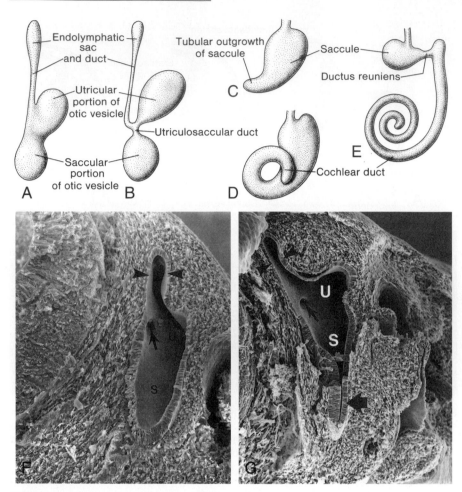

Figure 16.3 A and **B.** Development of the otocyst showing a dorsal utricular portion with the endolymphatic duct and a ventral saccular portion. **C** to **E.** Cochlear duct at 6, 7, and 8 weeks, respectively. Note formation of the ductus reuniens and the utriculosaccular duct. **F** and **G.** Scanning electron micrographs of mouse embryos showing similar stages of development of the otocyst as depicted in **A** and **B.** *Arrowheads,* endolymphatic duct; *S,* saccule; *small arrow,* opening of semicircular canal; *U,* utricle. **G** also shows initial stages of cochlear duct formation (*large arrow*).

acusticae, develop in the walls of the utricle and saccule. Impulses generated in sensory cells of the cristae and maculae as a result of a change in position of the body are carried to the brain by **vestibular fibers of cranial nerve VIII.**

During formation of the otic vesicle, a small group of cells breaks away from its wall and forms the **statoacoustic ganglion** (Fig. 16.2C). Other cells of this ganglion are derived from the neural crest. The ganglion subsequently

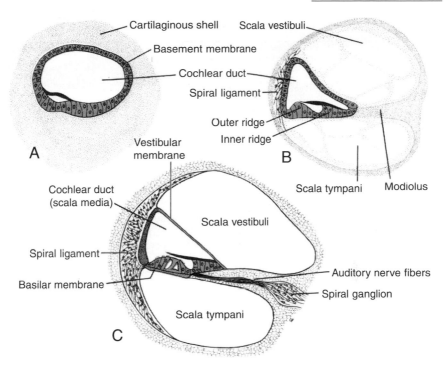

Figure 16.4 Development of the scala tympani and scala vestibuli. **A.** The cochlear duct is surrounded by a cartilaginous shell. **B.** During the 10th week large vacuoles appear in the cartilaginous shell. **C.** The cochlear duct (scala media) is separated from the scala tympani and the scala vestibuli by the basilar and vestibular membranes, respectively. Note the auditory nerve fibers and the spiral (cochlear) ganglion.

splits into **cochlear** and **vestibular** portions, which supply sensory cells of the organ of Corti and those of the saccule, utricle, and semicircular canals, respectively.

Middle Ear

TYMPANIC CAVITY AND AUDITORY TUBE

The **tympanic cavity**, which originates in the endoderm, is derived from the first pharyngeal pouch (Figs. 16.2 and 16.7). This pouch expands in a lateral direction and comes in contact with the floor of the first pharyngeal cleft. The distal part of the pouch, the **tubotympanic recess**, widens and gives rise to the primitive tympanic cavity, and the proximal part remains narrow and forms the **auditory tube (eustachian tube;** Fig. 16.7*B* and 16.8), through which the tympanic cavity communicates with the nasopharynx.

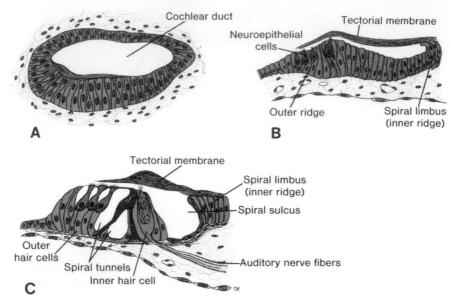

Figure 16.5 Development of the organ of Corti. **A.** 10 weeks. **B.** Approximately 5 months. **C.** Full-term infant. Note the appearance of the spiral tunnels in the organ of Corti.

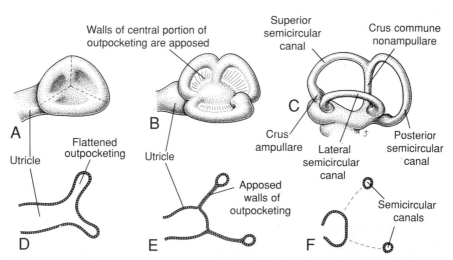

Figure 16.6 Development of the semicircular canals. **A.** 5 weeks. **C.** 6 weeks. **E.** 8 weeks. **B, D,** and **F.** Apposition, fusion, and disappearance, respectively, of the central portions of the walls of the semicircular outpocketings. Note the ampullae in the semicircular canals.

OSSICLES

The **malleus** and **incus** are derived from cartilage of the first pharyngeal arch, and the **stapes** is derived from that of the second arch (Fig. 16.9A). Although the ossicles appear during the first half of fetal life, they remain embedded in mesenchyme until the eighth month (Fig. 16.9B), when the surrounding tissue dissolves (Figs. 16.7, 16.8, and 16.9B). The endodermal epithelial lining of the primitive tympanic cavity then extends along the wall of the newly developing space. The tympanic cavity is now at least twice as large as before. When the ossicles are entirely free of surrounding mesenchyme, the endodermal epithelium connects them in a mesentery-like fashion to the wall of the cavity (Fig. 16.9B). The supporting ligaments of the ossicles develop later within these mesenteries.

Since the malleus is derived from the first pharyngeal arch, its muscle, the **tensor tympani,** is innervated by the **mandibular branch of the trigeminal nerve.** The **stapedius muscle,** which is attached to the stapes, is innervated by the **facial nerve,** the nerve to the second pharyngeal arch.

During late fetal life, the tympanic cavity expands dorsally by vacuolization of surrounding tissue to form the **tympanic antrum.** After birth, epithelium of the tympanic cavity invades bone of the developing **mastoid process,** and epithelium-lined air sacs are formed (**pneumatization**). Later, most of the mastoid air sacs come in contact with the antrum and tympanic cavity. Expansion of inflammations of the middle ear into the antrum and mastoid air cells is a common complication of middle ear infections.

External Ear

EXTERNAL AUDITORY MEATUS

The **external auditory meatus** develops from the dorsal portion of the first pharyngeal cleft (Fig. 16.7A). At the beginning of the third month, epithelial cells at the bottom of the meatus proliferate, forming a solid epithelial plate, the meatal plug (Fig. 16.7B). In the seventh month, this plug dissolves and the epithelial lining of the floor of the meatus participates in formation of the definitive eardrum. Occasionally the meatal plug persists until birth, resulting in congenital deafness.

EARDRUM OR TYMPANIC MEMBRANE

The eardrum is made up of (*a*) ectodermal epithelial lining at the bottom of the auditory meatus, (*b*) endodermal epithelial lining of the tympanic cavity, and (*c*) an intermediate layer of connective tissue (Fig. 16.9B) that forms the fibrous stratum. The major part of the eardrum is firmly attached to the handle of the malleus (Fig. 16.8 and 16.9B), and the remaining portion forms the separation between the external auditory meatus and the tympanic cavity.

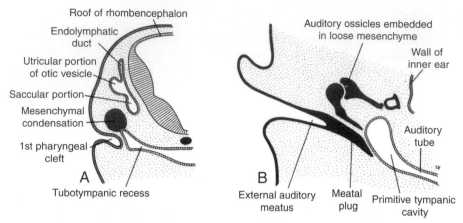

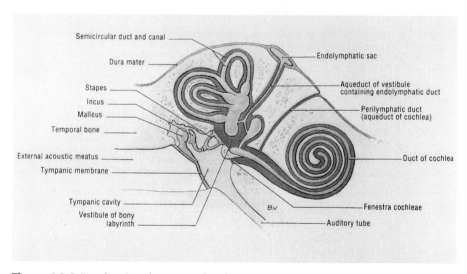

Figure 16.7 A. Transverse section of a 7-week embryo in the region of the rhomben-cephalon, showing the tubotympanic recess, the first pharyngeal cleft, and mesenchy-mal condensation, foreshadowing development of the ossicles. **B.** Middle ear showing the cartilaginous precursors of the auditory ossicles. *Thin yellow line* in mesenchyme indicates future expansion of the primitive tympanic cavity. Note the meatal plug ex-tending from the primitive auditory meatus to the tympanic cavity.

Figure 16.8 Ear showing the external auditory meatus, the middle ear with its ossicles, and the inner ear.

AURICLE

The **auricle** develops from six mesenchymal proliferations at the dorsal ends of the **first** and **second pharyngeal arches,** surrounding the first pharyngeal cleft (Fig. 16.10, *A* and *E*). These swellings (**auricular hillocks**), three on each side of the external meatus, later fuse and form the definitive auricle (Fig. 16.10,

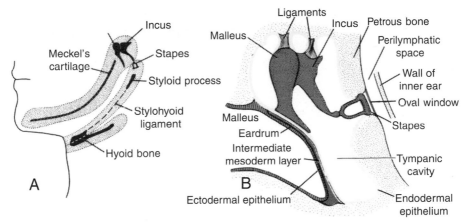

Figure 16.9 A. Derivatives of the first three pharyngeal arches. Note the malleus and incus at the dorsal tip of the first arch and the stapes at that of the second arch. **B**. Middle ear showing the handle of the malleus in contact with the eardrum. The stapes will establish contact with the membrane in the oval window. The wall of the tympanic cavity is lined with endodermal epithelium.

B–D and *G*). As fusion of the auricular hillocks is complicated, developmental abnormalities of the auricle are common. Initially, the external ears are in the lower neck region (Fig. 16.10*F*), but with development of the mandible, they ascend to the side of the head at the level of the eyes.

CLINICAL CORRELATES

Deafness and External Ear Abnormalities

Congenital deafness, usually associated with deaf-mutism, may be caused by abnormal development of the membranous and bony labyrinths or by malformations of the auditory ossicles and eardrum. In the most extreme cases the tympanic cavity and external meatus are absent.

Most forms of congenital deafness are caused by genetic factors, but environmental factors may also interfere with normal development of the internal and middle ear. Rubella virus, affecting the embryo in the seventh or eighth week, may cause severe damage to the organ of Corti. It has also been suggested that poliomyelitis, erythroblastosis fetalis, diabetes, hypothyroidism, and toxoplasmosis can cause congenital deafness.

External ear defects are common; they include minor and severe abnormalities (Fig. 16.11). They are significant from the standpoint of the psychological and emotional trauma they may cause and for the fact they are often associated with other malformations. Thus, they serve as clues to examine infants carefully for other abnormalities. **All of the frequently occurring chromosomal syndromes and most of the less common ones have ear anomalies as one of their characteristics.**

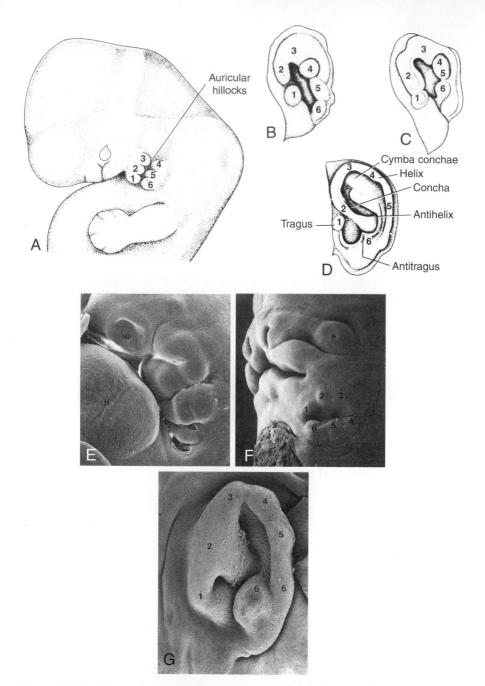

Figure 16.10 A. Lateral view of the head of an embryo showing the six auricular hillocks surrounding the dorsal end of the first pharyngeal cleft. **B** to **D**. Fusion and progressive development of the hillocks into the adult auricle. **E.** The six auricular hillocks from the first and second pharyngeal arches. *H*, heart; *NP*, nasal placode. **F.** The hillocks becoming more defined. Note the position of the ears with respect to the mouth and eyes (*e*). **G.** External ear nearly complete. Growth of the mandible and neck region places the ears in their permanent position.

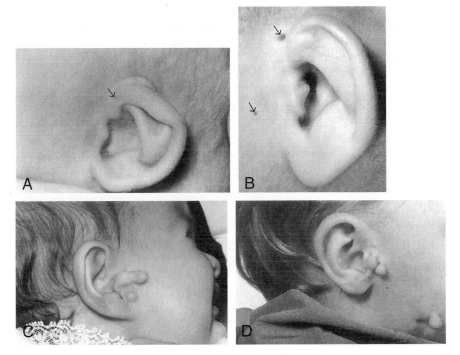

Figure 16.11 A. Microtia with preauricular pit (*arrow*). **B.** Preauricular pits (*arrows*). **C** and **D.** Preauricular appendages (skin tags). Note the low position of the tag in **D**.

Preauricular appendages and pits (Fig. 16.11) are skin tags and shallow depressions, respectively, anterior to the ear. Pits may indicate abnormal development of the auricular hillocks, whereas appendages may be due to accessory hillocks. Like other external ear defects, both are associated with other malformations.

Summary

The ear consists of three parts that have different origins, but that function as one unit. The **internal ear** originates from the **otic vesicle,** which in the fourth week of development detaches from surface ectoderm. This vesicle divides into a ventral component, which gives rise to the **saccule** and **cochlear duct,** and a dorsal component, which gives rise to the **utricle, semicircular canals,** and **endolymphatic duct** (Figs. 16.3, 16.6, and 16.8). The epithelial structures thus formed are known collectively as the **membranous labyrinth.** Except for the **cochlear duct,** which forms the **organ of Corti,** all structures derived from the membranous labyrinth are involved with equilibrium.

The **middle ear,** consisting of the **tympanic cavity** and **auditory tube,** is lined with epithelium of endodermal origin and is derived from the first pharyngeal pouch. The auditory tube extends between the tympanic cavity and nasopharynx. The **ossicles,** which transfer sound from the tympanic membrane to the oval window, are derived from the first (**malleus** and **incus**) and second (**stapes**) pharyngeal arches (Fig. 16.9).

The **external auditory meatus** develops from the first pharyngeal cleft and is separated from the tympanic cavity by the tympanic membrane (eardrum). The eardrum consists of (*a*) an ectodermal epithelial lining, (*b*) an intermediate layer of mesenchyme, and (*c*) an endodermal lining from the first pharyngeal pouch.

The **auricle** develops from six mesenchymal hillocks (Fig. 16.10) along the first and second pharyngeal arches. Defects in the auricle are often associated with other congenital malformations.

Problem to Solve

1. *A newborn has bilateral microtia. Should you be concerned about other malformations? What cell population might be involved in the embryological origin of the defect?*

SUGGESTED READING

Ars B: Organogenesis of the middle ear structures. *J. Laryngol Otol* 103:16, 1989.

Fritzsch B, Beisel KW: Evolution of the nervous system: evolution and development of the vertebrate ear. *Brain Res Bull* 55:711, 2001.

McPhee JR, Van De Water TR: Epithelial mesenchymal tissue interactions guiding otic capsule formation: the role of the otocyst. *J Embryol Exp Morphol* 97:1, 1986.

Michaels L: Evolution of the epidermoid formation and its role in the development of the middle ear and tympanic membrane during the first trimester. *J Otolaryngol* 17:22, 1988.

Michaels L, Soucek S: Auditory epithelial migration on the human tympanic membrane: 2. The existence of two discrete migratory pathways and their embryological correlates. *Am J Anat* 189:189, 1990.

O'Rahilly R: The early development of the otic vesicle in staged human embryos. *J Embryol Exp Morphol* 11:741, 1963.

Eye

Optic Cup and Lens Vesicle

The developing eye appears in the 22-day embryo as a pair of shallow grooves on the sides of the forebrain (Fig. 17.1). With closure of the neural tube, these grooves form outpocketings of the forebrain, the optic vesicles. These vesicles subsequently come in contact with the surface ectoderm and induce changes in the ectoderm necessary for lens formation (Fig. 17.1). Shortly thereafter the optic vesicle begins to invaginate and forms the double-walled optic cup (Figs. 17.1 and 17.2A). The inner and outer layers of this cup are initially separated by a lumen, the intraretinal space (Fig. 17.2B; see also Fig. 17.4A), but soon this lumen disappears, and the two layers appose each other (see Fig. 17.4). Invagination is not restricted to the central portion of the cup but also involves a part of the inferior surface (Fig. 17.2A) that forms the choroid fissure. Formation of this fissure allows the hyaloid artery to reach the inner chamber of the eye (Figs. 17.3 and 17.4; see also Fig. 17.8). During the seventh week, the lips of the choroid fissure fuse, and the mouth of the optic cup becomes a round opening, the future pupil.

During these events, cells of the surface ectoderm, initially in contact with the optic vesicle, begin to elongate and form the lens placode (Fig. 17.1). This placode subsequently invaginates and develops into the lens vesicle. During the fifth week, the lens vesicle loses contact with the surface ectoderm and lies in the mouth of the optic cup (Figs. 17.2C, 17.3, and 17.4).

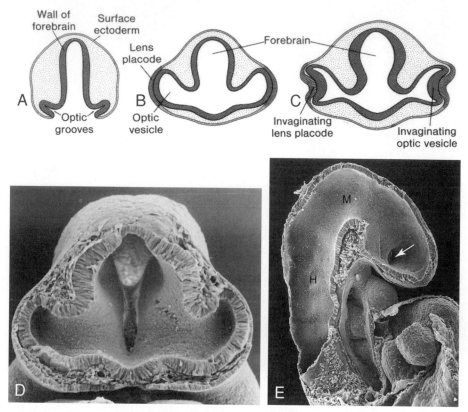

Figure 17.1 A. Transverse section through the forebrain of a 22-day embryo (approximately 14 somites), showing the optic grooves. **B.** Transverse section through the forebrain of a 4-week embryo showing the optic vesicles in contact with the surface ectoderm. Note the slight thickening of the ectoderm (lens placode). **C.** Transverse section through the forebrain of a 5-mm embryo showing invagination of the optic vesicle and the lens placode. **D.** Scanning electron micrograph showing a frontal view of a mouse embryo at a stage similar to that shown in **B. E.** Scanning electron micrograph of a mouse embryo during formation of the optic vesicle. The embryo has been cut sagittally to reveal the inside of the brain vesicles and outpocketing of the optic vesicle (*arrow*) from the forebrain. *H*, hindbrain; and *M*, midbrain.

Retina, Iris, and Ciliary Body

The outer layer of the optic cup, which is characterized by small pigment granules, is known as the **pigmented layer** of the retina (Figs. 17.3, 17.4, and 17.7). Development of the inner **(neural) layer** of the optic cup is more complicated. The posterior four-fifths, the **pars optica retinae,** contains cells bordering the intraretinal space (Fig. 17.3) that differentiate into light-receptive elements, **rods** and **cones** (Fig. 17.5). Adjacent to this photoreceptive layer is the mantle layer, which, as in the brain, gives rise to neurons and supporting cells, including the

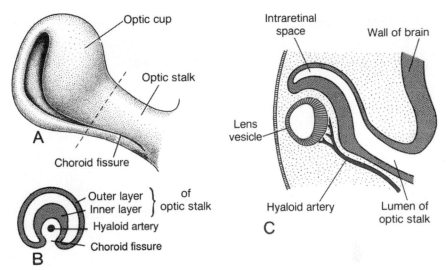

Figure 17.2 A. Ventrolateral view of the optic cup and optic stalk of a 6-week embryo. The choroid fissure on the undersurface of the optic stalk gradually tapers off. **B.** Transverse section through the optic stalk as indicated in **A,** showing the hyaloid artery in the choroid fissure. **C.** Section through the lens vesicle, the optic cup, and optic stalk at the plane of the choroid fissure.

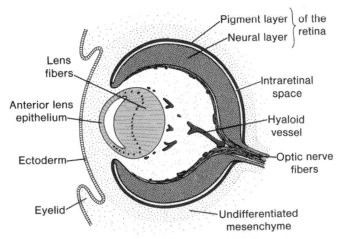

Figure 17.3 Section through the eye of a 7-week embryo. The eye primordium is completely embedded in mesenchyme. Fibers of the neural retina converge toward the optic nerve.

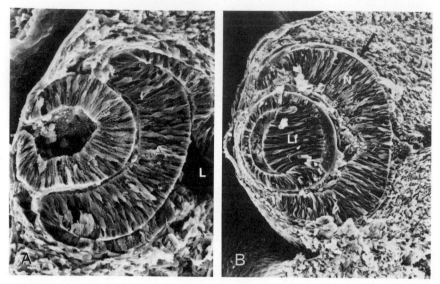

Figure 17.4 Scanning electron micrographs of sections through the eyes of mouse embryos at stages equivalent to **(A)** 6 weeks and **(B)** 7 weeks in the human. **A.** The forming lens vesicle, not entirely closed; the two layers of the optic cup; and the lumen (*L*) of the optic stalk. (Compare with Fig. 17.2**C**.) **B.** Lens fibers (*Lf*) and neural (*N*) and pigment layers (*arrow*) forming. (Compare with Fig. 17.3.)

outer nuclear layer, inner nuclear layer, and **ganglion cell layer** (Fig. 17.5). On the surface is a fibrous layer that contains axons of nerve cells of the deeper layers. Nerve fibers in this zone converge toward the optic stalk, which develops into the optic nerve (Figs. 17.3 and 17.5). Hence, light impulses pass through most layers of the retina before they reach the rods and cones.

The anterior fifth of the inner layer, the **pars ceca retinae,** remains one cell layer thick. It later divides into the **pars iridica retinae,** which forms the inner layer of the iris, and the **pars ciliaris retinae,** which participates in formation of the **ciliary body** (Fig. 17.6 and 17.7).

Meanwhile, the region between the optic cup and the overlying surface epithelium is filled with loose mesenchyme (Figs. 17.3, 17.4, and 17.7). The **sphincter** and **dilator pupillae** muscles form in this tissue (Fig. 17.6). These muscles develop from the underlying ectoderm of the optic cup. In the adult, the iris is formed by the pigment-containing external layer, the unpigmented internal layer of the optic cup, and a layer of richly vascularized connective tissue that contains the pupillary muscles (Fig. 17.6).

The **pars ciliaris retinae** is easily recognized by its marked folding (Figs. 17.6*B* and 17.7). Externally it is covered by a layer of mesenchyme that forms the **ciliary muscle;** on the inside it is connected to the lens by a network of elastic fibers, the **suspensory ligament** or zonula (Fig. 17.7). Contraction of the ciliary muscle changes tension in the ligament and controls curvature of the lens.

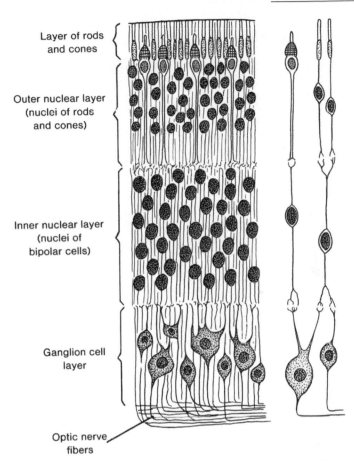

Layer of rods and cones

Outer nuclear layer (nuclei of rods and cones)

Inner nuclear layer (nuclei of bipolar cells)

Ganglion cell layer

Optic nerve fibers

Figure 17.5 Various layers of the pars optica retinae in a fetus of approximately 25 weeks.

Lens

Shortly after formation of the lens vesicle (Fig. 17.2*C*), cells of the posterior wall begin to elongate anteriorly and form long fibers that gradually fill the lumen of the vesicle (Figs. 17.3 and 17.4*B*). By the end of the seventh week, these **primary lens fibers** reach the anterior wall of the lens vesicle. Growth of the lens is not finished at this stage, however, since new (secondary) lens fibers are continuously added to the central core.

Choroid, Sclera, and Cornea

At the end of the fifth week, the eye primordium is completely surrounded by loose mesenchyme (Fig. 17.3). This tissue soon differentiates into an inner layer

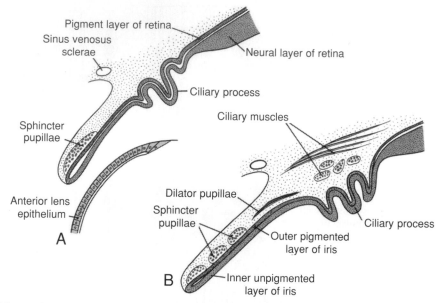

Figure 17.6 Development of the iris and ciliary body. The rim of the optic cup is covered by mesenchyme, in which the sphincter and dilator pupillae develop from the underlying ectoderm.

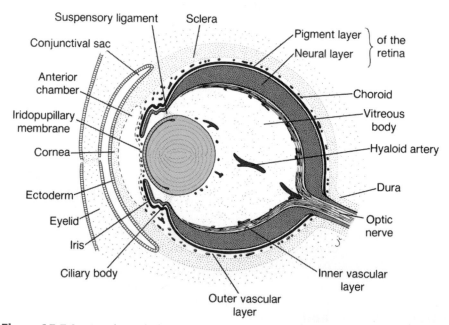

Figure 17.7 Section through the eye of a 15-week fetus showing the anterior chamber, iridopupillary membrane, inner and outer vascular layers, choroid, and sclera.

comparable with the pia mater of the brain and an outer layer comparable with the dura mater. The inner layer later forms a highly vascularized pigmented layer known as the **choroid;** the outer layer develops into the sclera and is continuous with the dura mater around the optic nerve (Fig. 17.7).

Differentiation of mesenchymal layers overlying the anterior aspect of the eye is different. The **anterior chamber** forms through vacuolization and splits the mesenchyme into an inner layer in front of the lens and iris, the **iridopupillary membrane,** and an outer layer continuous with the sclera, the **substantia propria** of the **cornea** (Fig. 17.7). The anterior chamber itself is lined by flattened mesenchymal cells. Hence, the cornea is formed by (*a*) an epithelial layer derived from the surface ectoderm, (*b*) the **substantia propria** or **stroma,** which is continuous with the sclera, and (*c*) an epithelial layer, which borders the anterior chamber. The iridopupillary membrane in front of the lens disappears completely, providing communication between the anterior and posterior eye chambers.

Vitreous Body

Mesenchyme not only surrounds the eye primordium from the outside but also invades the inside of the optic cup by way of the choroid fissure. Here it forms the hyaloid vessels, which during intrauterine life supply the lens and form the vascular layer on the inner surface of the retina (Fig. 17.7). In addition, it forms a delicate network of fibers between the lens and retina. The interstitial spaces of this network later fill with a transparent gelatinous substance, forming the **vitreous body** (Fig. 17.7). The hyaloid vessels in this region are obliterated and disappear during fetal life, leaving behind the hyaloid canal.

Optic Nerve

The optic cup is connected to the brain by the optic stalk, which has a groove, the **choroid fissure,** on its ventral surface (Figs. 17.2 and 17.3). In this groove are the hyaloid vessels. The nerve fibers of the retina returning to the brain lie among cells of the inner wall of the stalk (Fig. 17.8). During the seventh week, the choroid fissure closes, and a narrow tunnel forms inside the optic stalk (Fig. 17.8*B*). As a result of the continuously increasing number of nerve fibers, the inner wall of the stalk grows, and the inside and outside walls of the stalk fuse (Fig. 17.8*C*). Cells of the inner layer provide a network of neuroglia that support the optic nerve fibers.

The optic stalk is thus transformed into the **optic nerve.** Its center contains a portion of the hyaloid artery, later called the **central artery of the retina.** On the outside, a continuation of the choroid and sclera, the **pia arachnoid** and **dura** layer of the nerve, respectively, surround the optic nerve.

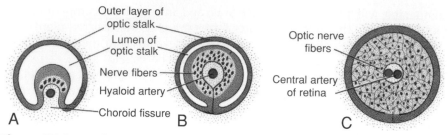

Figure 17.8 Transformation of the optic stalk into the optic nerve. **A.** Sixth week (9 mm). **B.** Seventh week (15 mm). **C.** Ninth week. Note the central artery of the retina in the optic nerve.

Molecular Regulation of Eye Development

PAX6 is the key regulatory gene for eye development. It is a member of the *PAX* (paired box) family of transcription factors and contains two DNA binding motifs that include a paired domain and a paired type homeodomain. Initially, this transcription factor is expressed in a band in the anterior neural ridge of the neural plate before neurulation begins. At this stage, there is a single eye field that later separates into two optic primordia. The signal for separation of this field is **sonic hedgehog** (**SHH**) expressed in the prechordal plate (see Fig. 19.32). *SHH* expression up-regulates *PAX2* in the center of the eye field and down-regulates *PAX6*. Later this pattern is maintained so that *PAX2* is expressed in the optic stalks and *PAX6* is expressed in the optic cup and overlying surface ectoderm that forms the lens. As development proceeds, it appears that *PAX6* is not essential for optic cup formation. Instead, this process is regulated by inter-active signals between the optic vesicle and surrounding mesenchyme and the overlying surface ectoderm in the lens-forming region (Fig. 17.9). Thus, fibro-blast growth factors (FGFs) from the surface ectoderm promote differentiation of the neural (inner layer) retina, while transforming growth factor β (TGF-β), se-creted by surrounding mesenchyne, directs formation of the pigmented (outer) retinal layer. Downstream from these gene products the transcription factors *MITF* and *CHX10* are expressed and direct differentiation of the pigmented and neural layer, respectively (Fig. 17.9). Thus, the lens ectoderm is essential for proper formation of the optic cup such that without a lens placode no cup invagination occurs.

Differentiation of the lens is dependent upon *PAX6*, although the gene is not responsible for inductive activity by the optic vesicle. Instead, *PAX6* acts au-tonomously in the surface ectoderm to regulate lens development (Fig. 17.10). The process begins with *PAX6* expression in the neural plate that upregulates the transcription factor *SOX2* and also maintains *PAX6* expression in the prospec-tive lens ectoderm. In turn, the optic vesicle secretes BMP-4, which also upreg-ulates and maintains *SOX2* expression as well as expression of *LMAF,* another transcription factor. Next, the expression of two homeobox genes, *SIX3* and

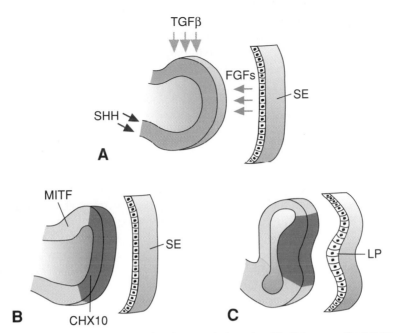

Figure 17.9 Drawing showing molecular regulation of epithelial-mesenchymal interactions responsible for patterning eye development. **A.** Once *PAX6* establishes the eye field, fibroblast growth factors (FGFs), secreted by surface ectoderm (SE) in the prospective lens-forming region overlying the optic vesicle, promote differentiation of the neural retinal layer; while members of the transforming growth factor β (TGF-β) family, secreted by surrounding mesenchyme, promote differentiation of the pigmented retinal layer. These external signals cause regionalization of the inner and outer layers of the optic cup and upregulate downstream genes, including *CHX10* and *MITF*, that regulate continued differentiation of these structures (**B** and **C**). In addition to its role in determining the eye fields, *PAX6* specifies the lens placode (LP) region (**B**) and is also important for retinal development.

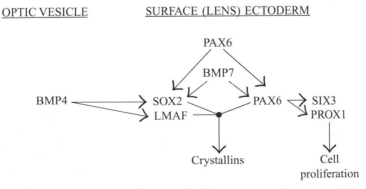

Figure 17.10 Schematic showing the cascade of gene expression responsible for early stages of lens development.

PROX1, is upregulated by *PAX6*, while *BMP-7* expression in the lens ectoderm is increased to maintain expression of *SOX2* and *PAX6*. Finally, the combined expression of *PAX6*, *SOX2*, and *LMAF* initiates expression of genes responsible for formation of lens crystalline proteins, while *PROX1* expression regulates genes controlling cell proliferation.

CLINICAL CORRELATES

Eye Abnormalities

Coloboma may occur if the choroid fissure fails to close. Normally this fissure closes during the seventh week of development (Fig. 17.8). When it does not, a cleft persists. Although such a cleft is usually in the iris only—**coloboma iridis** (Fig. 17.11 *A*)—it may extend into the ciliary body, the retina, the choroid, and the optic nerve. Coloboma is a common eye abnormality frequently associated with other eye defects. Colobomas (clefts) of the eyelids may also occur. Mutations in the **PAX2** gene have been linked with optic nerve colobomas and may play a role in the other types as well. Renal defects also occur with mutations in *PAX2* as part of the **renal coloboma syndrome** (see Chapter 14).

The **iridopupillary membrane** (Fig. 17.11 *B*) may persist instead of being resorbed during formation of the anterior chamber.

In **congenital cataracts** the lens becomes opaque during intrauterine life. Although this anomaly is usually genetically determined, many children of mothers who have had German measles (rubella) between the fourth and seventh weeks of pregnancy have cataracts. If the mother is infected after the seventh week of pregnancy, the lens escapes damage, but the child may be deaf as a result of abnormalities of the cochlea.

The **hyaloid artery** may persist to form a cord or cyst. Normally the distal portion of this vessel degenerates, leaving the proximal part to form the central artery of the retina.

In **microphthalmia** the eye is too small; the eyeball may be only two-thirds of its normal volume. Usually associated with other ocular abnormalities,

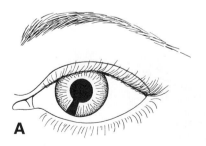
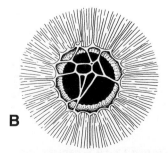

Figure 17.11 A. Coloboma iris. **B.** Persistence of the iridopupillary membrane.

microphthalmia frequently results from intrauterine infections such as cytomegalovirus and toxoplasmosis.

Anophthalmia is absence of the eye. In some cases histological analysis reveals some ocular tissue. The defect is usually accompanied by severe cranial abnormalities.

Congenital aphakia (absence of the lens) and **aniridia** (absence of the iris) are rare anomalies that are due to disturbances in induction and formation of tissues responsible for formation of these structures. Mutations in *PAX6* result in aniridia and may also contribute to anophthalmia and microphthalmia.

Cyclopia (single eye) and **synophthalmia** (fusion of the eyes) comprise a spectrum of defects in which the eyes are partially or completely fused (Fig. 17.12). The defects are due to a loss of midline tissue that may occur as early as days 19 to 21 of gestation or at later stages when facial development is initiated. This loss results in underdevelopment of the forebrain and frontonasal prominence. These defects are invariably associated with cranial defects, such as holoprosencephaly, in which the cerebral hemispheres are partially or completely merged into a single telencephalic vesicle. Factors affecting the midline include alcohol, mutations in *sonic hedgehog*, (*SHH*) and abnormalities in cholesterol metabolism that may disrupt *SHH* signaling (see Chapter 19).

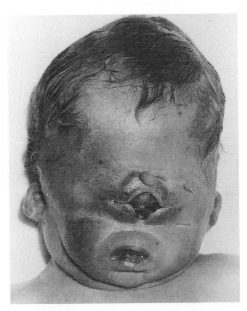

Figure 17.12 Synophthalmia. The eyes are fused because loss of midline structures prevented the eye fields from separating. Such babies also have severe cranial defects, including holoprosencephaly.

Summary

The eyes begin to develop as a pair of outpocketings that will become the **optic vesicles** on each side of the forebrain at the end of the fourth week of development. The optic vesicles contact the surface ectoderm and induce lens formation. When the optic vesicle begins to invaginate to form the pigment and neural layers of the retina, the lens placode invaginates to form the lens vesicle. Through a groove at the inferior aspect of the optic vesicle, the choroid fissure, the hyaloid artery (later the central artery of the retina) enters the eye (Figs. 17.2 and 17.3). Nerve fibers of the eye also occupy this groove to reach the optic areas of the brain. The cornea is formed by (*a*) a layer of surface ectoderm, (*b*) the stroma, which is continuous with the sclera, and (*c*) an epithelial layer bordering the anterior chamber (Fig. 17.7).

PAX6, the master gene for eye development, is expressed in the single eye field at the neural plate stage. The eye field is separated into two optic primordia by *SHH*, which up-regulates *PAX2* expression in the optic stalks while down-regulating *PAX6*, restricting this gene's expression to the optic cup and lens. Epithelial-mesenchymal interactions between prospective lens ectoderm, optic vesicle, and surrounding mesenchyme then regulate lens and optic cup differentiation (Figs. 17.9 and 17.10).

Problems to Solve

1. *A newborn has unilateral aphakia (absent lens). What is the embryological origin of this defect?*

2. *In taking a history of a young woman in her 10th week of gestation you become concerned that she may have contracted rubella sometime during the fourth to eighth weeks of her pregnancy. What types of defects might be produced in her offspring?*

3. *Physical examination of a newborn reveals clefts in the lower portion of the iris bilaterally. What is the embryological basis for this defect? What other structures might be involved?*

SUGGESTED READING

Ashery-Padau R, Gruss P: Pax6 lights up the way for eye development. *Curr Op Cell Biol* 13:706, 2001.

Li HS, et al.: A single morphogenetic field gives rise to two retina primordia under the influence of the prechordal plate. *Development* 124:603, 1997.

Macdonald R, et al.: Midline signaling is required for Pax gene regulation and patterning of the eyes. *Development* 121:3267, 1995.

O'Rahilly R: The timing and sequence of events in the development of the human eye and ear during the embryonic period proper. *Anat Embryol (Berl)* 168:87, 1983.

Saha MS, Spann C, Grainger RM: Embryonic lens induction: more than meets the optic vesicle. *Cell Differ Dev* 28:153, 1989.

Stromland K, Miller M, Cook C: Ocular teratology. *Surv Ophthalmol* 35:429, 1991.

chapter 18

Integumentary System

Skin

The skin has a dual origin: (*a*) A superficial layer, the **epidermis,** develops from the surface ectoderm. (*b*) A deep layer, the **dermis,** develops from the underlying mesenchyme.

EPIDERMIS

Initially, the embryo is covered by a single layer of ectodermal cells (Fig. 18.1*A*). In the beginning of the second month, this epithelium divides, and a layer of flattened cells, the **periderm,** or **epitrichium,** is laid down on the surface (Fig. 18.1*B*). With further proliferation of cells in the basal layer, a third, intermediate zone is formed (Fig. 18.1*C*). Finally, at the end of the fourth month, the epidermis acquires its definitive arrangement, and four layers can be distinguished (Fig. 18.1*D*):

The **basal layer,** or **germinative layer,** is responsible for production of new cells. This layer later forms ridges and hollows, which are reflected on the surface of the skin in the fingerprint.

A thick **spinous layer** consists of large polyhedral cells containing fine tonofibrils.

The **granular layer** contains small keratohyalin granules in its cells.

The **horny layer,** forming the tough scalelike surface of the epidermis, is made up of closely packed dead cells containing keratin. Cells of the periderm are usually cast off during the second part of intrauterine life and can be found in the amniotic fluid.

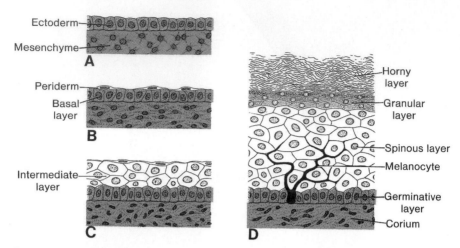

Figure 18.1 Formation of the skin at various stages of development. **A.** 5 weeks. **B.** 7 weeks. **C.** 4 months. **D.** Birth.

During the first 3 months of development, the epidermis is invaded by cells arising from the **neural crest.** These cells synthesize **melanin** pigment, which can be transferred to other cells of the epidermis by way of dendritic processes. After birth, these **melanocytes** cause pigmentation of the skin (Fig. 18.1*D*).

CLINICAL CORRELATES

Fingerprints

The epidermal ridges that produce typical patterns on the surface of the fingertips, palms of the hand, and soles of the feet are genetically determined. They form the basis for many studies in medical genetics and criminal investigations (**dermatoglyphics**). In children with chromosomal abnormalities, the epidermal pattern on the hand and fingers is sometimes used as a diagnostic tool.

DERMIS

The **dermis** is derived from lateral plate mesoderm and the dermatomes from somites. During the third and fourth months, this tissue, the **corium** (Fig. 18.1*D*), forms many irregular papillary structures, the **dermal papillae,** which project upward into the epidermis. Most of these papillae contain a small capillary or sensory nerve end organ. The deeper layer of the dermis, the **subcorium,** contains large amounts of fatty tissue.

At birth, the skin is covered by a whitish paste, the **vernix caseosa,** formed by secretions from sebaceous glands and degenerated epidermal cells and hairs. It protects the skin against the macerating action of amniotic fluid.

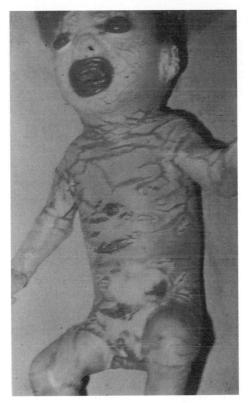

Figure 18.2 Ichthyosis in a harlequin fetus with massive thickening of the keratin layer, which cracks to form fissures between thickened plaques.

CLINICAL CORRELATES

Keratinization of the skin

Ichthyosis, excessive keratinization of the skin, is characteristic of a group of hereditary disorders that are usually inherited as an autosomal recessive trait but may also be X-linked. In severe cases, ichthyosis may result in a grotesque appearance, as in the case of a **harlequin fetus** (Fig. 18.2).

Hair

Hairs appear as solid epidermal proliferations penetrating the underlying dermis (Fig. 18.3*A*). At their terminal ends, hair buds invaginate. The invaginations, the **hair papillae,** are rapidly filled with mesoderm in which vessels and nerve endings develop (Fig. 18.3, *B* and *C*). Soon cells in the center of the hair buds become spindle-shaped and keratinized, forming the **hair shaft,** while

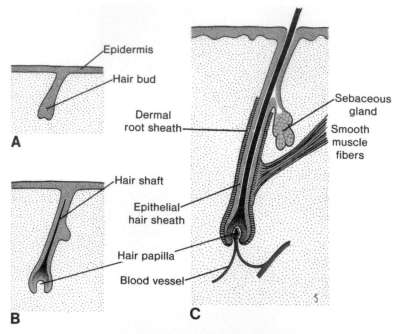

Epidermis

Hair bud

Dermal
root sheath

A

Hair shaft

Epithelial
hair sheath

Hair papilla

Blood vessel

B

C

Sebaceous
gland

Smooth
muscle
fibers

Figure 18.3 Development of a hair and a sebaceous gland. **A.** 4 months. **B.** 6 months. **C.** Newborn.

peripheral cells become cuboidal, giving rise to the **epithelial hair sheath** (Fig. 18.3, *B* and *C*).

The **dermal root sheath** is formed by the surrounding mesenchyme. A small smooth muscle, also derived from mesenchyme, is usually attached to the dermal root sheath. The muscle is the **arrector pili muscle.** Continuous proliferation of epithelial cells at the base of the shaft pushes the hair upward, and by the end of the third month the first hairs appear on the surface in the region of the eyebrow and upper lip. The first hair that appears, **lanugo hair,** is shed at about the time of birth and is later replaced by coarser hairs arising from new hair follicles.

The epithelial wall of the hair follicle usually shows a small bud penetrating the surrounding mesoderm (Fig. 18.3*C*). Cells from these buds form the **se-baceous glands.** Cells from the gland degenerate, forming a fatlike substance secreted into the hair follicle, and from there it reaches the skin.

CLINICAL CORRELATES

Abnormalities of Hair Distribution

Hypertrichosis (excessive hairiness) is caused by an unusual abundance of hair follicles. It may be localized to certain areas of the body, especially the lower lumbar region covering a spina bifida occulta defect or may cover the entire body.

Atrichia, the congenital absence of hair, is usually associated with abnormalities of other ectodermal derivatives, such as teeth and nails.

Mammary Glands

The first indication of mammary glands is found in the form of a bandlike thickening of the epidermis, the **mammary line** or **mammary ridge.** In a 7-week embryo, this line extends on each side of the body from the base of the forelimb to the region of the hindlimb (Fig. 18.4C). Although the major part of the mammary line disappears shortly after it forms, a small portion in the thoracic region persists and penetrates the underlying mesenchyme (Fig. 18.4A). Here it forms 16 to 24 sprouts, which in turn give rise to small, solid buds. By the end of prenatal life, the epithelial sprouts are canalized and form the **lactiferous ducts,** and the buds form small ducts and alveoli of the gland. Initially, the **lactiferous ducts** open into a small epithelial pit (Fig. 18.4B). Shortly after birth, this pit is transformed into the **nipple** by proliferation of the underlying mesenchyme.

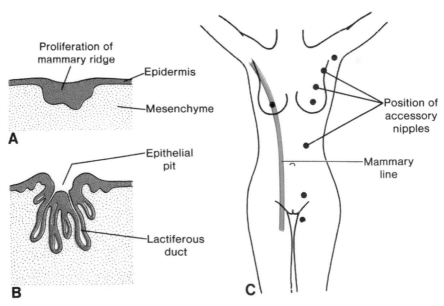

Figure 18.4 A and **B.** Sections through the developing mammary gland at the third and eighth months, respectively. **C.** Positions of accessory nipples (*blue line,* mammary line).

CLINICAL CORRELATES

Mammary Gland Abnormalities

Polythelia is a condition where accessory nipples have formed due to the persistence of fragments of the mammary line (Fig. 18.4C). Accessory nipples may develop anywhere along the original mammary line, but usually appear in the axillary region.

Polymastia occurs when a remnant of the mammary line develops into a complete breast.

Inverted nipple is a condition in which the lactiferous ducts open into the original epithelial pit that has failed to evert.

Summary

The skin and its associated structures, hair, nails, and glands, are derived from surface ectoderm. **Melanocytes,** which give the skin its color, are derived from **neural crest cells,** which migrate into the epidermis. The production of new cells occurs in the **germinative** layer. After moving to the surface, cells are sloughed off in the horny layer (Fig. 18.1). The dermis, the deep layer of the skin, is derived from lateral plate mesoderm and from dermatomes of the somites.

Hairs develop from downgrowth of epidermal cells into the underlying dermis. By about 20 weeks, the fetus is covered by downy hair, **lanugo hair,** which is shed at the time of birth. **Sebaceous glands, sweat glands,** and **mammary glands** all develop from epidermal proliferations. Supernumerary nipples (**polythelia**) and breasts (**polymastia**) are relatively common (see Fig. 18.4).

Problem to Solve

1. *A woman appears to have accessory nipples in her axilla and on her abdomen bilaterally. What is the embryological basis for these additional nipples, and why do they occur in these locations?*

SUGGESTED READING

Beller F: Development and anatomy of the breast. *In* Mitchell GW Jr, Bassett LW (eds): *The Female Breast and Its Disorders.* Baltimore, Williams & Wilkins, 1990.

Fuchs E: Epidermal differentiation: the bare essential. *J Cell Biol* 111:2807, 1990.

Hirschhorn K: Dermatoglyphics. *In* Behrman RE (ed): *Nelson Textbook of Pediatrics.* 14th ed. Philadelphia, WB Saunders, 1992.

Newman M: Supernumerary nipples. *Am Fam Physician* 38:183, 1988.

Nordlund JJ, Abdel-Malek ZA, Boissy R, Rheins LA: Pigment cell biology: an historical review. *J Invest Dermatol* 92:53S, 1989.

Opitz JM: Pathogenetic analysis of certain developmental and genetic ectodermal defects. *Birth Defects* 24:75, 1988.

Smith LT, Holbrook KA: Embryogenesis of the dermis in human skin. *Pediatr Dermatol* 3:271, 1986.

chapter 19

Central Nervous System

The central nervous system (CNS) appears at the beginning of the third week as a slipper-shaped plate of thickened ectoderm, **the neural plate,** in the middorsal region in front of the **primitive node.** Its lateral edges soon elevate to form the **neural folds** (Fig. 19.1).

With further development, the neural folds continue to elevate, approach each other in the midline, and finally fuse, forming the **neural tube** (Figs. 19.2 and 19.3). Fusion begins in the cervical region and proceeds in cephalic and caudal directions (Fig. 19.3A). Once fusion is initiated, the open ends of the neural tube form the **cranial** and **caudal neuropores** that communicate with the overlying amniotic cavity (Fig. 19.3B). Closure of the cranial neuropore proceeds cranially from the initial closure site in the cervical region (19.3A) and from a site in the forebrain that forms later. This later site proceeds cranially, to close the rostralmost region of the neural tube, and caudally to meet advancing closure from the cervical site (19.3B). Final closure of the cranial neuropore occurs at the 18- to 20-somite stage (25th day); closure of the caudal neuropore occurs approximately 2 days later.

The cephalic end of the neural tube shows three dilations, the **primary brain vesicles:** (a) the **prosencephalon,** or forebrain; (b) the **mesencephalon,** or midbrain; and (c) the **rhombencephalon,** or hindbrain (Fig. 19.4). Simultaneously it forms two flexures: (a) the **cervical flexure**

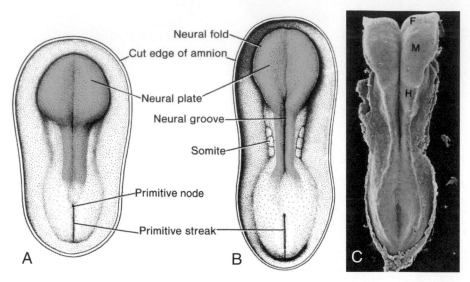

Figure 19.1 A. Dorsal view of a late presomite embryo at approximately 18 days. The amnion has been removed, and the neural plate is clearly visible. **B.** Dorsal view at approximately 20 days. Note the somites and the neural groove and neural folds. **C.** Scanning electron micrograph of a mouse embryo at a stage similar to that in **B.** *F*, forebrain; *M*, midbrain; *H*, hindbrain.

at the junction of the hindbrain and the spinal cord and (*b*) the **cephalic flexure** in the midbrain region (Fig. 19.4).

When the embryo is 5 weeks old, the prosencephalon consists of two parts: (*a*) the **telencephalon,** formed by a midportion and two lateral outpocketings, the **primitive cerebral hemispheres,** and (*b*) the **diencephalon,** characterized by outgrowth of the optic vesicles (Fig. 19.5). A deep furrow, the **rhombencephalic isthmus,** separates the mesencephalon from the rhombencephalon.

The rhombencephalon also consists of two parts: (*a*) the **metencephalon,** which later forms the **pons** and **cerebellum,** and (*b*) the **myelencephalon.** The boundary between these two portions is marked by the **pontine flexure** (Fig.19.5).

The lumen of the spinal cord, the **central canal,** is continuous with that of the brain vesicles. The cavity of the rhombencephalon is the **fourth ventricle,** that of the diencephalon is the **third ventricle,** and those of the cerebral hemispheres are the **lateral ventricles** (Fig. 19.5). The lumen of the mesencephalon connects the third and fourth ventricles. This lumen becomes very narrow and is then known as the **aqueduct of Sylvius.** The lateral ventricles communicate with the third ventricle through the **interventricular foramina of Monro** (Fig. 19.5).

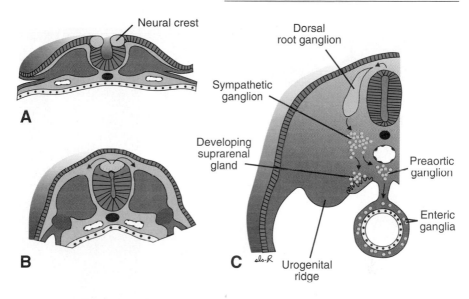

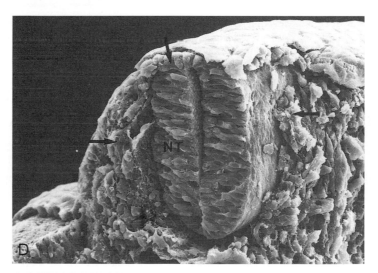

Figure 19.2 A–C. Transverse sections through successively older embryos showing formation of the neural groove, neural tube, and neural crest. Cells of the neural crest, migrate from the edges of the neural folds and develop into spinal and cranial sensory ganglia **(A–C). D.** Scanning electron micrograph of a mouse embryo showing the neural tube (*NT*) and neural crest cells (*arrows*) migrating from the dorsal region (compare with **B** and **C**).

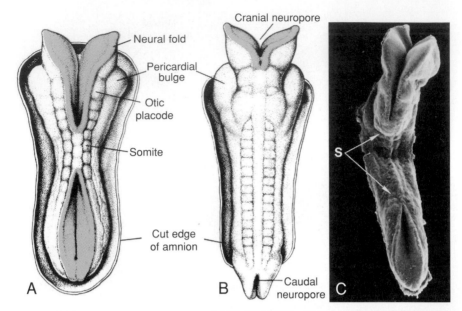

Figure 19.3 A. Dorsal view of a human embryo at approximately day 22. Seven distinct somites are visible on each side of the neural tube. **B.** Dorsal view of a human embryo at approximately day 23. The nervous system is in connection with the amniotic cavity through the cranial and caudal neuropores. **C.** Scanning electron micrograph of a mouse embryo at a stage similar to that in **A.** *S,* somites.

Spinal Cord

NEUROEPITHELIAL, MANTLE, AND MARGINAL LAYERS

The wall of a recently closed neural tube consists of **neuroepithelial cells.** These cells extend over the entire thickness of the wall and form a thick pseudostratified epithelium (Fig. 19.6). Junctional complexes at the lumen connect them. During the neural groove stage and immediately after closure of the tube, they divide rapidly, producing more and more neuroepithelial cells. Collectively they constitute the **neuroepithelial layer** or **neuroepithelium.**

Once the neural tube closes, neuroepithelial cells begin to give rise to another cell type characterized by a large round nucleus with pale nucleoplasm and a dark-staining nucleolus. These are the primitive nerve cells, or **neuroblasts** (Fig. 19.7). They form the **mantle layer,** a zone around the neuroepithelial layer (Fig. 19.8). The mantle layer later forms the **gray matter of the spinal cord.**

The outermost layer of the spinal cord, the **marginal layer,** contains nerve fibers emerging from neuroblasts in the mantle layer. As a result of myelination of nerve fibers, this layer takes on a white appearance and therefore is called the **white matter of the spinal cord** (Fig. 19.8).

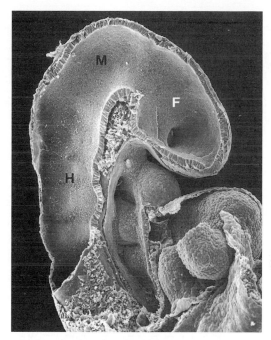

Figure 19.4 Scanning electron micrograph of a sagittal section through a mouse embryo at approximately 27 days of human development. Three brain vesicles represent the forebrain (*F*), midbrain (*M*), and hindbrain (*H*).

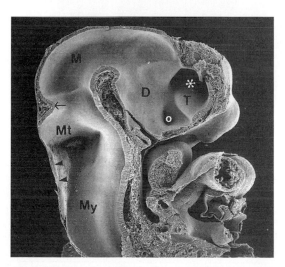

Figure 19.5 Scanning electron micrograph of a sagittal section through a mouse embryo at approximately 32 days of human development. The three brain vesicles have segregated into the telencephalon (*T*), diencephalon (*D*), mesencephalon (*M*), metencephalon (*Mt*), and myelencephalon (*My*). *Asterisk*, outpocketing of the telencephalon; *arrow*, rhombencephalic isthmus; *arrowheads*, roof of the fourth ventricle; *o*, optic stalk.

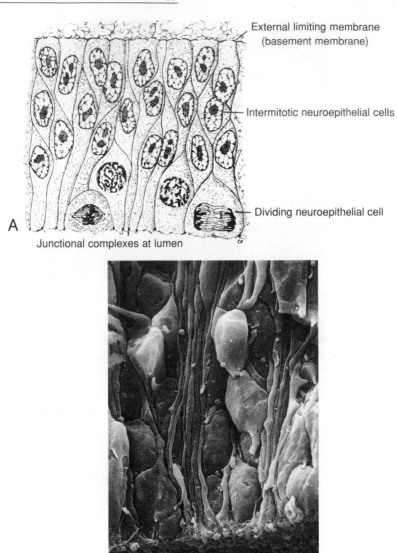

External limiting membrane
(basement membrane)

Intermitotic neuroepithelial cells

Dividing neuroepithelial cell

Junctional complexes at lumen

A

B

Figure 19.6 A. Section of the wall of the recently closed neural tube showing neuroepithelial cells, which form a pseudostratified epithelium extending over the full width of the wall. Note the dividing cells at the lumen of the tube. **B.** Scanning electron micrograph of a section of the neural tube of a mouse embryo similar to that in **A.**

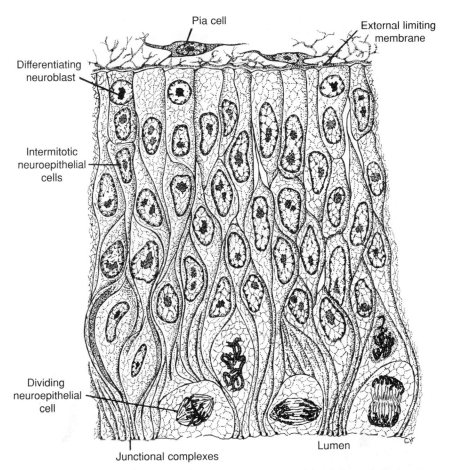

Figure 19.7 Section of the neural tube at a slightly more advanced stage than in Figure 19.6. The major portion of the wall consists of neuroepithelial cells. On the periphery, immediately adjacent to the external limiting membrane, neuroblasts form. These cells, which are produced by the neuroepithelial cells in ever-increasing numbers, will form the mantle layer.

BASAL, ALAR, ROOF, AND FLOOR PLATES

As a result of continuous addition of neuroblasts to the mantle layer, each side of the neural tube shows a ventral and a dorsal thickening. The ventral thickenings, the **basal plates,** which contain ventral motor horn cells, form the motor areas of the spinal cord; the dorsal thickenings, the **alar plates,** form the **sensory areas** (Fig. 19.8*A*). A longitudinal groove, the **sulcus limitans,** marks the boundary between the two. The dorsal and ventral midline portions of the neural tube, known as the **roof** and **floor plates,** respectively, do not contain neuroblasts; they serve primarily as pathways for nerve fibers crossing from one side to the other.

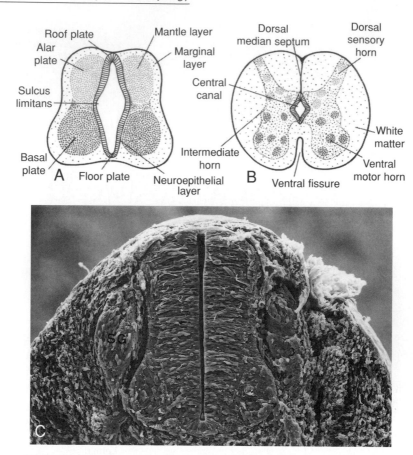

Figure 19.8 A and **B.** Two successive stages in the development of the spinal cord. Note formation of ventral motor and dorsal sensory horns and the intermediate column. **C.** Scanning electron micrograph of a section through the spinal cord of a mouse embryo showing a stage similar to that in **A.** *SG,* spinal ganglion.

In addition to the ventral motor horn and the dorsal sensory horn, a group of neurons accumulates between the two areas and forms a small **intermediate horn** (Fig. 19.8*B*). This horn, containing neurons of the sympathetic portion of the autonomic nervous system, is present only at thoracic (T1–T12) and upper lumbar levels (L2 or L3) of the spinal cord.

HISTOLOGICAL DIFFERENTIATION

Nerve Cells

Neuroblasts, or primitive nerve cells, arise exclusively by division of the neuroepithelial cells. Initially they have a central process extending to the lumen (**transient dendrite**), but when they migrate into the mantle layer, this process disappears, and neuroblasts are temporarily round and **apolar** (Fig. 19.9 *A*). With further differentiation, two new cytoplasmic processes appear on opposite sides of the cell body, forming a **bipolar neuroblast** (Fig. 19.9 *B*). The process at one end of the cell elongates rapidly to form the **primitive axon,** and the process at the other end shows a number of cytoplasmic arborizations, the **primitive dendrites** (Fig. 19.9*C*). The cell is then known as a **multipolar neuroblast** and with further development becomes the adult nerve cell or **neuron.** Once neuroblasts form, they lose their ability to divide. Axons of neurons in the basal plate break through the marginal zone and become visible on the ventral aspect of the cord. Known collectively as the **ventral motor root of the spinal nerve,** they conduct motor impulses from the spinal cord to the muscles (Fig. 19.10). Axons of neurons in the dorsal sensory

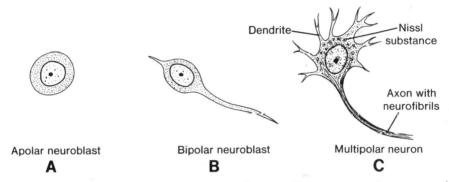

Figure 19.9 Various stages of development of a neuroblast. A neuron is a structural and functional unit consisting of the cell body and all its processes.

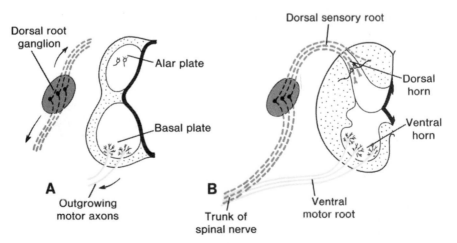

Figure 19.10 A. Motor axons growing out from neurons in the basal plate and centrally and peripherally growing fibers of nerve cells in the dorsal root ganglion. **B.** Nerve fibers of the ventral motor and dorsal sensory roots join to form the trunk of the spinal nerve.

horn (alar plate) behave differently from those in the ventral horn. They penetrate into the marginal layer of the cord, where they ascend to either higher or lower levels to form **association neurons.**

Glial Cells

The majority of primitive supporting cells, the **gliablasts,** are formed by neuroepithelial cells after production of neuroblasts ceases. Gliablasts migrate from the neuroepithelial layer to the mantle and marginal layers. In the mantle layer, they differentiate into **protoplasmic astrocytes** and **fibrillar astrocytes** (Fig. 19.11).

Another type of supporting cell possibly derived from gliablasts is the **oligodendroglial cell.** This cell, which is found primarily in the marginal layer, forms myelin sheaths around the ascending and descending axons in the marginal layer.

In the second half of development, a third type of supporting cell, the **microglial cell,** appears in the CNS. This highly phagocytic cell type is derived from mesenchyme (Fig. 19.11). When neuroepithelial cells cease to produce neuroblasts and gliablasts, they differentiate into ependymal cells lining the central canal of the spinal cord.

Neural Crest Cells

During elevation of the neural plate, a group of cells appears along each edge (the crest) of the neural folds (Fig. 19.2). These **neural crest cells** are ectodermal

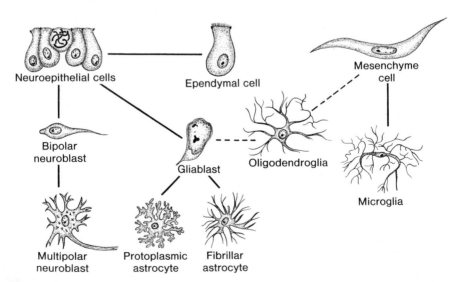

Figure 19.11 Origin of the nerve cell and the various types of glial cells. Neuroblasts, fibrillar and protoplasmic astrocytes, and ependymal cells originate from neuroepithelial cells. Microglia develop from mesenchyme cells. The origin of the oligodendroglia is not clear.

in origin and extend throughout the length of the neural tube. Crest cells migrate laterally and give rise to **sensory ganglia (dorsal root ganglia)** of the spinal nerves and other cell types (Fig. 19.2).

During further development, neuroblasts of the sensory ganglia form two processes (Fig. 19.10*A*). The centrally growing processes penetrate the dorsal portion of the neural tube. In the spinal cord, they either end in the dorsal horn or ascend through the marginal layer to one of the higher brain centers. These processes are known collectively as the **dorsal sensory root of the spinal nerve** (Fig. 19.10*B*). The peripherally growing processes join fibers of the ventral motor roots and thus participate in formation of the trunk of the spinal nerve. Eventually these processes terminate in the sensory receptor organs. Hence, neuroblasts of the sensory ganglia derived from neural crest cells give rise to the **dorsal root neurons.**

In addition to forming sensory ganglia, cells of the neural crest differentiate into sympathetic neuroblasts, Schwann cells, pigment cells, odontoblasts, meninges, and mesenchyme of the pharyngeal arches (see Chapter 5).

Spinal Nerves

Motor nerve fibers begin to appear in the fourth week, arising from nerve cells in the basal plates (ventral horns) of the spinal cord. These fibers collect into bundles known as **ventral nerve roots** (Fig. 19.10). **Dorsal nerve roots** form as collections of fibers originating from cells in **dorsal root ganglia (spinal ganglia)**. Central processes from these ganglia form bundles that grow into the spinal cord opposite the dorsal horns. Distal processes join the ventral nerve roots to form a **spinal nerve** (Fig. 19.10). Almost immediately, spinal nerves divide into **dorsal** and **ventral primary rami.** Dorsal primary rami innervate dorsal axial musculature, vertebral joints, and the skin of the back. Ventral primary rami innervate the limbs and ventral body wall and form the major nerve plexuses (brachial and lumbosacral).

Myelination

Schwann cells myelinate the peripheral nerves. These cells originate from neural crest, migrate peripherally, and wrap themselves around axons, forming the **neurilemma sheath** (Fig. 19.12). Beginning at the fourth month of fetal life, many nerve fibers take on a whitish appearance as a result of deposition of **myelin,** which is formed by repeated coiling of the Schwann cell membrane around the axon (Fig. 19.12*C*).

The myelin sheath surrounding nerve fibers in the spinal cord has a completely different origin, the **oligodendroglial cells** (Fig. 19.12, *B* and *C*). Although myelination of nerve fibers in the spinal cord begins in approximately the fourth month of intrauterine life, some of the motor fibers descending from higher brain centers to the spinal cord do not become myelinated until the first year of postnatal life. Tracts in the nervous system become myelinated at about the time they start to function.

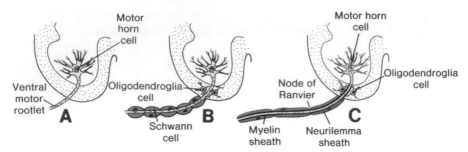

Figure 19.12 A. Motor horn cell with naked rootlet. **B.** In the spinal cord oligodendroglia cells surround the ventral rootlet; outside the spinal cord, Schwann cells begin to surround the rootlet. **C.** In the spinal cord the myelin sheath is formed by oligodendroglia cells; outside the spinal cord the sheath is formed by Schwann cells.

POSITIONAL CHANGES OF THE CORD

In the third month of development the spinal cord extends the entire length of the embryo, and spinal nerves pass through the intervertebral foramina at their level of origin (Fig. 19.13*A*). With increasing age, however, the vertebral column and dura lengthen more rapidly than the neural tube, and the terminal end of the spinal cord gradually shifts to a higher level. At birth, this end is at the level of the third lumbar vertebra (Fig. 19.13*C*). As a result of this disproportionate growth, spinal nerves run obliquely from their segment of origin in the spinal cord to the corresponding level of the vertebral column. The dura remains attached to the vertebral column at the coccygeal level.

In the adult, the spinal cord terminates at the level of L2 to L3, whereas the dural sac and subarachnoid space extend to S2. Below L2 to L3, a threadlike extension of the pia mater forms the **filum terminale,** which is attached to the periosteum of the first coccygeal vertebra and which marks the tract of regression of the spinal cord. Nerve fibers below the terminal end of the cord collectively constitute the **cauda equina.** When cerebrospinal fluid is tapped during a **lumbar puncture,** the needle is inserted at the lower lumbar level, avoiding the lower end of the cord.

Molecular Regulation of Spinal Cord Development

At the neural plate stage in the spinal cord region, the entire plate expresses the transcription factors *PAX3, PAX7, MSX1,* and *MSX2,* all of which contain a homeodomain. This expression pattern is altered by *sonic hedgehog* (*SHH*) expressed in the notochord and **bone morphogenetic proteins 4** and 7 (**BMP4** and **BMP7**) expressed in the nonneural ectoderm at the border of the neural plate (Fig. 19.14 *A*). The *SHH* signal represses expression of *PAX3* and *PAX7* and *MSX1* and *MSX2*. Thus, SHH **ventralizes** the neural tube. This ventral region then acquires the capacity to form a **floor plate,** which also expresses *SHH,* and **motor neurons** in the **basal plate.** *BMP4* and *BMP7* expression maintains

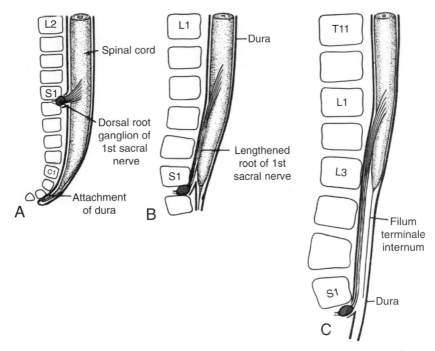

Figure 19.13 Terminal end of the spinal cord in relation to that of the vertebral column at various stages of development. **A.** Approximately the third month. **B.** End of the fifth month. **C.** Newborn.

and up-regulates *PAX3* and *PAX7* in the dorsal half of the neural tube, where **sensory neurons** in the **alar plate** will form (Fig. 19.14B). These two genes are required for formation of neural crest cells in the top of the neural folds, but their roles and those of the *MSX* genes in differentiation of sensory neurons and interneurons is not clear. However, their expression throughout the neural plate at earlier stages is essential for formation of ventral cell types, despite the fact that their expression is excluded from ventral regions by SHH at later stages. Thus, they confer on ventral cell types competence to respond appropriately to SHH and other ventralizing signals. Yet another *PAX* gene, *PAX6,* is expressed throughout the elevating neural folds except in the midline, and this pattern is maintained after fold closure. However, the role of this gene has not been determined (Fig. 19.14 *B*).

CLINICAL CORRELATES

Neural Tube Defects

Most defects of the spinal cord result from abnormal closure of the neural folds in the third and fourth weeks of development. The resulting abnormalities,

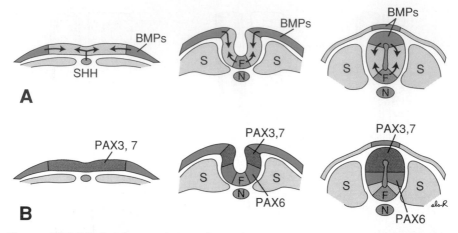

Figure 19.14 Molecular regulation of spinal cord development. **A.** Sonic hedgehog (SHH), secreted by the notochord, ventralizes the neural tube and induces a floor plate region (*F*) that also expresses this gene. Bone morphogenetic proteins 4 and 7 are secreted by the nonneural ectoderm and contribute to differentiation of the roof and alar plates. **B.** Initially, *PAX3* and 7 and *MSX 1* and 2 are expressed uniformly throughout the neural plate. SHH represses expression of these genes in the ventral half of the neural tube that will become the floor and basal plates. Simultaneously, BMPs upregulate and maintain expression of *PAX 3* and *7* in the dorsal half of the neural tube that will form the roof and alar plates. *PAX 6* begins expression throughout the neural ectoderm as the neural folds elevate and close. The exact roles of the *PAX* and *MSX* genes in differentiation of these regions have not been determined.

neural tube defects (NTDs), may involve the meninges, vertebrae, muscles, and skin. Severe NTDs involving neural and non-neural structures occur in approximately 1 in 1000 births, but the incidence varies among different populations and may be as high as 1 in 100 births in some areas, such as Northern China.

Spina bifida is a general term for NTDs affecting the spinal region. It consists of a splitting of the vertebral arches and may or may not involve underlying neural tissue. Two different types of spina bifida occur:

1) Spina bifida occulta is a defect in the vertebral arches that is covered by skin and usually does not involve underlying neural tissue (Fig. 19.15*A*). It occurs in the lumbosacral region (L4 to S1) and is usually marked by a patch of hair overlying the affected region. The defect, which is due to a lack of fusion of the vertebral arches, affects about 10% of otherwise normal people.

2) Spina bifida cystica is a severe NTD in which neural tissue and/or meninges protrude through a defect in the vertebral arches and skin to form a cystlike sac (Fig. 19.15). Most lie in the lumbosacral region and result in neurological deficits, but they are usually not associated with mental retardation. In some cases only fluid-filled meninges protrude through the defect (spina bifida with **meningocele**) (Fig. 19.15*B*); in others neural tissue is included

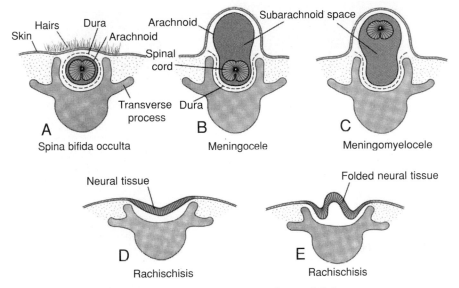

Figure 19.15 Various types of spina bifida.

in the sac (spina bifida with **meningomyelocele**) (Figs. 19.15C and 19.16A). Occasionally the neural folds do not elevate but remain as a flattened mass of neural tissue (spina bifida with **myeloschisis** or **rachischisis**) (Figs. 19.15, *D* and *E*, and 19.16B). **Hydrocephaly** develops in virtually every case of spina bifida cystica because the spinal cord is tethered to the vertebral column. As the vertebral column lengthens, tethering pulls the cerebellum into the foramen magnum, cutting off the flow of cerebrospinal fluid.

Spina bifida cystica can be diagnosed prenatally by ultrasound and by determination of α-fetoprotein (AFP) levels in maternal serum and amniotic fluid. The vertebra can be visualized by 12 weeks of gestation, and defects in closure of the vertebral arches can be detected. A new treatment for the defect is to perform surgery in utero at approximately 28 weeks of gestation. The baby is exposed by cesarean section, the defect is repaired, and the infant is placed back in the uterus. Preliminary results indicate that this approach reduces the incidence of hydrocephalus, improves bladder and bowel control, and increases motor development to the lower limbs.

Hyperthermia, valproic acid, and hypervitaminosis A produce NTDs, as do a large number of other teratogens. The origin of most NTDs is multifactorial, and the likelihood of having a child with such a defect increases significantly once one affected offspring is born. Recent evidence proves that **folic acid (folate)** reduces the incidence of NTDs by as much as 70% if 400 μg is taken daily beginning 2 months prior to conception and continuing throughout gestation.

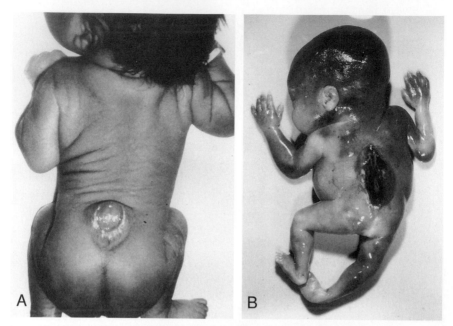

Figure 19.16 Lumbosacral region of patients with neural tube defects. **A.** Patient with a large meningomyelocele. **B.** Patient with a severe defect in which the neural folds failed to elevate throughout the lower thoracic and lumbosacral regions, resulting in rachischisis.

Brain

Distinct **basal** and **alar plates,** representing motor and sensory areas, respectively, are found on each side of the midline in the rhombencephalon and mesencephalon. In the prosencephalon, however, the alar plates are accentuated and the basal plates regress.

RHOMBENCEPHALON: HINDBRAIN

The rhombencephalon consists of the **myelencephalon,** the most caudal of the brain vesicles, and the **metencephalon,** which extends from the pontine flexure to the rhombencephalic isthmus (Figs. 19.5 and 19.17).

Myelencephalon

The myelencephalon is a brain vesicle that gives rise to the **medulla oblongata.** It differs from the spinal cord in that its lateral walls are everted (Fig. 19.18, *B* and *C*). Alar and basal plates separated by the sulcus limitans can be clearly distinguished. The basal plate, similar to that of the spinal cord, contains motor nuclei. These nuclei are divided into three groups: (*a*) a medial **somatic efferent**

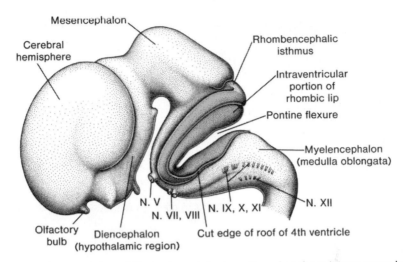

Mesencephalon

Cerebral hemisphere

Rhombencephalic isthmus

Intraventricular portion of rhombic lip

Pontine flexure

Myelencephalon (medulla oblongata)

N. V

N. VII, VIII

N. IX, X, XI

N. XII

Olfactory bulb

Diencephalon (hypothalamic region)

Cut edge of roof of 4th ventricle

Figure 19.17 Lateral view of the brain vesicles in an 8-week embryo (crown-rump length approximately 27 mm). The roof plate of the rhombencephalon has been removed to show the intraventricular portion of the rhombic lip. Note the origin of the cranial nerves.

group, (*b*) an intermediate **special visceral efferent** group, and (*c*) a lateral **general visceral efferent** group (Fig. 19.18*C*).

The first group contains motor neurons, which form the **cephalic continuation of the anterior horn cells.** Since this somatic efferent group continues rostrally into the mesencephalon, it is called the **somatic efferent motor column.** In the myelencephalon it includes neurons of the **hypoglossal nerve** that supply the tongue musculature. In the metencephalon and the mesencephalon, the column contains neurons of the **abducens** (Fig. 19.19), **trochlear,** and **oculomotor nerves** (see Fig. 19.23), respectively. These nerves supply the eye musculature.

The **special visceral efferent** group extends into the metencephalon, forming the **special visceral efferent motor column.** Its motor neurons supply **striated muscles** of the pharyngeal arches. In the myelencephalon the column is represented by neurons of the **accessory, vagus, and glossopharyngeal nerves.**

The **general visceral efferent** group contains motor neurons that supply **involuntary musculature** of the respiratory tract, intestinal tract, and heart.

The alar plate contains three groups of **sensory relay nuclei** (see Fig. 19.18*C*). The most lateral of these, the **somatic afferent** (sensory) group, receives impulses from the ear and surface of the head by way of the **vestibulocochlear** and **trigeminal nerves.** The intermediate, or **special visceral afferent,** group receives impulses from taste buds of the tongue and from the palate, oropharynx, and epiglottis. The medial, or **general visceral afferent,** group receives interoceptive information from the gastrointestinal tract and heart.

The roof plate of the myelencephalon consists of a single layer of ependymal cells covered by vascular mesenchyme, the **pia mater** (Figs. 19.5 and

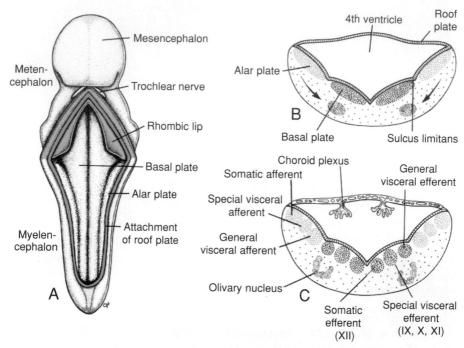

Figure 19.18 A. Dorsal view of the floor of the fourth ventricle in a 6-week embryo after removal of the roof plate. Note the alar and basal plates in the myelencephalon. The rhombic lip is visible in the metencephalon. **B** and **C.** Position and differentiation of the basal and alar plates of the myelencephalon at different stages of development. Note formation of the nuclear groups in the basal and alar plates. *Arrows,* path followed by cells of the alar plate to the olivary nuclear complex. The choroid plexus produces cerebrospinal fluid.

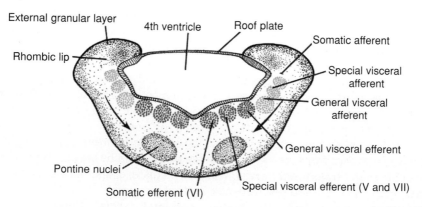

Figure 19.19 Transverse section through the caudal part of the metencephalon. Note the differentiation of the various motor and sensory nuclear areas in the basal and alar plates, respectively, and the position of the rhombic lips, which project partly into the lumen of the fourth ventricle and partly above the attachment of the roof plate. *Arrows,* direction of migration of the pontine nuclei.

19.18*B*). The two combined are known as the **tela choroidea**. Because of active proliferation of the vascular mesenchyme, a number of saclike invaginations project into the underlying ventricular cavity (Figs. 19.18*C* and 19.20*D*). These tuftlike invaginations form the **choroid plexus,** which produces cerebrospinal fluid.

Metencephalon

The metencephalon, similar to the myelencephalon, is characterized by basal and alar plates (Fig. 19.19). Two new components form: (*a*) the **cerebellum,** a coordination center for posture and movement (Fig. 19.20), and (*b*) the **pons,** the pathway for nerve fibers between the spinal cord and the cerebral and cerebellar cortices.

Each basal plate of the metencephalon (Fig. 19.19) contains three groups of motor neurons: (*a*) the medial **somatic efferent** group, which gives rise to

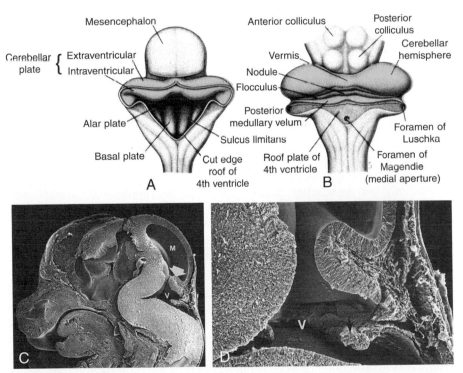

Figure 19.20 A. Dorsal view of the mesencephalon and rhombencephalon in an 8-week embryo. The roof of the fourth ventricle has been removed, allowing a view of its floor. **B.** Similar view in a 4-month embryo. Note the choroidal fissure and the lateral and medial apertures in the roof of the fourth ventricle. **C.** Scanning electron micrograph of a mouse embryo at a slightly younger stage than in **A,** showing the cerebellar primordium (*arrow*) extending into the fourth ventricle (*V*). M, mesencephalon. **D.** High magnification of the cerebellar region in **C.** Choroid plexus (*arrow*) in the roof of the fourth ventricle (*V*).

the nucleus of the **abducens nerve;** (*b*) the **special visceral efferent** group, containing nuclei of the **trigeminal** and **facial nerves,** which innervate the musculature of the first and second pharyngeal arches; and (*c*) the **general visceral efferent** group, whose axons supply the submandibular and sublingual glands.

The marginal layer of the basal plates of the metencephalon expands as it makes a bridge for nerve fibers connecting the cerebral cortex and cerebellar cortex with the spinal cord. Hence this portion of the metencephalon is known as the **pons** (bridge). In addition to nerve fibers, the pons contains the **pontine nuclei,** which originate in the alar plates of the metencephalon and myelencephalon (*arrows,* Fig. 19.19).

The alar plates of the metencephalon contain three groups of sensory nuclei: (*a*) a lateral **somatic afferent** group, which contains neurons **of the trigeminal nerve** and a small portion of the **vestibulocochlear complex,** (*b*) the **special visceral afferent** group, and (*c*) the **general visceral afferent** group (Fig. 19.19).

Cerebellum

The dorsolateral parts of the alar plates bend medially and form the **rhombic lips** (Fig. 19.18). In the caudal portion of the metencephalon, the rhombic lips are widely separated, but immediately below the mesencephalon they approach each other in the midline (Fig. 19.20). As a result of a further deepening of the pontine flexure, the rhombic lips compress cephalocaudally and form the **cerebellar plate** (Fig. 19.20). In a 12-week embryo, this plate shows a small midline portion, the **vermis,** and two lateral portions, the **hemispheres.** A transverse fissure soon separates the **nodule** from the vermis and the lateral **flocculus** from the hemispheres (Fig. 19.20*B*). This **flocculonodular** lobe is phylogenetically the most primitive part of the cerebellum.

Initially, the **cerebellar plate** consists of neuroepithelial, mantle, and marginal layers (Fig. 19.21*A*). During further development, a number of cells formed by the neuroepithelium migrate to the surface of the cerebellum to form the **external granular layer.** Cells of this layer retain their ability to divide and form a proliferative zone on the surface of the cerebellum (Fig. 19.21, *B* and *C*).

In the sixth month of development, the external granular layer gives rise to various cell types. These cells migrate toward the differentiating Purkinje cells (Fig. 19.22) and give rise to **granule cells. Basket** and **stellate cells** are produced by proliferating cells in the cerebellar white matter. The cortex of the cerebellum, consisting of Purkinje cells, Golgi II neurons, and neurons produced by the external granular layer, reaches its definitive size after birth (Fig. 19.22*B*). The deep cerebellar nuclei, such as the **dentate nucleus,** reach their final position before birth (Fig. 19.21*D*).

MESENCEPHALON: MIDBRAIN

In the mesencephalon (Fig. 19.23), each basal plate contains two groups of motor nuclei: (*a*) a medial **somatic efferent** group, represented by the

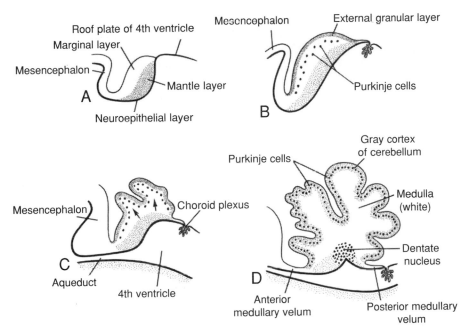

Figure 19.21 Sagittal sections through the roof of the metencephalon showing development of the cerebellum. **A.** 8 weeks (approximately 30 mm). **B.** 12 weeks (70 mm). **C.** 13 weeks. **D.** 15 weeks. Note formation of the external granular layer on the surface of the cerebellar plate (**B** and **C**). During later stages, cells of the external granular layer migrate inward to mingle with Purkinje cells and form the definitive cortex of the cerebellum. The dentate nucleus is one of the deep cerebellar nuclei. Note the anterior and posterior velum.

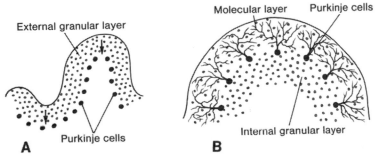

Figure 19.22 Stages in development of the cerebellar cortex. **A.** The external granular layer on the surface of the cerebellum forms a proliferative layer from which granule cells arise. They migrate inward from the surface (*arrows*). Basket and stellate cells derive from proliferating cells in the cerebellar white matter. **B.** Postnatal cerebellar cortex showing differentiated Purkinje cells, the molecular layer on the surface, and the internal granular layer beneath the Purkinje cells.

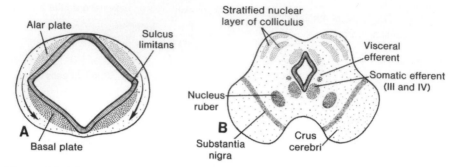

Figure 19.23 A and **B.** Position and differentiation of the basal and alar plates in the mesencephalon at various stages of development. *Arrows* in **A** indicate the path followed by cells of the alar plate to form the nucleus ruber and substantia nigra. Note the various motor nuclei in the basal plate.

oculomotor and **trochlear nerves,** which innervate the eye musculature, and (*b*) a small **general visceral efferent** group, represented by the **nucleus of Edinger-Westphal,** which innervates the **sphincter pupillary muscle** (Fig. 19.23*B*). The marginal layer of each basal plate enlarges and forms the **crus cerebri.** These crura serve as pathways for nerve fibers descending from the cerebral cortex to lower centers in the pons and spinal cord. Initially the alar plates of the mesencephalon appear as two longitudinal elevations separated by a shallow midline depression (Fig. 19.23). With further development, a transverse groove divides each elevation into an **anterior** (superior) and a **posterior** (inferior) **colliculus** (Fig. 19.23*B*). The posterior colliculi serve as synaptic relay stations for auditory reflexes; the anterior colliculi function as correlation and reflex centers for visual impulses. The colliculi are formed by waves of neuroblasts migrating into the overlying marginal zone. Here they are arranged in layers (Fig. 19.23*B*).

PROSENCEPHALON: FOREBRAIN

The **prosencephalon** consists of the **telencephalon,** which forms the cerebral hemispheres, and the **diencephalon,** which forms the optic cup and stalk, pituitary, thalamus, hypothalamus, and epiphysis.

Diencephalon

Roof Plate and Epiphysis. The diencephalon, which develops from the median portion of the prosencephalon (Figs. 19.5 and 19.17), is thought to consist of a roof plate and two alar plates but to lack floor and basal plates (interestingly, *sonic hedgehog,* a ventral midline marker, is expressed in the floor of the diencephalon, suggesting that a floor plate does exist). The roof plate of the diencephalon consists of a single layer of ependymal cells covered by

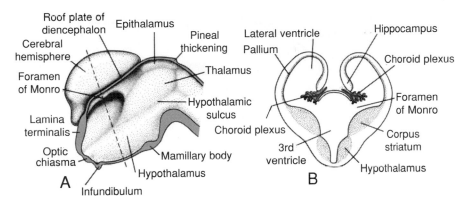

Figure 19.24 A. Medial surface of the right half of the prosencephalon in a 7-week embryo. **B.** Transverse section through the prosencephalon at the level of the *broken line* in **A.** The corpus striatum bulges out in the floor of the lateral ventricle and the foramen of Monro.

vascular mesenchyme. Together these layers give rise to the **choroid plexus** of the third ventricle (see Fig. 19.30). The most caudal part of the roof plate develops into the **pineal body,** or **epiphysis.** This body initially appears as an epithelial thickening in the midline, but by the seventh week it begins to evaginate (Figs. 19.24 and 19.25). Eventually it becomes a solid organ on the roof of the mesencephalon (see Fig. 19.30) that serves as a channel through which light and darkness affect endocrine and behavioral rhythms. In the adult, calcium is frequently deposited in the epiphysis and then serves as a landmark on radiographs of the skull.

Alar Plate, Thalamus, and Hypothalamus. The alar plates form the lateral walls of the diencephalon. A groove, the **hypothalamic sulcus,** divides the plate into a dorsal and a ventral region, the **thalamus** and **hypothalamus,** respectively (Figs. 19.24 and 19.25).

As a result of proliferative activity, the thalamus gradually projects into the lumen of the diencephalon. Frequently this expansion is so great that thalamic regions from the right and left sides fuse in the midline, forming the **massa intermedia,** or **interthalamic connexus.**

The hypothalamus, forming the lower portion of the alar plate, differentiates into a number of nuclear areas that regulate the visceral functions, including sleep, digestion, body temperature, and emotional behavior. One of these groups, the **mamillary body,** forms a distinct protuberance on the ventral surface of the hypothalamus on each side of the midline (Figs. 19.24*A* and 19.25*A*).

Hypophysis or Pituitary Gland. The hypophysis, or pituitary gland, develops from two completely different parts: (*a*) an ectodermal outpocketing of the stomodeum immediately in front of the buccopharyngeal membrane, known

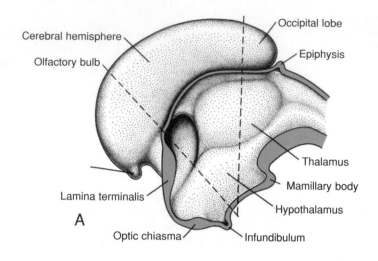

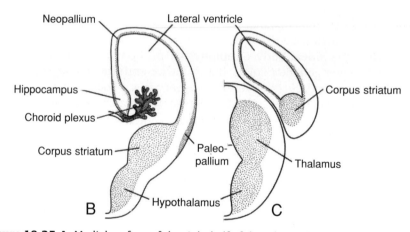

Figure 19.25 A. Medial surface of the right half of the telencephalon and diencephalon in an 8-week embryo. **B** and **C.** Transverse sections through the right half of the telencephalon and diencephalon at the level of the *broken lines* in **A.**

as **Rathke's pouch,** and (*b*) a downward extension of the diencephalon, the **infundibulum** (Fig. 19.26, *A* and *D*).

When the embryo is approximately 3 weeks old, Rathke's pouch appears as an evagination of the oral cavity and subsequently grows dorsally toward the infundibulum. By the end of the second month it loses its connection with the oral cavity and is then in close contact with the infundibulum.

During further development, cells in the anterior wall of Rathke's pouch increase rapidly in number and form the **anterior lobe of the hypophysis,** or **adenohypophysis** (Fig. 19.26*B*). A small extension of this lobe,

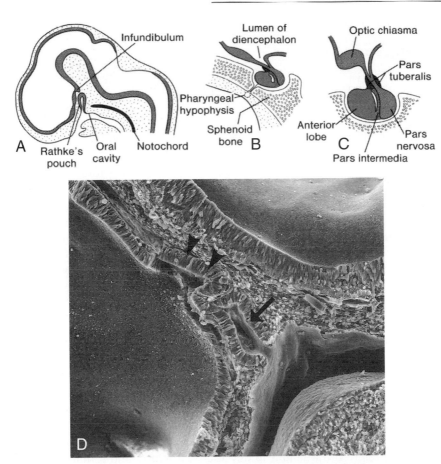

Figure 19.26 A. Sagittal section through the cephalic part of a 6-week embryo showing Rathke's pouch as a dorsal outpocketing of the oral cavity and the infundibulum as a thickening in the floor of the diencephalon. **B** and **C.** Sagittal sections through the developing hypophysis in the 11th and 16th weeks of development, respectively. Note formation of the pars tuberalis encircling the stalk of the pars nervosa. **D.** High-magnification scanning electron micrograph of the region of the developing hypophysis similar to that in **A.** Rathke's pouch (*arrow*) and the infundibulum (*arrowheads*) are visible.

the **pars tuberalis,** grows along the stalk of the infundibulum and eventually surrounds it (Fig. 19.26*C*). The posterior wall of Rathke's pouch develops into the **pars intermedia,** which in humans seems to have little significance.

The infundibulum gives rise to the **stalk** and the **pars nervosa,** or **posterior lobe of the hypophysis** (neurohypophysis) (Fig. 19.26*C*). It is composed of neuroglial cells. In addition, it contains a number of nerve fibers from the hypothalamic area.

CLINICAL CORRELATES

Hypophyseal Defects

Occasionally a small portion of Rathke's pouch persists in the roof of the pharynx as a **pharyngeal hypophysis. Craniopharyngiomas** arise from remnants of Rathke's pouch. They may form within the sella turcica or along the stalk of the pituitary but usually lie above the sella. They may cause hydrocephalus and pituitary dysfunction (e.g., diabetes insipidus, growth failure).

Telencephalon

The telencephalon, the most rostral of the brain vesicles, consists of two lateral outpocketings, the **cerebral hemispheres,** and a median portion, the **lamina terminales** (Figs. 19.4, 19.5, 19.24, and 19.25). The cavities of the hemispheres, the **lateral ventricles,** communicate with the lumen of the diencephalon through the **interventricular foramina of Monro** (Fig. 19.24).

Cerebral Hemispheres. The cerebral hemispheres arise at the beginning of the fifth week of development as bilateral evaginations of the lateral wall of the prosencephalon (Fig. 19.24). By the middle of the second month the basal part of the hemispheres (i.e., the part that initially formed the forward extension of the thalamus) (Fig. 19.24*A*) begins to grow and bulges into the lumen of the lateral ventricle and into the floor of the foramen of Monro (Figs. 19.24*B* and 19.25, *A* and *B*). In transverse sections, the rapidly growing region has a striated appearance and is therefore known as the **corpus striatum** (Fig. 19.25*B*).

In the region where the wall of the hemisphere is attached to the roof of the diencephalon, the wall fails to develop neuroblasts and remains very thin (Fig. 19.24*B*). Here the hemisphere wall consists of a single layer of ependymal cells covered by vascular mesenchyme, and together they form the **choroid plexus.** The choroid plexus should have formed the roof of the hemisphere, but as a result of the disproportionate growth of the various parts of the hemisphere, it protrudes into the lateral ventricle along the **choroidal fissure** (Figs. 19.25 and 19.27). Immediately above the choroidal fissure, the wall of the hemisphere thickens, forming the **hippocampus** (Figs. 19.24*B* and 19.25*B*). This structure, whose primary function is olfaction, bulges into the lateral ventricle.

With further expansion, the hemispheres cover the lateral aspect of the diencephalon, mesencephalon, and cephalic portion of the metencephalon (Figs. 19.27 and 19.28). The corpus striatum (Fig. 19.24*B*), being a part of the wall of the hemisphere, likewise expands posteriorly and is divided into two parts: (*a*) a dorsomedial portion, the **caudate nucleus,** and (*b*) a ventrolateral portion, the **lentiform nucleus** (Fig. 19.27*B*). This division is accomplished by axons passing to and from the cortex of the hemisphere and breaking through the nuclear mass of the corpus striatum. The fiber bundle thus formed is known as the **internal capsule** (Fig. 19.27*B*). At the same time, the medial wall of the

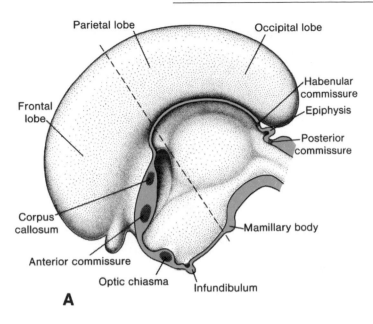

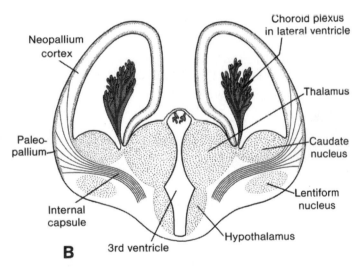

Figure 19.27 A. Medial surface of the right half of the telencephalon and diencephalon in a 10-week embryo. **B.** Transverse section through the hemisphere and diencephalon at the level of the *broken line* in **A.**

hemisphere and the lateral wall of the diencephalon fuse, and the caudate nucleus and thalamus come into close contact (Fig. 19.27*B*).

Continuous growth of the cerebral hemispheres in anterior, dorsal, and inferior directions results in the formation of frontal, temporal, and occipital lobes, respectively. As growth in the region overlying the corpus striatum slows,

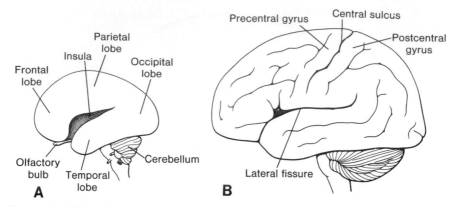

Figure 19.28 Development of gyri and sulci on the lateral surface of the cerebral hemisphere. **A.** 7 months. **B.** 9 months.

however, the area between the frontal and temporal lobes becomes depressed and is known as the **insula** (Fig. 19.28*A*). This region is later overgrown by the adjacent lobes and at the time of birth is almost completely covered. During the final part of fetal life, the surface of the cerebral hemispheres grows so rapidly that a great many convolutions (**gyri**) separated by fissures and sulci appear on its surface (Fig. 19.28*B*).

Cortex Development. The cerebral cortex develops from the pallium (Fig. 19.24), which has two regions: (*a*) the **paleopallium,** or **archipallium,** immediately lateral to the corpus striatum (Fig. 19.25*B*), and (*b*) the **neopallium,** between the hippocampus and the paleopallium (Figs. 19.25*B* and 19.27*B*).

In the neopallium, waves of neuroblasts migrate to a subpial position and then differentiate into fully mature neurons. When the next wave of neuroblasts arrives, they migrate through the earlier-formed layers of cells until they reach the subpial position. Hence the early-formed neuroblasts obtain a deep position in the cortex, while those formed later obtain a more superficial position.

At birth the cortex has a stratified appearance due to differentiation of the cells in layers. The motor cortex contains a large number of **pyramidal cells,** and the sensory areas are characterized by **granular cells.**

Olfactory Bulbs. Differentiation of the olfactory system is dependent upon epithelial-mesenchymal interactions. These occur between neural crest cells and ectoderm of the frontonasal prominence to form the **olfactory placodes** (Fig. 19.29) and between these same crest cells and the floor of the telencephalon to form the **olfactory bulbs.** Cells in the nasal placodes differentiate into primary sensory neurons of the nasal epithelium whose axons grow and make contact with secondary neurons in the developing olfactory bulbs (Fig. 19.29). By the seventh week, these contacts are well established. As growth of the brain continues, the olfactory bulbs and the olfactory tracts of the secondary neurons lengthen, and together they constitute the olfactory nerve (Fig. 19.30).

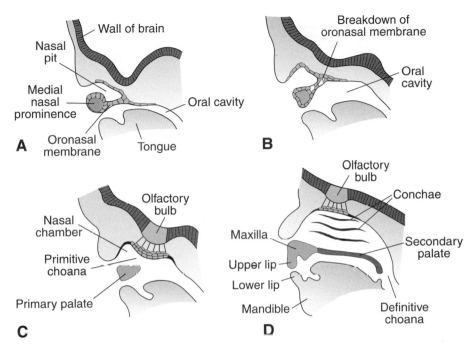

Figure 19.29 A. Sagittal section through the nasal pit and lower rim of the medial nasal prominence of a 6-week embryo. The primitive nasal cavity is separated from the oral cavity by the oronasal membrane. **B.** Similar section as in **A** toward the end of the sixth week showing breakdown of the oronasal membrane. **C.** At 7 weeks, neurons in the nasal epithelium have extended processes that contact the floor of the telencephalon in the region of the developing olfactory bulbs. **D.** By 9 weeks, definitive oronasal structures have formed, neurons in the nasal epithelium are well differentiated, and secondary neurons from the olfactory bulbs to the brain begin to lengthen. Togther, the olfactory bulbs and tracts of the secondary neurons constitute the olfactory nerve (see Fig. 19.30).

Commissures. In the adult, a number of fiber bundles, the **commissures**, which cross the midline, connect the right and left halves of the hemispheres. The most important fiber bundles make use of the **lamina terminalis** (Figs. 19.24, 19.27, and 19.30). The first of the crossing bundles to appear is the **anterior commissure.** It consists of fibers connecting the olfactory bulb and related brain areas of one hemisphere to those of the opposite side (Figs. 19.27 and 19.30).

The second commissure to appear is the **hippocampal commissure,** or **fornix commissure.** Its fibers arise in the hippocampus and converge on the lamina terminalis close to the roof plate of the diencephalon. From here the fibers continue, forming an arching system immediately outside the choroid fissure, to the mamillary body and the hypothalamus.

The most important commissure is the **corpus callosum.** It appears by the 10th week of development and connects the nonolfactory areas of the right and the left cerebral cortex. Initially, it forms a small bundle in the lamina terminalis. As a result of continuous expansion of the neopallium, however, it extends first

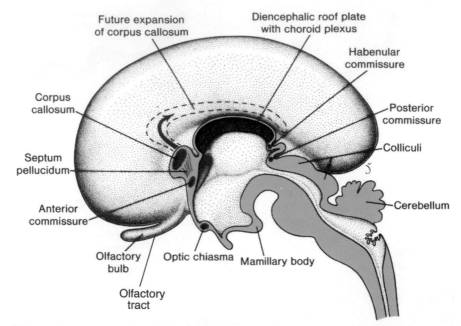

Figure 19.30 Medial surface of the right half of the brain in a 4-month embryo showing the various commissures. *Broken line,* future site of the corpus callosum. The hippocampal commissure is not indicated.

anteriorly and then posteriorly, arching over the thin roof of the diencephalon (Fig. 19.30).

In addition to the three commissures developing in the lamina terminalis, three more appear. Two of these, the **posterior** and **habenular commissures,** are just below and rostral to the stalk of the pineal gland. The third, the **optic chiasma,** which appears in the rostral wall of the diencephalon, contains fibers from the medial halves of the retinae (Fig. 19.30).

Molecular Regulation of Brain Development

Anteroposterior (craniocaudal) patterning of the central nervous system begins early in development, during gastrulation and neural induction (see Chapters 4 and 5). Once the neural plate is established, signals for segregation of the brain into forebrain, midbrain, and hindbrain regions are derived from **homeobox** genes expressed in the notochord, prechordal plate, and neural plate. The hindbrain has eight segments, the **rhombomeres,** that have variable expression patterns of the *Antennapedia* class of homeobox genes, the ***HOX*** genes (see Chapter 5). These genes are expressed in overlapping patterns, with genes at the most 3' end of a cluster having more anterior boundaries and paralogous genes having identical expression domains (Fig. 19.31). Genes at the 3' end are also expressed earlier than those at the 5' end, so that a temporal relation

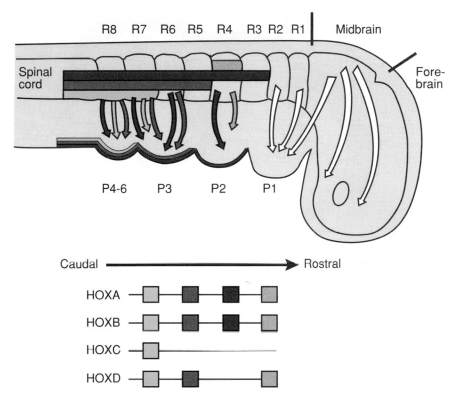

Figure 19.31 Patterns of *HOX* gene expression in the hindbrain and the pattern of neural crest cell migration into the pharyngeal arches. *HOX* genes are expressed in overlapping patterns ending at specific rhombomere boundaries. Genes at the 3' end of a cluster have the most anterior boundaries, and paralogous genes have identical expression domains. These genes confer positional value along the anterior-posterior axis of the hindbrain, determine the identity of the rhombomeres, and specify their derivatives.

to the expression pattern is established. These genes, then, confer positional value along the anteroposterior axis of the hindbrain, determine the identity of the rhombomeres, and specify their derivatives. How this regulation occurs is not clear, although the **retinoids (retinoic acid)** play a critical role in regulating *HOX* expression. For example, excess retinoic acid shifts *HOX* gene expression anteriorly and causes more cranial rhombomeres to differentiate into more caudal types. Retinoic acid deficiency results in a small hindbrain. There is also a differential response to retinoic acid by the *HOX* genes; those at the 3' end of the cluster are more sensitive than those at the 5' end.

Specification of the forebrain and midbrain areas is also regulated by genes containing a homeodomain. However, these genes are not of the *Antennapedia* class, whose most anterior boundary of expression stops at rhombomere 3. Thus, new genes have assumed the patterning role for these regions of the

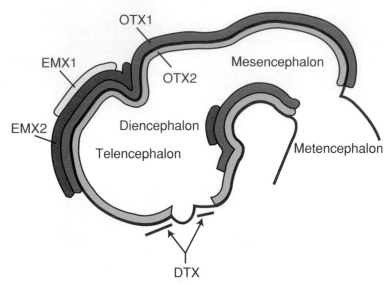

OTX1

EMX1

OTX2

Mesencephalon

EMX2

Diencephalon

Telencephalon

Metencephalon

DTX

Figure 19.32 Overlapping expression patterns of homeobox genes that specify the identities of the forebrain and midbrain regions.

brain, which evolutionarily constitute the "new head." At the neural plate stage, ***LIM1,*** expressed in the prechordal plate, and ***OTX2,*** expressed in the neural plate, are important for designating the forebrain and midbrain areas, with *LIM1* supporting *OTX2* expression. (These genes are also expressed at the earliest stages of gastrulation, and they assist in specifying the entire cranial region of the epiblast.) Once the neural folds and pharyngeal arches appear, additional **homeobox** genes, including *OTX1, EMX1,* and *EMX2* are expressed in specific and in overlapping (nested) patterns (Fig. 19.32) that specify the identity of the forebrain and midbrain regions. Once these boundaries are established, two additional organizing centers appear: the **anterior neural ridge (ANR)** at the junction of the cranial border of the neural plate and nonneural ectoderm (Fig. 19.33) and the **isthmus** (Fig. 19.34) between the hindbrain and midbrain. In both locations, **fibroblast growth factor-8 (FGF-8)** is the key signaling molecule, inducing subsequent gene expression that regulates differentiation. In the ANR at the four-somite stage, FGF-8 induces expression of ***brain factor 1*** (***BF1***; Fig. 19.33). *BF1* then regulates development of the telencephalon (cerebral hemispheres) and regional specification within the forebrain, including the basal telencephalon and the retina. In the isthmus at the junction between the midbrain and hindbrain territories, *FGF-8* is expressed in a ring around the circumference of this location (Fig. 19.34). FGF-8 induces expression of ***engrailed 1*** and ***2*** (***EN1*** and ***EN2***), two homeobox-containing genes, expressed in gradients radiating anteriorly and posteriorly from the isthmus. *EN1* regulates development throughout its expression domain, including the dorsal midbrain (tectum) and anterior hindbrain (cerebellum), whereas *EN2* is involved only in cerebellar development. FGF-8 also induces ***WNT1*** expression in a

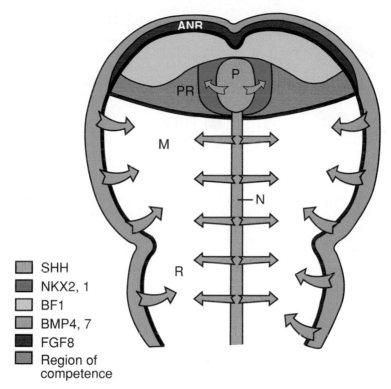

Figure 19.33 Diagram illustrating the organizing center known as the anterior neural ridge (*ANR*). This area lies in the most anterior region of the neural plate and secretes FGF-8, which induces expression of *brain factor 1* (*BF1*) in adjacent neurectoderm. *BF1* regulates development of the telencephalon (cerebral hemispheres) and regional specification within the prosencephalon (*PR*). Sonic hedgehog (SHH), secreted by the prechordal plate *(P)* and notochord *(N)*, ventralizes the brain and induces expression of *NKX2.1*, which regulates development of the hypothalamus. Bone morphogenetic proteins 4 and 7, secreted by the adjacent nonneural ectoderm, control dorsal patterning of the brain. *M*, mesencephalon; *R*, rhombencephalon.

circumferential band anterior to the region of *FGF-8* expression (Fig. 19.34). *WNT1* interacts with *EN1* and *EN2* to regulate development of this region, including the cerebellum. In fact, WNT1 may assist in early specification of the midbrain area since it is expressed in this region at the neural plate stage. *FGF-8* is also expressed at this early time in mesoderm underlying the midbrain-hindbrain junction and may therefore regulate *WNT1* expression and initial patterning of this region. The constriction for the isthmus is slightly posterior to the actual midbrain-hindbrain junction, which lies at the caudal limit of *OTX2* expression (Fig. 19.32).

Dorsoventral (mediolateral) patterning also occurs in the forebrain and midbrain areas. Ventral patterning is controlled by **SHH** just as it is throughout the remainder of the central nervous system. SHH, secreted by the prechordal plate, induces expression of **NKX2.1,** a homeodomain-containing gene that regulates

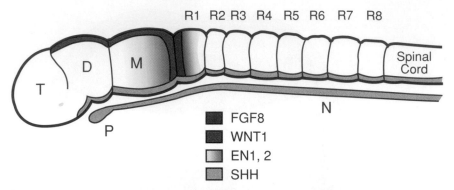

Figure 19.34 Organizing center in the isthmus at the boundaries between the midbrain and hindbrain. This region secretes FGF-8 in a circumferential ring that induces expression of *engrailed 1* and *2(EN1* and *EN2)* in gradients anteriorly and posteriorly from this area. *EN1* regulates development of the dorsal midbrain, and both genes participate in formation of the cerebellum. *WNT1,* another gene induced by FGF-8, also assists in development of the cerebellum. *N,* notochord; *P,* prechordal plate.

development of the hypothalamus. Interestingly, SHH signaling requires cleavage of the protein. The carboxy terminal portion executes this process, which also includes covalent linkage of **cholesterol** to the carboxy terminus of the amino terminal product. The amino terminal portion retains all of the signaling properties of SHH, and its association with cholesterol assists in its distribution.

Dorsal (lateral) patterning of the neural tube is controlled by **bone morphogenetic proteins 4** and **7** (**BMP4** and **BMP7**) expressed in the nonneural ectoderm adjacent to the neural plate. These proteins induce expression of *MSX1* in the midline and repress expression of *BF1* (Fig. 19.33).

Expression patterns of genes regulating anterior-posterior (craniocaudal) and dorsoventral (mediolateral) patterning of the brain overlap and interact at the borders of these regions. Furthermore, various brain regions are competent to respond to specific signals and not to others. For example, only the cranial part of the neural plate expresses *NKX2.1* in response to SHH. Likewise, only the anterior neural plate produces BF1 in response to FGF-8; midbrain levels express EN2 in response to the same FGF-8 signal. Thus, a **competence to respond** also assists in specifying regional differences.

CLINICAL CORRELATES

Cranial Defects

Holoprosencephaly (HPE) refers to a spectrum of abnormalities in which a loss of midline structures results in malformations of the brain and face. In severe cases, the lateral ventricles merge into a single **telencephalic vesicle (alobar HPE),** the eyes are fused, and there is a single nasal chamber along

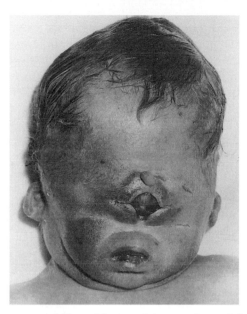

Figure 19.35 Holoprosencephaly and fusion of the eyes (synophthalmia). A loss of the midline in the brain causes the lateral ventricles to merge into a single chamber and the eye fields to fail to separate. Mutations in the gene *sonic hedgehog* (*SHH*), which specifies the midline of the central nervous system at neural plate stages, is one cause for this spectrum of abnormalities.

with other midline facial defects (Fig. 19.35). In less severe cases, some division of the prosencephalon into two cerebral hemispheres occurs, but there is incomplete development of midline structures. Usually the olfactory bulbs and tracts and the corpus callosum are hypoplastic or absent. In very mild cases, sometimes the only indication that some degree of HPE has occurred is the presence of a single central incisor. HPE occurs in 1 in 15,000 live births, but is present in 1 in 250 pregnancies that end in early miscarriage. Mutations in **SHH,** the gene that regulates establishment of the ventral midline in the CNS, result in some forms of holoprosencephaly. Another cause is defective **cholesterol biosynthesis** leading to **Smith-Lemli-Opitz syndrome.** These children have craniofacial and limb defects, and 5 % have holoprosencephaly. Smith-Lemli-Opitz syndrome is due to abnormalities in **7-dehydrocholesterol reductase,** which metabolizes 7-dehydrocholesterol to cholesterol. Many of the defects, including those of the limbs and brain, may be due to abnormal *SHH* signaling, since cholesterol is necessary for this gene to exert its effects (see page 466). Other genetic causes include mutations in the transcription factors **sine occulis homeobox3 (SIX3), TG interacting factor (TGIF)** and the **zinc finger protein ZIC2.** Yet another cause of holoprosencephaly is alcohol abuse, which at early stages of development selectively kills midline cells.

Schizencephaly is a rare disorder in which large clefts occur in the cerebral hemispheres, sometimes causing a loss of brain tissue. Mutations in the homeobox gene *EMX2* appear to account for some of these cases.

Meningocele, meningoencephalocele, and **meningohydroencephalocele** are all caused by an ossification defect in the bones of the skull. The most frequently affected bone is the squamous part of the occipital bone, which may be partially or totally lacking. If the opening of the occipital bone is small, only meninges bulge through it (**meningocele**), but if the defect is large, part of the brain and even part of the ventricle may penetrate through the opening into the meningeal sac (Figs. 19.36 and 19.37). The latter two malformations are known as **meningoencephalocele** and **meningohydroencephalocele,** respectively. These defects occur in 1/2000 births.

Exencephaly is characterized by failure of the cephalic part of the neural tube to close. As a result, the vault of the skull does not form, leaving the malformed brain exposed. Later this tissue degenerates, leaving a mass of necrotic tissue. This defect is called **anencephaly,** although the brainstem remains intact (Fig. 19.38, *A* and *B*). Since the fetus lacks the mechanism for swallowing, the last 2 months of pregnancy are characterized by **hydramnios.** The abnormality can be recognized on a radiograph, since the vault of the skull is absent. Anencephaly is a common abnormality (1/1500) that occurs 4 times more often in females than in males. Like spina bifida, up to 70% of these cases can be prevented by having women take 400 μg of folic acid per day before and during pregnancy.

Hydrocephalus is characterized by an abnormal accumulation of cerebrospinal fluid within the ventricular system. In most cases, hydrocephalus in the newborn is due to an obstruction of the **aqueduct of Sylvius (aqueductal stenosis).** This prevents the cerebrospinal fluid of the lateral and third ventricles from passing into the fourth ventricle and from there into the subarachnoid space, where it would be resorbed. As a result, fluid accumulates

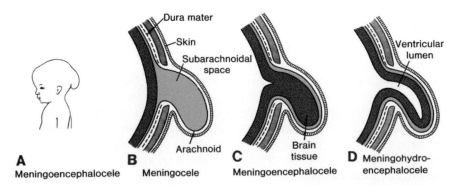

Figure 19.36 A–D. Various types of brain herniation due to abnormal ossification of the skull.

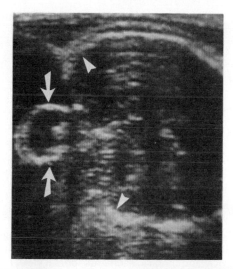

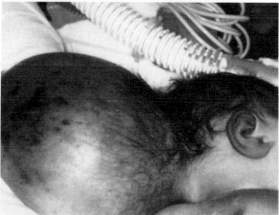

Figure 19.37 Ultrasonogram (**top**) and photograph (**bottom**) of a child with a meningoencephalocele. The defect was detected by ultrasound in the seventh month of gestation and repaired after birth. Ultrasound shows brain tissue (*arrows*) extending through the bony defect in the skull (*arrowheads*).

in the lateral ventricles and presses on the brain and bones of the skull. Since the cranial sutures have not yet fused, spaces between them widen as the head expands. In extreme cases, brain tissue and bones become thin and the head may be very large (Fig. 19.39).

The **Arnold-Chiari malformation** is caudal displacement and herniation of cerebellar structures through the foramen magnum. Arnold-Chiari malformation occurs in virtually every case of spina bifida cystica and is usually accompanied by hydrocephalus.

Microcephaly describes a cranial vault that is smaller than normal (Fig. 19.40). Since the size of the cranium depends on growth of the brain, the

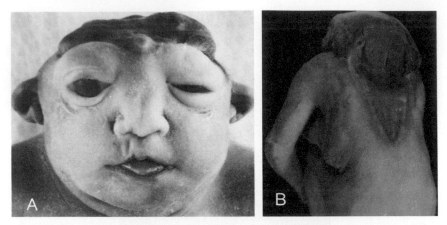

Figure 19.38 A. Anencephalic child, ventral view. This abnormality occurs frequently (1/1500 births). Usually the child dies a few days after birth. **B.** Anencephalic child with spina bifida in the cervical and thoracic segments, dorsal view.

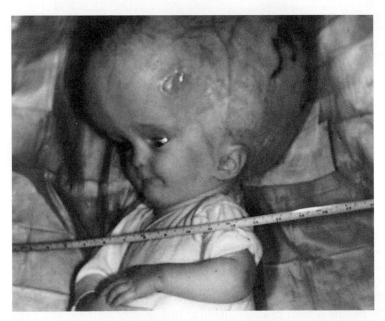

Figure 19.39 Child with severe hydrocephalus. Since the cranial sutures had not closed, pressure from the accumulated cerebrospinal fluid enlarged the head, thinning the bones of the skull and cerebral cortex.

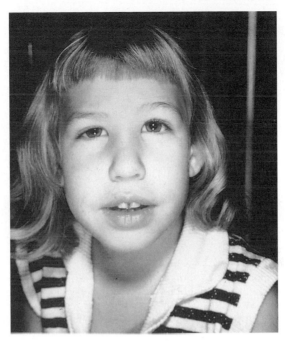

Figure 19.40 Child with microcephaly. This abnormality, due to poor growth of the brain, is frequently associated with mental retardation.

underlying defect is in brain development. Causation of the abnormality is varied; it may be genetic (autosomal recessive) or due to prenatal insults such as infection or exposure to drugs or other teratogens. Impaired mental development occurs in more than half of cases.

Fetal infection by toxoplasmosis may result in cerebral calcification, mental retardation, hydrocephalus, or microcephaly. Likewise, exposure to radiation during the early stages of development may produce microcephaly. Hyperthermia produced by maternal infection or by sauna baths may cause spina bifida and exencephaly.

The aforementioned abnormalities are the most serious ones, and they may be incompatible with life. A great many other defects of the CNS may occur without much external manifestation. For example, the **corpus callosum** may be partially or completely absent without much functional disturbance. Likewise, partial or complete absence of the cerebellum may result in only a slight disturbance of coordination. On the other hand, cases of severe **mental retardation** may not be associated with morphologically detectable brain abnormalities. Mental retardation may result from genetic abnormalities (e.g., Down and Klinefelter syndromes) or from exposures to teratogens, including infectious agents (rubella, cytomegalovirus, toxoplasmosis). The leading cause of mental retardation is, however, **maternal alcohol abuse.**

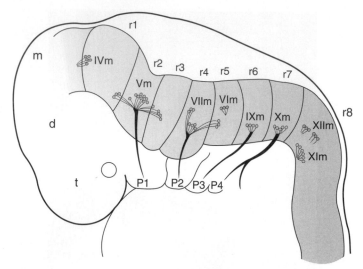

Figure 19.41 Segmentation patterns in the brain and mesoderm that appear by the 25th day of development. The hindbrain (*coarse stipple*) is divided into 8 rhombomeres (*r1* to *r8*), and these structures give rise to the cranial motor nerves *(m)*. *P1–P4*, pharyngeal (branchial) arches; *t*, telencephalon; *d*, diencephalon; *m*, mesencephalon.

Cranial Nerves

By the fourth week of development, nuclei for all 12 cranial nerves are present. All except the olfactory (I) and optic (II) nerves arise from the brainstem, and of these only the oculomotor (III) arises outside the region of the hindbrain. In the hindbrain, proliferation centers in the neuroepithelium establish eight distinct segments, the rhombomeres. These rhombomeres give rise to motor nuclei of cranial nerves IV, V, VI, VII, IX, X, XI, and XII (Figs. 19.17 and 19.41). Establishment of this segmental pattern appears to be directed by mesoderm collected into somitomeres beneath the overlying neuroepithelium.

Motor neurons for cranial nuclei are within the brainstem, while sensory ganglia are outside of the brain. Thus the organization of cranial nerves is homologous to that of spinal nerves, although not all cranial nerves contain both motor and sensory fibers (Table 19.1, p. 473–474).

Cranial nerve sensory ganglia originate from **ectodermal placodes** and **neural crest cells.** Ectodermal placodes include the nasal, otic, and four **epibranchial placodes** represented by ectodermal thickenings dorsal to the pharyngeal (branchial) arches (Table 19.2, p. 475; see Fig. 15.2). Epibranchial placodes contribute to ganglia for nerves of the pharyngeal arches (V, VII, IX, and X). Parasympathetic (visceral efferent) ganglia are derived from neural crest cells, and their fibers are carried by cranial nerves III, VII, IX, and X (Table 19.1).

TABLE 19.1 **Origins of Cranial Nerves and Their Composition**

Cranial Nerve	Brain Region	Type[a]	Innervation
Olfactory (I)	Telencephalon	SVA	Nasal epithelium (smell)
Optic (II)	Diencephalon	SSA	Retina (vision)
Oculomotor (III)	Mesencephalon	GSE	Sup., inf., med. Rectus, inf. oblique, levator palpebrae sup. m.
		GVE (ciliary ganglion)	sphincter pupillae, ciliary m.
Trochlear (IV)	Metencephalon (exits mesencephalon)	GSE	Sup. oblique m.
Trigeminal (V)	Metencephalon	GSA (trigeminal ganglion)	Skin, mouth, facial m., teeth, ant. two thirds of tongue
		GVA (trigeminal ganglion)	proprioception: skin, muscles, joints
		SVE (branchio-motor)	M. of mastication, mylohyoid, ant. belly of digastric, tensor velipalatini, post. belly of diagastric m.
Abducens (VI)	Metencephalon	GSE	Lateral rectus m.
Facial (VII)	Metencephalon	SVA (geniculate ganglion)	Taste ant. two thirds of tongue
		GSA (geniculate ganglion)	Skin ext. auditory meatus
		GVA (geniculate ganglion)	Ant. two thirds of tongue
		SVE (branchiomotor)	M. of facial expression, stapeduis, stylohyoid, post. belly of digastric
		GVE	Submandibular, sublingual, and lacrimal glands
Vestibulo-cochlear (VIII)	Metencephalon	SSA (vestibular and spiral ganglia)	Semicircular canals, utricle, saccule (balance) spiral organ of Corti (hearing)
Glossopharyn-geal (IX)	Myelencephalon	SVA (inferior ganglion)	Post. one third of tongue (taste)
		GVA (superior ganglion)	Parotid gland, carotid body and sinus, middle ear
		GSA (inferior ganglion)	External ear
		SVE (branchiomotor)	Stylopharyngeus
		GVE (otic ganglion)	Parotid gland

(Continued)

TABLE 19.1 (*Continued*)

Cranial Nerve	Brain Region	Type	Innervation
Vagus (X)	Myelencephalon	SVA (inferior ganglion)	Palate and epiglottis (taste)
		GVA (superior ganglion)	Base of tongue, pharynx, larynx, trachea, heart, esophagus, stomach, intestines
		GSA (superior ganglion)	External auditory meatus
		SVE (branchiomotor)	Constrictor m. pharynx, intrinsic m. larynx, sup. two thirds esophagus
		GVE (ganglia at or near viscera)	Trachea, bronchi, digestive tract, heart
Spinal Accessory (XI)	Myelencephalon	SVE (branchiomotor)	Sternocleidomastoid, trapezius m.
		GSE	Solf palate, pharynx (with X)
Hypoglossal (XII)	Myelencephalon	GSE	M. of tongue (except palatoglassus)

a SVA, Special Visceral Affarent; SSA, Special Somatic Affarent; SVE, Special Visceral Efferent; GVE, General Visceral Efferent; GSE, General Somatic Efferent; GSA, General Somatic Affarent; GVA, General Visceral Affarent.

Autonomic Nervous System

Functionally the autonomic nervous system can be divided into two parts: a **sympathetic** portion in the thoracolumbar region and a **parasympathetic** portion in the cephalic and sacral regions.

SYMPATHETIC NERVOUS SYSTEM

In the fifth week, cells originating in the **neural crest** of the thoracic region migrate on each side of the spinal cord toward the region immediately behind the dorsal aorta (Fig. 19.42). Here they form a bilateral chain of segmentally arranged sympathetic ganglia interconnected by longitudinal nerve fibers. Together they form the sympathetic chains on each side of the vertebral column. From their position in the thorax, neuroblasts migrate toward the cervical and lumbosacral regions, extending the sympathetic chains to their full length. Although initially the ganglia are arranged segmentally, this arrangement is later obscured, particularly in the cervical region, by fusion of the ganglia.

Some sympathetic neuroblasts migrate in front of the aorta to form **preaortic ganglia,** such as the **celiac** and **mesenteric** ganglia. Other sympathetic cells migrate to the heart, lungs, and gastrointestinal tract, where they give rise to **sympathetic organ plexuses** (Fig. 19.42).

TABLE 19.2 Contributions of Neural Crest Cells and Placodes to Ganglia of the Cranial Nerves

Nerve	Ganglion	Origin
Oculomotor (III)	Ciliary (visceral efferent)	Neural crest at forebrain-midbrain junction
Trigeminal (V)	Trigeminal (general afferent)	Neural crest at forebrain-midbrain junction, trigeminal placode
Facial (VII)	Superior (general and special afferent)	Hindbrain neural crest, first epibranchial placode
	Inferior (geniculate) (general and special afferent)	First epibranchial placode
	Sphenopalatine (visceral efferent)	Hindbrain neural crest
	Submandibular (visceral efferent)	Hindbrain neural crest
Vestibulocochlear (VIII)	Acoustic (cochlear) (special afferent)	Otic placode
	Vestibular (special afferent)	Otic placode, hindbrain neural crest
Glossopharyngeal (IX)	Superior (general and special afferent)	Hindbrain neural crest
	Inferior (petrosal) (general and special afferent)	Second epibranchial placode
	Otic (visceral efferent)	Hindbrain neural crest
Vagus (X)	Superior (general afferent)	Hindbrain neural crest
	Inferior (nodose) (general and special afferent)	Hindbrain neural crest; third, fourth epibranchial placodes
	Vagal parasympathetic (visceral efferent)	Hindbrain neural crest

Once the sympathetic chains have been established, nerve fibers originating in the **visceroefferent column (intermediate horn)** of the thoracolumbar segments (T1-L1,2) of the spinal cord penetrate the ganglia of the chain (Fig. 19.43). Some of these nerve fibers synapse at the same levels in the sympathetic chains or pass through the chains to **preaortic** or **collateral ganglia** (Fig. 19.43). They are known as **preganglionic fibers,** have a myelin sheath, and stimulate the sympathetic ganglion cells. Passing from spinal nerves to the sympathetic ganglia, they form the **white communicating rami.** Since the visceroefferent column extends only from the first thoracic to the second or third lumbar segment of the spinal cord, white rami are found only at these levels.

Axons of the sympathetic ganglion cells, the **postganglionic fibers,** have no myelin sheath. They either pass to other levels of the sympathetic chain or extend to the heart, lungs, and intestinal tract (*broken lines* in Fig. 19.43).

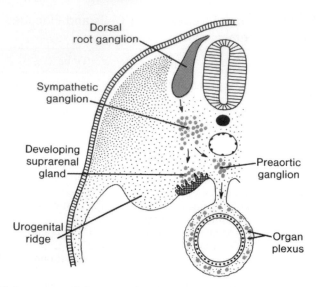

Figure 19.42 Formation of the sympathetic ganglia. A portion of the sympathetic neuroblasts migrates toward the proliferating mesothelium to form the medulla of the suprarenal gland.

Other fibers, the **gray communicating rami,** pass from the sympathetic chain to spinal nerves and from there to peripheral blood vessels, hair, and sweat glands. Gray communicating rami are found at all levels of the spinal cord.

Suprarenal Gland

The suprarenal gland develops from two components: (*a*) a mesodermal portion, which forms the **cortex,** and (*b*) an ectodermal portion, which forms the **medulla.** During the fifth week of development, mesothelial cells between the root of the mesentery and the developing gonad begin to proliferate and penetrate the underlying mesenchyme (Fig. 19.42). Here they differentiate into large acidophilic organs, which form the **fetal cortex,** or **primitive cortex,** of the suprarenal gland (Fig. 19.44*A*). Shortly afterward a second wave of cells from the mesothelium penetrates the mesenchyme and surrounds the original acidophilic cell mass. These cells, smaller than those of the first wave, later form the **definitive cortex** of the gland (Fig. 19.44, *A* and *B*). After birth the fetal cortex regresses rapidly except for its outermost layer, which differentiates into the reticular zone. The adult structure of the cortex is not achieved until puberty.

While the fetal cortex is being formed, cells originating in the sympathetic system (**neural crest cells**) invade its medial aspect, where they are arranged in cords and clusters. These cells give rise to the medulla of the suprarenal gland. They stain yellow-brown with chrome salts and hence are called **chromaffin**

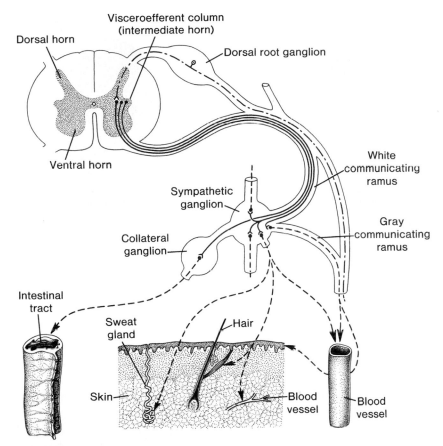

Figure 19.43 Relation of the preganglionic and postganglionic nerve fibers of the sympathetic nervous system to the spinal nerves. Note the origin of preganglionic fibers in the visceroefferent column of the spinal cord.

cells (Fig. 19.44). During embryonic life, chromaffin cells are scattered widely throughout the embryo, but in the adult the only persisting group is in the medulla of the adrenal glands.

PARASYMPATHETIC NERVOUS SYSTEM

Neurons in the brainstem and the sacral region of the spinal cord give rise to **preganglionic parasympathetic fibers.** Fibers from nuclei in the brainstem travel via the **oculomotor (III), facial (VII), glossopharyngeal (IX),** and **vagus (X) nerves.** **Postganglionic fibers** arise from neurons (ganglia) derived from **neural crest cells** and pass to the structures they innervate (e.g., pupil of the eye, salivary glands, viscera).

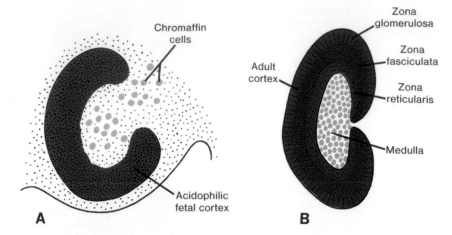

Figure 19.44 A. Chromaffin (sympathetic) cells penetrating the fetal cortex of the suprarenal gland. **B.** Later in development the definitive cortex surrounds the medulla almost completely.

CLINICAL CORRELATES

Congenital Megacolon (Hirschsprung Disease)

Congenital megacolon (Hirschsprung disease) results from a failure of parasympathetic ganglia to form in the wall of part or all of the colon and rectum because the neural crest cells fail to migrate. Most familial cases of Hirschsprung disease are due to mutations in the ***RET gene,*** which codes for a cell membrane **tyrosine kinase receptor.** This gene on chromosome 10q11 is essential for crest cell migration. The ligand for the receptor is **glial cell–derived neurotrophic growth factor** (GDNF) secreted by mesenchyme cells through which crest cells migrate. Receptor ligand interactions then regulate crest cell migration. Consequently, if there are abnormalities in the receptor, migration is inhibited and no parasympathetic ganglia form in affected areas. The rectum is involved in nearly all cases, and the rectum and sigmoid are involved in 80% of affected infants. The transverse and ascending portions of the colon are involved in only 10 to 20%. The colon is dilated above the affected region, which has a small diameter because of tonic contraction of noninnervated musculature.

Summary

The CNS **originates in the ectoderm** and appears as the **neural plate** at the middle of the third week (Fig. 19.1). After the edges of the plate fold, the **neural folds** approach each other in the midline to fuse into the **neural tube** (Figs. 19.2 and 19.3). The cranial end closes approximately at day 25, and the caudal end closes at day 27. The CNS then forms a tubular structure with a broad cephalic portion, the **brain,** and a long caudal portion, the **spinal cord.** Failure of the neural tube to close results in defects such as **spina bifida** (Figs. 19.15 and 19.16) and **anencephaly** (Fig. 19.38), defects that can be prevented by folic acid.

The **spinal cord,** which forms the caudal end of the CNS, is characterized by the **basal plate** containing the **motor neurons,** the **alar plate** for the **sensory neurons,** and a **floor plate** and a **roof plate** as connecting plates between the two sides (Fig. 19.8). *SHH* ventralizes the neural tube in the spinal cord region and induces the floor and basal plates. **Bone morphogenetic proteins 4 and 7,** expressed in nonneural ectoderm, maintain and up-regulate expression of *PAX3* and *PAX7* in the alar and roof plates.

The **brain,** which forms the cranial part of the CNS, consists originally of three vesicles: the rhombencephalon (hindbrain), mesencephalon (midbrain), and prosencephalon (forebrain).

The **rhombencephalon** is divided into (*a*) the myelencephalon, which forms the **medulla oblongata** (this region has a basal plate for somatic and visceral efferent neurons and an alar plate for somatic and visceral afferent neurons) (Fig. 19.18), and (*b*) the **metencephalon,** with its typical basal (efferent) and alar (afferent) plates (Fig. 19.19). This brain vesicle is also characterized by formation of the **cerebellum** (Fig. 19.20), a coordination center for posture and movement, and the **pons,** the pathway for nerve fibers between the spinal cord and the cerebral and the cerebellar cortices (Fig. 19.19).

The **mesencephalon,** or **midbrain,** resembles the spinal cord with its basal efferent and alar afferent plates. The mesencephalon's alar plates form the anterior and posterior colliculi as relay stations for visual and auditory reflex centers, respectively (Fig. 19.23).

The **diencephalon,** the posterior portion of the forebrain, consists of a thin roof plate and a thick alar plate in which the **thalamus** and **hypothalamus** develop (Figs. 19.24 and 19.25). It participates in formation of the pituitary gland, which also develops from Rathke's pouch (Fig. 19.26). Rathke's pouch forms the **adenohypophysis,** the **intermediate lobe,** and **pars tuberalis,** and the diencephalon forms the **posterior lobe,** the **neurohypophysis,** which contains neuroglia and receives nerve fibers from the hypothalamus.

The **telencephalon,** the most rostral of the brain vesicles, consists of two lateral outpocketings, the **cerebral hemispheres,** and a median portion, the **lamina terminalis** (Fig. 19.27). The lamina terminalis is used by the commissures as a connection pathway for fiber bundles between the right and left hemispheres (Fig. 19.30). The cerebral hemispheres, originally two small outpocketings (Figs. 19.24 and 19.25), expand and cover the lateral aspect of the diencephalon, mesencephalon, and metencephalon (Figs. 19.26–19.28). Eventually, nuclear regions of the telencephalon come in close contact with those of the diencephalon (Fig. 19.27).

The ventricular system, containing cerebrospinal fluid, extends from the lumen in the spinal cord to the fourth ventricle in the rhombencephalon, through the narrow duct in the mesencephalon, and to the third ventricle in the diencephalon. By way of the foramina of Monro, the ventricular system extends from the third ventricle into the lateral ventricles of the cerebral hemispheres. Cerebrospinal fluid is produced in the choroid plexus of the third, fourth, and lateral ventricles. Blockage of cerebrospinal fluid in the ventricular system or subarachnoid space may lead to hydrocephalus.

The brain is patterned along the anteroposterior (craniocaudal) and dorsoventral (mediolateral) axes. *HOX* **genes** pattern the anteroposterior axis in the hindbrain and specify rhombomere identity. Other transcription factors containing a homeodomain pattern the anteroposterior axis in the forebrain and midbrain regions, including *LIM1* and *OTX2*. Two other organizing centers, the **anterior neural ridge** and the **rhombencephalic isthmus,** secrete **FGF-8,** which serves as the inducing signal for these areas. In response to this growth factor, the cranial end of the forebrain expresses *BF1,* which regulates development of the telencephalon, and the isthmus expresses *engrailed genes* that regulate differentiation of the cerebellum and the roof of the midbrain. As it does throughout the central nervous system, **SHH,** secreted by the prechordal plate and notochord, ventralizes the forebrain and midbrain areas. **Bone morphogenetic proteins 4 and 7,** secreted by nonneural ectoderm, induce and maintain expression of dorsalizing genes.

Problems to Solve

1. *How are cranial nerves and spinal nerves similar? How are they different?*

2. *At what level is a spinal tap performed? From an embryological standpoint, why is this possible?*

3. *What is the embryological basis for most neural tube defects? Can they be diagnosed prenatally? Are there any means of prevention?*

4. *Prenatal ultrasound reveals an infant with an enlarged head and expansion of both lateral ventricles. What is this condition called, and what might have caused it?*

SUGGESTED READING

Chiang C, et al: Cyclopia and defective axial patterning in mice lacking *sonic hedgehog* gene function. *Nature* 383:407, 1996.

Cohen MM, Sulik KK: Perspectives on holoprosencephaly: Part II. Central nervous system, craniofacial anatomy, syndrome commentary, diagnostic approach, and experimental studies. *J Craniofac Genet Dev Biol* 12:196, 1992.

Cordes SP: Molecular genetics of cranial nerve development in mouse. *Nat Rev Neurosci* 2:611, 2001.

Dehart DB, Lanoue L, Tint GS, Sulik KK: Pathogenesis of malformations in a rodent model for Smith-Lemli-Opitz syndrome. *Am J Med Genet* 68:328, 1997.

Gavalis A, Krumlauf R: Retinoid signaling and hindbrain patterning. *Curr Op Genet Dev* 10:380, 2000.

Geelen JAG, Langman J: Closure of the neural tube in the cephalic region of the mouse embryo. *Anat Rec* 189:625, 1977.

Hinrichsen K, Mestres P, Jacob HJ: Morphological aspects of the pharyngeal hypophysis in human embryos. *Acta Morphol Neerl Scand* 24:235, 1986.

Hu D, Helms JA: The role of sonic hedgehog in normal and abnormal craniofacial morphogenesis. *Dev* 126:4873, 1999.

LeDouarin N, Fontaine-Perus J, Couly G: Cephalic ectodermal placodes and neurogenesis. *Trends Neurosci* 9:175, 1986.

LeDouarin N, Smith J: Development of the peripheral nervous system from the neural crest. *Annu Rev Cell Biol* 4:375, 1988.

Le Mantia AS, Bhasin N, Rhodes K, Heemskerk J: Mesenchymal epithelial induction mediates olfactory pathway formation. *Neuron* 28:411, 2000.

Loggie JMH: Growth and development of the autonomic nervous system. *In* Davis JA, Dobbing J (eds): *Scientific Foundations of Pediatrics*. Philadelphia, WB Saunders, 1974.

Lumsden A, Krumlauf R: Patterning the vertebrate neuraxis. *Science* 274:1109, 1996.

Lumsden A, Sprawson N, Graham A: Segmental origin and migration of neural crest cells in the hindbrain region of the chick embryo. *Development* 113:1281, 1991.

Muller F, O'Rahilly R: The development of the human brain and the closure of the rostral neuropore at stage 11. *Anat Embryol* 175:205, 1986.

Muller F, O'Rahilly R: The development of the human brain from a closed neural tube at stage 13. *Anat Embryol* 177:203, 1986.

O'Rahilly R, Muller F: The meninges in human development. *J Neuropathol Exp Neurol* 45:588, 1986.

Rodier PM, Reynolds SS, Roberts WN: Behavioral consequences of interference with CNS development in the early fetal period. *Teratology* 19:327, 1979.

Roessler E, et al.: Mutations in the human *sonic hedgehog* gene cause holoprosencephaly. *Nat Genet* 14:357, 1996.

Rubenstein JLR, Beachy PA: Patterning of the embryonic forebrain. *Curr Opin Neurobiol* 8:18, 1998.

Sakai Y: Neurulation in the mouse: 1. The ontogenesis of neural segments and the determination of topographical regions in a central nervous system. *Anat Rec* 218:450, 1987.

Schoenwolf G: On the morphogenesis of the early rudiments of the developing central nervous system. *Scanning Electron Microsc* 1:289, 1982.

Schoenwolf G, Smith JL: Mechanisms of neurulation: traditional viewpoint and recent advances. *Development* 109:243, 1990.

Shimamura K, Rubenstein JLR: Inductive interactions direct early regionalization of the mouse forebrain. *Development* 124:2709, 1997.

Tanabe Y, Jessell TM: Diversity and patterning in the developing spinal cord. *Science* 274:1115, 1996.

Watkins-Chow DE, Camper SA: How many homeobox genes does it take to make a pituitary gland? *Trends Genet* 14:284, 1998.

Wilkie AOM, Morriss-Kay GM: Genetics of craniofacial development and malformation. *Nat Rev Genet* 2:458, 2001.

part three

Appendix

appendix

Answers to Problems

CHAPTER 1

1. The most common cause for abnormal chromosome number is nondisjunction during either meiosis or mitosis. For unknown reasons chromosomes fail to separate during cell division. Nondisjunction during meiosis I or II results in half of the gametes having no copy and half having two copies of a chromosome. If fertilization occurs between a gamete lacking a chromosome and a normal one, monosomy results; if it occurs between a gamete with two copies and a normal one, trisomy results. Trisomy 21 (Down syndrome), the most common numerical abnormality resulting in birth defects (mental retardation, abnormal facies, heart malformations), is usually due to nondisjunction in the mother and occurs most frequently in children of women more than 35 years of age, reflecting the fact that the risk of meiotic nondisjunction increases with increasing maternal age. Other trisomies that result in syndromes of abnormal development involve chromosomes 8, 9, 13, and 18. Monosomies involving autosomal chromosomes are fatal, but monosomy of the X chromosome (Turner syndrome) is compatible with life. This condition is usually (80%) a result of nondisjunction during meiosis of paternal chromosomes and is characterized by infertility, short stature, webbing of the neck, and other defects. Karyotyping of embryonic cells obtained by amniocentesis or chorionic villus biopsy (see Clinical Correlates in Chapter 7) can detect chromosome abnormalities prenatally.

2. Chromosomes sometimes break and the pieces may create partial monosomies or trisomies or become attached (translocated) to other chromosomes. Translocation of part of chromosome 21 onto chromosome 14, for example, accounts for approximately 4% of cases of Down syndrome. Chromosomes may also be altered by mutations in single genes. The risk of chromosomal abnormalities is increased by maternal and paternal age over 35 years.

3. Mosaicism occurs when an individual has two or more cell lines that are derived from a single zygote but that have different genetic characteristics. The different cell lines may arise by mutation or by mitotic nondisjunction during cleavage, as in some cases of Down syndrome.

CHAPTER 2

1. Infertility occurs in approximately 20% of married couples. A major cause of infertility in women is blockage of the uterine (fallopian) tubes due to scarring from repeated pelvic inflammatory disease; in men the primary cause is low sperm count. In vitro fertilization (IVF) techniques can circumvent these problems, although the success rate (approximately 20%) is low.

2. Pelvic inflammatory diseases, such as gonorrhea, are a major cause of occluded oviducts (uterine tubes). Although the patient may be cured, scarring closes the lumen of the tubes and prevents passage of sperm to the oocyte and of oocytes to the uterine cavity. IVF can overcome the difficulty by fertilizing the woman's oocytes in culture and transferring them to her uterus for implantation. The alternative programs, gamete intrafallopian transfer (GIFT) and zygote intrafallopian transfer (ZIFT), are not possibilities, since both techniques require patent uterine tubes.

CHAPTER 3

1. The second week is known as the week of twos because the trophoblast differentiates into two layers, the syncytiotrophoblast and cytotrophoblast; the embryoblast differentiates into two layers, the epiblast and hypoblast; the extraembryonic mesoderm splits into two layers, the splanchnopleure and somatopleure; and two cavities, the amniotic and yolk sac cavities, form.

2. It is not clear why the conceptus is not rejected by the maternal system. Recent evidence suggests that secretion of immunosuppressive molecules, such as cytokines and proteins, and expression of unrecognizable antigens of the major histocompatibility complex protect the conceptus from rejection. In some cases maternal immunological responses do adversely affect pregnancy, as in some cases of autoimmune disease. Thus patients with systemic lupus erythematosus have poor reproductive outcomes and histories of multiple spontaneous abortions. It has not been conclusively shown that maternal antibodies can cause birth defects.

3. In some cases, trophoblastic tissue is the only tissue in the uterus, and embryo-derived cells are either absent or present in small numbers. This condition is termed a hydatidiform mole, which, because of its trophoblastic origin, secretes human chorionic gonadotropin (hCG) and mimics the initial stages of pregnancy. Most moles are aborted early in pregnancy, but those containing remnants of an embryo may remain into the second trimester. If pieces of tro-phoblast are left behind following spontaneous abortion or surgical removal of a mole, cells may continue to proliferate and form tumors known as invasive moles, or choriocarcinoma. Since early trophoblast development is controlled by paternal genes, it is thought that the origin of moles may be from fertilization of an ovum without a nucleus.

4. The most likely diagnosis is an ectopic pregnancy in the uterine tube, which can be confirmed by ultrasound. Implantation in a uterine tube is due to poor transport of the zygote and may be a result of scarring. As with Down syndrome, the frequency of ectopic pregnancy increases with maternal age over 35.

CHAPTER 4

1. Unfortunately, consuming large quantities of alcohol at any stage during pregnancy may adversely affect embryonic development. In this case the woman has exposed the embryo during the third week of gestation (assuming that fertilization occurred at the midpoint of the menstrual cycle), at the time of gastrulation. This stage is particularly vulnerable to insult by alcohol and may result in fetal alcohol syndrome (mental retardation, abnormal facies) (see Chap-ter 7). Although fetal alcohol syndrome is most common in offspring of alcoholic mothers, no *safe* levels of blood alcohol concentration have been established for embryogenesis. Therefore, since alcohol causes birth defects and is the leading cause of mental retardation, it is recommended that women who are planning a pregnancy or who are already pregnant refrain from use of any alcohol.

2. Such a mass is probably a sacrococcygeal teratoma. These tumors arise as remnants of the primitive streak, usually in the sacral region. The term *teratoma* refers to the fact that the tumor contains different types of tissues. Since it is derived from the streak, which contains cells for all three germ layers, it may contain tissues of ectoderm, mesoderm, or endoderm origin. Such tumors are 3 times as common in females as in males.

3. The baby has a severe form of caudal dysgenesis called sirenomelia (mermaid-like). Sirenomelia, which occurs in varying degrees, is probably due to abnormalities in gastrulation in caudal segments. It was initially termed caudal regression, but it is clear that structures do not regress; they simply do not form. Also known as caudal agenesis and sacral agenesis, sirenomelia is characterized by varying degrees of flexion, inversion, lateral rotation, and oc-casional fusion of the lower limbs; defects in lumbar and sacral vertebrae; renal agenesis; imperforate anus; and agenesis of internal genital structures except the testes and ovaries. Its cause is unknown. It occurs sporadically but is most frequently observed among infants of diabetic mothers.

4. This patient has left-sided laterality sequence and should be evaluated for additional defects. Sidedness is established at the time of primitive streak for-mation (gastrulation) and is regulated by genes, such as *sonic hedgehog* and *nodal,* that become restricted in their expression. Partial reversal of left-right asymmetry is more often associated with other defects than complete asymmetry (situs inversus).

CHAPTER 5

1. Development during the third to the eighth week is critical because this is when cell populations responsible for organ formation are established and when organ primordia are being formed. Early in the third week, gastrulation begins to provide cells that constitute the three germ layers responsible for organogenesis. Late in the third week, differentiation of the central nervous system is initiated, and over the next 5 weeks, all of the primordia for the major organ systems will be established. At these times, cells are rapidly proliferating, and critical cell-cell signals are occurring. These phenomena are particularly sensitive to disruption by outside factors, such as environmental hazards, pharmaceutical agents, and drugs of abuse. Thus exposure to such factors may result in abnormalities known as birth defects or congenital malformations.

CHAPTER 6

1. An excess of amniotic fluid is called hydramnios or polyhydramnios, and many times (35%) the cause is unknown (idiopathic). A high incidence (25%) is also associated with maternal diabetes and with birth defects that interfere with fetal swallowing, such as esophageal atresia and anencephaly.

2. No. She is not correct. The placenta does not act as a complete barrier, and many compounds cross freely, especially lipophilic substances, such as toluene and alcohol. Furthermore, early in pregnancy the placenta is not completely developed, and the embryo is particularly vulnerable. These early weeks are also very sensitive to insult by compounds such as toluene, which causes the toluene embryopathy.

CHAPTER 7

1. Neural tube defects, such as spina bifida and anencephaly, produce elevated α-fetoprotein (AFP) levels, as do abdominal defects, such as gastroschisis and omphalocele. Maternal serum AFP levels are also elevated, so that they may be used as a screen to be confirmed by amniocentesis. Ultrasonography is used to confirm the diagnosis.

2. Since Down syndrome is a chromosomal abnormality resulting most commonly from trisomy 21 (see Chapter 1), cells for chromosomal analysis can be collected by amniocentesis or chorionic villus biopsy (CVS). CVS has the advantage that sufficient cells can be obtained immediately to do the analysis, whereas cells collected by amniocentesis, which is not usually done prior to

14 weeks' gestation, must be cultured for approximately 2 weeks to obtain sufficient numbers. The risk of fetal loss following CVS is 1%, which is about twice as high as that of amniocentesis.

3. Status of the fetus is critical for managing pregnancy, delivery, and postnatal care. Size, age, and position are important for determining the time and mode of delivery. Knowing whether birth defects are present is important for planning postnatal care. Tests for determining fetal status are dictated by maternal history and factors that increase risk, such as exposure to teratogens, chromosome abnormalities in either parent, advanced maternal age, or the birth of a previous infant with a birth defect.

4. Factors that influence the action of a teratogen are (*a*) genotype of the mother and conceptus, (*b*) dose and duration of exposure to the agent, and (*c*) stage of embryogenesis when exposure occurs. Most major malformations are produced during the embryonic period (teratogenic period), the third to the eighth week of gestation. However, stages prior to this time, including the preimplantation period, and after the eighth week (fetal period) remain susceptible. The brain, for example, remains sensitive to insult throughout the fetal period. No stage of pregnancy is free of risk from teratogenic insult.

5. The woman is correct that drugs may be teratogenic. Severe hyperthermia such as this, however, is known to cause neural tube defects (spina bifida and anencephaly at this stage of gestation). Therefore, one must weigh the risk of teratogenicity of an antipyretic agent with a low teratogenic potential, such as low-dose aspirin, against the risk of hyperthermia. Interestingly, malformations have been associated with sauna-induced hyperthermia. No information about exercise-induced hyperthermia and birth defects is available, but strenuous physical activity (running marathons) raises body temperature significantly and probably should be avoided during pregnancy.

6. Since more than 50% of pregnancies are unplanned, all women of childbearing age should consume 400 μg of folic acid daily as a supplement to prevent neural tube defects. If a woman has not been taking folate and is planning a pregnancy, she should begin the supplement 2 months prior to conception and continue throughout gestation. Folic acid is nontoxic even at high doses, can prevent up to 70% of neural tube defects, and may prevent conotruncal heart defects and facial clefts.

7. The woman's concerns are valid, since infants of insulin-dependent diabetic mothers have an increased incidence of birth defects, including a broad spectrum of minor and major anomalies. Placing the mother under strict metabolic control using multiple insulin injections prior to conception, however, significantly reduces the incidence of abnormalities and affords the greatest opportunity for a normal pregnancy. A similar scenario occurs with women who have phenylketonuria (PKU). Strict management of these patients' disease prior to conception virtually eliminates the risk of congenital defects in the offspring.

Both situations stress the need for planning pregnancies and for avoiding potential teratogenic exposures, especially during the first 8 weeks of gestation, when most defects are produced.

CHAPTER 8

1. Cranial sutures are fibrous regions between flat bones of the skull. Membranous regions between the flat bones are known as fontanelles, the largest of which is the anterior fontanelle (soft spot). These sutures and fontanelles permit (*a*) molding of the head as it passes through the birth canal and (*b*) growth of the brain. Growth of the skull, which continues after birth as the brain enlarges, is greatest during the first 2 years of life. Premature closure of one or more sutures (craniosynostosis) results in deformities in the shape of the head, depending on which sutures are involved. Craniosynostosis is often associated with other skeletal defects, and evidence suggests that genetic factors are important in the causation (see Table 8.1, p. 181).

2. Defects of the long bones and digits are often associated with other malformations and should prompt a thorough examination of all systems. Clusters of defects that occur simultaneously with a common cause are called syndromes, and limb anomalies, especially of the radius and digits, are common components of such clusters. Diagnosis of syndromes is important in determining recurrence risks and thus in counseling parents about subsequent pregnancies.

3. Formation of the vertebrae is a complex process involving growth and fusion of the caudal portion of one sclerotome with the cranial portion of an adjacent one. Not surprisingly, mistakes occur, and they result in fusions and increases and decreases in the number of vertebrae (Klippel-Feil syndrome). In some cases, only half a vertebra forms (hemivertebra), resulting in asymmetry and lateral curvature of the spine (scoliosis). *HOX* (homeobox) genes that pattern the vertebra may have mutations that cause part of one not to form properly. Scoliosis may also be caused by weakness of muscles of the back.

CHAPTER 9

1. Muscle cells are derived from the dorsolateral and dorsomedial regions of the somite. Dorsolateral cells express *MyoD* and migrate early to form the hypomeric muscles. These muscles include those of the limb and body wall. Dorsomedial cells express *Myf5*, migrate beneath cells forming the dermatome, and form epimeric muscles. These are the extensor muscles of the vertebral column.

2. Partial or complete absence of the pectoralis major muscle, the defect known as Poland anomaly, is the most likely diagnosis. Poland anomaly is often associated with shortness of the middle digits (brachydactyly) and digital fusion (syndactyly). Loss of the pectoralis major muscle produces little or no loss of function, since other muscles compensate.

3. Patterning for muscles depends on connective tissue that forms from fibroblasts. In the head, with its complicated pattern of muscles of facial expression, neural crest cells direct patterning; in cervical and occipital regions, connective tissue from somites directs it; and in the body wall and limbs, somatic mesoderm directs it.

4. Innervation for muscles is derived from the vertebral level from which the muscle cells originate, and this relation is maintained regardless of where the muscle cells migrate. Thus myoblasts forming the diaphragm originate from cervical segments 3, 4, and 5, migrate to the thoracic region, and carry their nerves with them.

CHAPTER 10

1. Failure of the left pleuroperitoneal membrane to close the pericardioperitoneal canal on that side is responsible for the defect. This canal is larger on the left than on the right, closes later, and therefore may be more susceptible to abnormalities. The degree of hypoplasia of the lungs resulting from compression by abdominal viscera determines the fate of the infant. Treatment requires surgical repair of the defect, and attempts to correct the malformation in utero have been made.

2. The defect is gastroschisis. It occurs because of a weakness in the body wall caused by regression of the right umbilical vein. Since the bowel is not covered by the amnion, it may become necrotic because of exposure to the amniotic fluid. It is also possible for the bowel loops to twist around themselves (volvulus), cutting off their blood supply and producing an infarction. Gastroschisis is not associated with genetic abnormalities or with other malformations. Therefore, if damage to the bowel is not too extensive, survival rates are good.

CHAPTER 11

1. A four-chambered view is sought in ultrasound scans of the heart. The chambers are divided by the atrial septum superiorly, the ventricular septum inferiorly, and the endocardial cushions surrounding the atrioventricular canals laterally. Together these structures form a cross whose integrity is readily visualized by ultrasound. In this case, however, the fetus probably has a ventricular septal defect, the most commonly occurring heart malformation, in the membranous portion of the septum. The integrity of the great vessels should also be checked carefully, since the conotruncal septum dividing the aortic and pulmonary channels must come into contact with the membranous portion of the interventricular septum for this structure to develop normally.

2. Since neural crest cells contribute to much of the development of the face and to the truncal portion of the conotruncal septum, these cells have probably been disrupted. Crest cells may have failed to migrate to these regions, failed to proliferate, or been killed. Retinoic acid (vitamin A) is a potent teratogen that targets neural crest cells among other cell populations. Since retinoids are effective in treating acne and since acne is common in young women of childbearing age, great care should be employed before prescribing the drug to this cohort.

3. Endocardial cushion tissue is essential for proper development of these structures. In the common atrioventricular canal, the superior, the inferior, and two lateral endocardial cushions divide the opening and contribute to the mitral and tricuspid valves in the left and right atrioventricular canals. In addition, the superior and inferior cushions are essential for complete septation of the atria by fusion with the septum primum and of the ventricles by forming the membranous part of the interventricular septum. Cushion tissue in the conus and truncus forms the conotruncal septum, which spirals down to separate the aorta and pulmonary channels and to fuse with the inferior endocardial cushion to complete the interventricular septum. Therefore, any abnormality of cushion tissue may result in a number of cardiac defects, including atrial and ventricular septal defects, transposition of the great vessels, and other abnormalities of the outflow tract.

4. In the development of the vascular system for the head and neck, a series of arterial arches forms around the pharynx. Most of these arches undergo alterations, including regression, as the original patterns are modified. Two such alterations that produce difficulty swallowing are (*a*) double aortic arch, in which a portion of the right dorsal aorta (that normally regresses) persists between the seventh intersegmental artery and its junction with the left dorsal aorta, creating a vascular ring around the esophagus, and (*b*) right aortic arch, in which the ascending aorta and the arch form on the right. If in such cases the ligamentum arteriosum remains on the left, it passes behind the esophagus and may constrict it.

CHAPTER 12

1. This infant most likely has some type of tracheoesophageal atresia with or without a tracheoesophageal fistula. The baby cannot swallow, and this results in polyhydramnios. The defect is caused by abnormal partitioning of the trachea and esophagus by the tracheoesophageal septum. These defects are often associated with other malformations, including a constellation of vertebral anomalies, anal atresia, cardiac defects, renal anomalies, and limb defects known as the VACTERL association.

2. Babies born prior to 7 months of gestation do not produce sufficient amounts of surfactant to reduce surface tension in the alveoli to permit normal lung function. Consequently, alveoli collapse, resulting in respiratory distress syndrome. Recent improvements in artificial surfactants have improved the prognosis for these infants.

CHAPTER 13

1. The baby most likely has some type of esophageal atresia and/or tracheo-esophageal fistula. In 90% of these cases the proximal part of the esophagus ends in a blind pouch, and a fistula connects the distal part with the trachea. Polyhydramnios results because the baby cannot swallow amniotic fluid. Aspiration of fluids at birth may cause pneumonia. The defect is caused by an abnormal partitioning of the respiratory diverticulum from the foregut by the tracheoesophageal septum.

2. The most likely diagnosis is an omphalocele resulting from a failure of herniated bowel to return to the abdominal cavity at 10 to 12 weeks of gestation. Since the bowel normally herniates into the umbilical cord, it is covered by amnion. This situation is in contrast to gastroschisis, in which loops of bowel herniate through an abdominal wall defect and are not covered by amnion. The prognosis is not good, since 25% of infants with omphalocele die before birth, 40 to 88% have associated anomalies, and approximately 50% show chromosomal abnormalities. If no other defects are present, surgical repair is possible, and in experienced hands, survival is 100%.

3. This infant has an imperforate anus with a rectovaginal fistula, part of an anorectal atresia complex. She appears to have a high anorectal atresia, since the fistula connects the rectum to the vagina, accounting for meconium (intestinal contents) in this structure. The defect was probably caused by a cloaca that was too small, so that the cloacal membrane was shortened posteriorly. This condition causes the opening of the hindgut to shift anteriorly. The smaller the cloaca posteriorly, the farther anteriorly the hindgut opening shifts, resulting in a higher defect.

CHAPTER 14

1. The three systems to form are the pronephros, mesonephros, and metanephros, all derivatives of the intermediate mesoderm. They form in succession in cranial to caudal sequence. Thus, the pronephros forms in cervical segments at the end of the third week but is rudimentary and rapidly regresses. The mesonephros, which begins early in the fourth week, extends from thoracic to upper lumbar regions. It is segmented in only its upper portion and contains excretory tubules that connect to the mesonephric (wolffian) duct. This kidney also regresses but may function for a short time. It is more important because the tubules and collecting duct contribute to the genital ducts in the male. Collecting ducts near the testis form the efferent ductules, whereas the mesonephric duct forms the epididymis, ductus deferens, and ejaculatory duct. In the female these tubules and ducts degenerate, since maintaining them depends on testosterone production. The metanephros lies in the pelvic region as a mass of unsegmented mesoderm (metanephric blastema) that forms the definitive kidneys. Ureteric buds grow from the mesonephric ducts and, on contact with the metanephric blastema, induce it to differentiate. The ureteric buds form collecting ducts and ureters, while the metanephric blastema forms nephrons (excretory units), each of which consists of a glomerulus (capillaries) and renal tubules.

2. Both the ovaries and testes develop in the abdominal cavity from intermediate mesoderm along the urogenital ridge. Both also descend by similar mechanisms from their original position, but the uterus prevents migration of the ovary out of the abdominal cavity. In the male, however, a mesenchymal condensation, the gubernaculum (which also forms in females but attaches to the uterus), attaches the caudal pole of the testis, first to the inguinal region and then to the scrotal swellings. Growth and retraction of the gubernaculum, together with increasing intra-abdominal pressure, cause the testis to descend. Failure of these processes causes undescended testes, known as cryptorchism. Approximately 2 to 3% of term male infants have an undescended testicle, and in 25% of these the condition is bilateral. In many cases the undescended testis descends by age 1. If it does not, testosterone administration (since this hormone is thought to play a role in descent) or surgery may be necessary. Fertility may be affected if the condition is bilateral.

3. Male and female external genitalia pass through an indifferent stage during which it is impossible to differentiate between the two sexes. Under the influence of testosterone, these structures assume a masculine appearance, but the derivatives are homologous between males and females. These homologies include (*a*) the clitoris and penis, derived from the genital tubercle; (*b*) the labia majora and scrotum, derived from the genital swellings that fuse in the male; and (*c*) the labia minora and penile urethra, derived from the urethral folds that fuse in the male. During early stages the genital tubercle is larger in the female than in the male, and this has led to misidentification of sex by ultrasound.

4. The uterus is formed by fusion of the lower portions of the paramesonephric (müllerian) ducts. Numerous abnormalities have been described; the most common consists of two uterine horns (bicornuate uterus). Complications of this defect include difficulties in becoming pregnant, high incidence of spontaneous abortion, and abnormal fetal presentations. In some cases a part of the uterus has a blind end (rudimentary horn), causing problems with menstruation and abdominal pain.

CHAPTER 15

1. Neural crest cells are important for craniofacial development because they contribute to so many structures in this region. They form all of the bones of the face and the anterior part of the cranial vault, and the connective tissue that provides patterning of the facial muscles. They also contribute to cranial nerve ganglia, meninges, dermis, odontoblasts, and stroma for glands derived from pharyngeal pouches. In addition, crest cells from the hindbrain region of the neural folds migrate ventrally to participate in septation of the conotruncal region of the heart into aortic and pulmonary vessels. Unfortunately, crest cells appear to be vulnerable to a number of compounds, including alcohol and retinoids, perhaps because they lack catalase and superoxide dismutase enzymes that scavenge toxic free radicals. Many craniofacial defects are due to insults on neural crest cells and may be associated with cardiac abnormalities because of the contribution of these cells to heart morphogenesis.

2. The child may have DiGeorge anomaly, which is characterized by these types of craniofacial defects and partial or complete absence of thymic tissue. It is the loss of thymic tissue that compromises the immune system, resulting in numerous infections. Damage to neural crest cells is the most likely cause of the sequence, since these cells contribute to development of all of these structures, including the stroma of the thymus. Teratogens, such as alcohol, have been shown to cause these defects experimentally.

3. Children with midline clefts of the lip often have mental retardation. Median clefts are associated with loss of other midline structures, including those in the brain. In its extreme form, the entire cranial midline is lost, and the lateral ventricles of the cerebral hemispheres are fused into a single ventricle, a condition called holoprosencephaly. Midline clefts, induced as the cranial neural folds begin to form (approximately days 19 to 21), result from the loss of midline tissue in the prechordal plate region.

4. The child most likely has a thyroglossal cyst that results from incomplete regression of the thyroglossal duct. These cysts may form anywhere along the line of descent of the thyroid gland as it migrates from the region of the foramen cecum of the tongue to its position in the neck. A cyst must be differentiated from ectopic glandular tissue, which may also remain along this pathway.

CHAPTER 16

1. Microtia involves defects of the external ear that range from small but well-formed ears to absence of the ear (anotia). Other defects occur in 20 to 40% of children with microtia and/or anotia, including the oculoauriculovertebral spectrum (hemifacial microsomia), in which case the craniofacial defects may be asymmetrical. Since the external ear is derived from hillocks on the first two pharyngeal arches, which are largely formed by neural crest cells, this cell population plays a role in most ear malformations.

CHAPTER 17

1. The lens forms from a thickening of ectoderm (lens placode) adjacent to the optic cup. Lens induction may begin very early, but contact with the optic cup plays a role in this process as well as in maintenance and differentiation of the lens. Therefore, if the optic cup fails to contact the ectoderm or if the molecular and cellular signals essential for lens development are disrupted, a lens will not form.

2. Rubella is known to cause cataracts, microphthalmia, congenital deafness, and cardiac malformations. Exposure during the fourth to the eighth week places the offspring at risk for one or more of these birth defects.

3. As the optic cup reaches the surface ectoderm, it invaginates, and along its ventral surface it forms a fissure that extends along the optic stalk. It is through this fissure that the hyaloid artery reaches the inner chamber of the eye. Normally the distal portion of the hyaloid artery degenerates, and the choroid fissure closes by fusion of its ridges. If this fusion does not occur, colobomas occur. These defects (clefts) may occur anywhere along the length of the fissure. If they occur distally, they form colobomas of the iris; if they occur more proximally, they form colobomas of the retina, choroid, and optic nerve, depending on their extent. Mutations in *PAX2* can cause optic nerve colobomas and may be responsible for other types as well. Also, mutations in this gene have been linked to renal defects and renal coloboma syndrome.

CHAPTER 18

1. Mammary gland formation begins as budding of epidermis into the under-lying mesenchyme. These buds normally form in the pectoral region along a thickened ridge of ectoderm, the mammary or milk line. This line or ridge extends from the axilla into the thigh on both sides of the body. Occasionally, accessory sites of epidermal growth occur, so that extra nipples (polythelia) and extra breasts (polymastia) appear. These accessory structures always occur along the milk line and usually in the axillary region. Similar conditions also occur in males.

CHAPTER 19

1. Cranial and spinal nerves are homologues, but they differ in that cranial nerves are much less consistent in their composition. Motor neurons for both lie in basal plates of the central nervous system, and sensory ganglia, derived from the neural crest, lie outside the central nervous system. Fibers from sensory neurons synapse on neurons in the alar plates of the spinal cord and brain. Three cranial nerves (I, II, and VIII) are entirely sensory; four (IV, VI, XI, and XII) are entirely motor; three (VII, IX, and X) have motor, sensory, and parasympathetic fibers; and one (III) has only motor and parasympathetic components. In contrast, each spinal nerve has motor and sensory fibers.

2. A spinal tap is performed between vertebra L4 and vertebra L5, since the spinal cord ends at the L2 to L3 level. Thus it is possible to obtain cerebrospinal fluid at this level without damaging the cord. The space is created because after the third month, the cord, which initially extended the entire length of the vertebral column, does not lengthen as rapidly as the dura and vertebral column do, so that in the adult the spinal cord ends at the L2 to L3 level.

3. The embryological basis for most neural tube defects is inhibition of closure of the neural folds at the cranial and caudal neuropores. In turn, defects occur in surrounding structures, resulting in anencephaly, some types of encephaloceles, and spina bifida cystica. Severe neurological deficits accompany abnormalities in these regions. Neural tube defects, which occur in approximately 1/1000 births, may be diagnosed prenatally by ultrasound and findings of elevated levels of α-fetoprotein in maternal serum and amniotic fluid. Recent evidence has shown that daily supplements of 400 μg of folic acid started 2 months prior to conception prevent up to 70% of these defects.

4. This condition, hydrocephalus, results from a blockage in the flow of cerebrospinal fluid from the lateral ventricles through the foramina of Monro and the cerebral aqueduct into the fourth ventricle and out into the subarachnoid space, where it would be resorbed. In most cases, blockage occurs in the cerebral aqueduct in the midbrain. It may result from genetic causes (X-linked recessive) or viral infection (toxoplasmosis, cytomegalovirus).

Figure Credits

Figure 1.6. Reprinted with permission from Gelehrter TD, Collins FS, Ginsburg D. *Principles of Medical Genetics.* 2nd ed. Baltimore: Williams & Wilkins, 1998:166.

Figure 1.7. Courtesy of Dr. Kathleen Rao, Department of Pediatrics, University of North Carolina.

Figure 1.11. Reprinted with permission from McKusick VA. Klinefelter and Turner's syndromes. *Journal of Chronic Disease* 12:50, 1960.

Figure 1.12. Reprinted with permission from McKusick VA. Klinefelter and Turner's syndromes. *Journal of Chronic Disease* 12:52, 1960.

Figure 1.13. Courtesy of Dr. R. J. Gorlin, Department of Oral Pathology and Genetics, University of Minnesota.

Figure 1.14. Courtesy of Dr. R. J. Gorlin, Department of Oral Pathology and Genetics, University of Minnesota.

Figure 1.15. Courtesy of D. L. Van Dyke and A. Wiktor, Henry Ford Health Sciences Center.

Figure 1.19C. Reprinted with permission from Ross MH, Romrell LJ, Kaye GI. *Histology: A Text and Atlas.* 3rd ed. Baltimore: Williams & Wilkins, 1995:684.

Figure 1.22. Adapted from Fawcett DW. *Bloom and Fawcett: A Textbook of Histology.* Philadelphia: WB Saunders, 1986.

Figure 1.24. Adapted from Clermont Y. The cycle of the seminiferous epithelium in man. *American Journal of Anatomy* 112:35, 1963.

Figure 2.3, A and B. Reprinted with permission from Van Blerkom J, Motta P. *The Cellular Basis of Mammalian Reproduction.* Baltimore: Urban & Schwarzenberg, 1979.

Figure 2.5A. Courtesy of Dr. P. Motta.

Figure 2.7A. Courtesy of Drs. L. Dickmann and R. Noyes, Vanderbilt University.

Figure 2.7B. Reprinted with permission from Hertig AT, Rock J. Two human ova of the previllous stage, having a developmental age of about seven and nine days, respectively. *Contributions in Embryology* 31:65, 1945. Courtesy of Carnegie Institution of Washington, Washington, DC.

Figure 2.9, A and B. Courtesy of Dr. Caroline Ziomeck, Genzyme Transgenics Corporation.

Figure 2.10A. Reprinted with permission from Hertig AT, Rock J, Adams EC. A description of 34 human ova within the first 17 days of development. *American Journal of Anatomy* 98:435, 1956. Courtesy of Carnegie Institution of Washington, Washington, DC.

Figure 3.2. Reprinted with permission from Hertig AT, Rock J. Two human ova of the previllous stage, having a developmental age of about seven and nine days, respectively. *Contributions in Embryology* 31:65, 1945. Courtesy of Carnegie Institution of Washington, Washington, DC.

Figure 3.5. Reprinted with permission from Hertig AT, Rock J. Two human ova of the previllous stage, having a developmental age of 11 and 12 days, respectively. *Contributions in Embryology* 29:127, 1941. Courtesy of Carnegie Institution of Washington, Washington, DC.

Figure 3.7. Reprinted with permission from Hertig AT, Rock J, Adams EC. A description of 34 human ova within the first 17 days of development. *American Journal of Anatomy* 98:345, 1956. Courtesy of Carnegie Institution of Washington, Washington, DC.

Figure 3.8. Modified from Hamilton WJ, Mossman HW. *Human Embryology.* Baltimore: Williams & Wilkins, 1972.

Figure 4.2B. Reprinted with permission from Heuser CH. A presomite embryo with a definite chorda canal. *Contributions in Embryology* 23:253, 1932. Courtesy of Carnegie Institution of Washington, Washington, DC.

Figure 4.3, C and D. Courtesy of Dr. K. K. Sulik, Department of Cell Biology and Anatomy, University of North Carolina.

Figure 4.4, B, D, and F. Courtesy of Dr. K. K. Sulik, Department of Cell Biology and Anatomy, University of North Carolina.

Figure 4.6. Courtesy of Dr. Michael R. Kuehn, National Cancer Institute, Bethesda, MD.

Figure 4.7. Reprinted with permission from Niehrs C, Keller R, Cho KWY, DeRobertis EM. The homeobox gene goosecoid controls cell migration in Xenopus embryos. Cell 72:491–503, 1993.

Figure 4.8. Reprinted with permission from Herrmann BG. Expression pattern of the Brachyury gene in whole mount Twis/Twis mutant embryos. Development 113:913–917, 1991.

Figure 4.10. Courtesy of Dr. Michael R. Kuehn, National Cancer Institute, Bethesda, MD.

Figure 4.11. Reprinted with permission from Smith JL, Gestland KM, Schoenwolf GC. Prospective fate map of the mouse primitive streak at 7.5 days of gestation. *Developmental Dynamics* 201:279, 1994.

Figure 4.12. Courtesy of Dr. K. K. Sulik, Department of Cell Biology and Anatomy, University of North Carolina.

Figure 4.14. Courtesy of Dr. Don Nakayama, Department of Surgery, University of North Carolina.

Figure 4.18. Reprinted with permission from King BF, Mias JJ. Developmental changes in rhesus monkey placental villi and cell columns. *Anatomy and Embryology* 165:361–376, 1982.

Figure 5.1C. Reprinted with permission from Heuser CH. A presomite embryo with a definite chorda canal. *Contributions in Embryology* 23:253, 1932. Courtesy of Carnegie Institution of Washington, Washington, DC.

Figure 5.2A. Modified after Davis.

Figure 5.2B. Modified after Ingalls.

Figure 5.2C. Courtesy of Dr. K. K. Sulik, Department of Cell Biology and Anatomy, University of North Carolina.

Figure 5.3, D and E. Courtesy of Dr. K. K. Sulik, Department of Cell Biology and Anatomy, University of North Carolina.

Figure 5.4. Courtesy of Dr. K. K. Sulik, Department of Cell Biology and Anatomy, University of North Carolina.

Figure 5.5A. Modified after Payne.

Figure 5.5B. Modified after Corner.

Figure 5.6, A and B. Courtesy of Dr. K. K. Sulik, Department of Cell Biology and Anatomy, University of North Carolina.

Figure 5.7. Reprinted with permission from Blechschmidt E. *The Stages of Human Development Before Birth.* Philadelphia: WB Saunders, 1961.

Figure 5.8, A and B. Modified after Streeter GL. Developmental horizons in human embryos: age group XI, 13–20 somites, and age group XII, 21–29 somites. *Contributions in Embryology* 30:211, 1942.

Figure 5.10. Courtesy of Dr. K. K. Sulik, Department of Cell Biology and Anatomy, University of North Carolina.

Figure 5.12. Reprinted with permission from Cossu G, Tajbakhsh S, Buckingham M. How is myogenesis initiated in the embryo? *Trends in Genetics* 12:218–223, 1996.

Figure 5.14. Modified from Gilbert SF. *Developmental Biology.* Sunderland, MA: Sinauer, 2000.

Figure 5.19. Reprinted with permission from Blechschmidt E. *The Stages of Human Development Before Birth.* Philadelphia: WB Saunders, 1961.

Figure 5.20, A and B. Reprinted with permission from Streeter GL. Developmental horizons in human embryos: age groups XV, XVI, XVII, and XVIII [the third issue of a survey of the Carnegie Collection]. *Contributions in Embryology* 32:133, 1948. Courtesy of Carnegie Institution of Washington, Washington, DC.

Figure 5.21. Reprinted with permission from Blechschmidt E. *The Stages of Human Development Before Birth.* Philadelphia: WB Saunders, 1961.

Figure 5.22. Reprinted with permission from Coletta PL, Shimeld SM, Sharpe P. The molecular anatomy of Hox gene expression. *Journal of Anatomy* 184: 15, 1994.

Figure 5.23. Reprinted with permission from Hamilton WJ, Mossman HW. *Human Embryology.* Baltimore: Williams & Wilkins, 1972.

Figure 5.24. Reprinted with permission from Starck D. *Embryologie.* Stuttgart: Georg Thieme, 1965. Courtesy of Dietrich Starck, Professor of Anatomy, University of Frankfurt am Main.

Figure 6.7. Modified after von Ortmann.

Figure 6.13. Modified after Ramsey EM. The placenta and fetal membranes. *In* Greenhill JP (ed): *Obstetrics.* Philadelphia: WB Saunders, 1965; and Hamilton WJ, Boyd JD. Trophoblastin human uteroplacental arteries. *Nature* 212:906, 1966.

Figure 6.19. Reprinted with permission from Stevenson RE, Hall, JG, Goodman RM (eds). *Human Malformations and Related Anomalies.* New York: Oxford University Press, 1993.

Figure 6.20. Reprinted with permission from Stevenson RE, Hall, JG, Goodman RM (eds). *Human Malformations and Related Anomalies.* New York: Oxford University Press, 1993.

Figure 6.22. Reprinted with permission from Stevenson RE, Hall, JG, Goodman RM (eds). *Human Malformations and Related Anomalies.* New York: Oxford University Press, 1993.

Figure 7.2A. Reprinted with permission from Streissguth AP, Little RE. Unit 5: *Alcohol, Pregnancy, and the Fetal Alcohol Syndrome.* 2nd ed. Project Cork Institute Medical School Curriculum [slide lecture series] on Biomedical Education: Alcohol Use and Its Medical Consequences. Produced by Dartmouth Medical School, 1994.

Figure 7.3, A–D. Courtesy of Dr. Hytham Imseis, Department of Obstetrics and Gynecology, Mountain Area Health Education Center, Asheville, NC.

Figure 7.4, A and B. Courtesy of Dr. Hytham Imseis, Department of Obstetrics and Gynecology, Mountain Area Health Education Center, Asheville, NC.

Figure 7.5, A–D. Courtesy of Dr. Hytham Imseis, Department of Obstetrics and Gynecology, Mountain Area Health Education Center, Asheville, NC.

Figure 8.3. Modified from Noden DM. Interactions and fates of avarian craniofacial mesenchyme. *Development* 103:121–140, 1988.

Figure 8.7, A and B. Courtesy of Dr. J. Warkany. Reprinted with permission from Warkany J. *Congenital Malformations.* Chicago: Year Book Medical Publishers, 1971.

Figure 8.9. Reprinted with permission from Muenke M, Schell U. Fibroblast growth factor receptor mutations in human skeletal disorders. *Trends in Genetics* 2:308–313, 1995.

Figure 8.11. Reprinted with permission from Stevenson RE, Hall JG, Goodman RM (eds). *Human Malformations and Related Anomalies.* New York: Oxford University Press, 1993.

Figure 8.14, A–C. Courtesy of Dr. K. K. Sulik, Department of Cell and Developmental Biology, University of North Carolina.

Figure 8.15. Modified from Gilbert SF. *Developmental Biology.* Sunderland, MA: Sinaver, 2000.

Figure 8.16. Shubin N, Tabin C, Carroll S. Fossils, genes and the evolution of animal limbs. *Nature* 388:639–648, 1997.

Figure 8.17. Reprinted with permission from Honig LS, Summerbell D. Maps of strength of positional signaling activity in the developing chick wing bud. *Journal of Embryology and Experimental Morphology* 87:163–174, 1985.

Figure 8.19. Reprinted with permission from Stevenson RE, Hall JG, Goodman RM (eds). *Human Malformations and Related Anomalies.* New York: Oxford University Press, 1993.

Figure 8.20. Courtesy of Dr. A. Aylsworth, Department of Pediatrics, University of North Carolina.

Figure 8.22. Courtesy of Dr. Nancy Chescheir, Department of Obstetrics and Gynecology, University of North Carolina.

Figure 9.2. Reprinted with permission from Cossu G, Tajbakhsh S, Buckingham M. How is myogenesis initiated in the embryo? *Trends in Genetics* 12:218–223, 1996.

Figure 9.5, A and B. Reprinted with permission from Langman J, Woerdeman MW. *Atlas of Medical Anatomy.* Philadelphia: WB Saunders, 1978.

Figure 9.6. Courtesy of Dr. K. K. Sulik, Department of Cell and Developmental Biology, University of North Carolina.

Figure 9.7. Courtesy of Dr. D. Nakayama, Department of Surgery, University of North Carolina.

Figure 10.2, D and E. Courtesy of Jennifer Burgoon, Department of Cell and Developmental Biology, University of North Carolina.

Figure 10.3. Courtesy of Dr. S. Lacey, Department of Surgery, University of North Carolina.

Figure 11.1A. Courtesy of Dr. K. K. Sulik, Department of Cell and Developmental Biology, University of North Carolina.

Figure 11.2E. Courtesy of Jennifer Burgoon, Department of Cell and Developmental Biology, University of North Carolina.

Figure 11.4. Courtesy of Dr. K. K. Sulik, Department of Cell and Developmental Biology, University of North Carolina.

Figure 11.6, A–C. Modified from Kramer TC. The partitioning of the truncus and conus and the formation of the membranous portion of the interventricular septum in the human heart. *American Journal of Anatomy* 71:343, 1942.

Figure 11.6, D and E. Courtesy of Dr. K. K. Sulik, Department of Cell and Developmental Biology, University of North Carolina.

Figure 11.7, A and B. Modified from Kramer TC. The partitioning of the truncus and conus and the formation of the membranous portion of the interventricular septum in the human heart. *American Journal of Anatomy* 71:343, 1942.

Figure 11.7C. Courtesy of Dr. K. K. Sulik, Department of Cell and Developmental Biology, University of North Carolina.

Figure 11.9. Modified from Marvin MJ, di Rocco J, Gardiner A, Bush SM, Lassar AB. Inhibition of Wnt activity induces heart formation from posterior mesoderm. *Genes in Development* 15:316, 2001.

Figure 11.12, B and D. Courtesy of Dr. K. K. Sulik, Department of Cell and Developmental Biology, University of North Carolina.

Figure 11.16, B and C. Courtesy of Dr. K. K. Sulik, Department of Cell and Developmental Biology, University of North Carolina.

Figure 11.17, B. Courtesy of Dr. K. K. Sulik, Department of Cell and Developmental Biology, University of North Carolina.

Figure 11.22, A–C. Courtesy of Dr. K. K. Sulik, Department of Cell and Developmental Biology, University of North Carolina.

Figure 11.26, A–C. After Kramer TC. The partitioning of the truncus and conus and the formation of the membranous portion of the interventricular septum in the human heart. *American Journal of Anatomy* 71:343, 1942.

Figure 11.29, C and D. Courtesy of Dr. Nancy Chescheir, Department of Obstetrics and Gynecology, University of North Carolina.

Figure 13.3, D and E. Courtesy of Jennifer Burgoon, Department of Cell and Developmental Biology, University of North Carolina.

Figure 13.16. Reprinted with permission from Agur AMR. *Grant's Atlas of Anatomy.* 10th ed. Baltimore: Lippincott Williams & Wilkins, 1999:107.

Figure 13.19. Modified from Gilbert SF. *Developmental Biology.* Sunderland, MA: Sinaver, 2000.

Figure 13.26, B and C. Courtesy of Jennifer Burgoon, Department of Cell and Developmental Biology, University of North Carolina.

Figure 13.29. Reprinted with permission from Agur AMR. *Grant's Atlas of Anatomy.* 9th ed. Baltimore: Williams & Wilkins, 1991:123.

Figure 13.31C. Courtesy of Dr. S. Lacey, Department of Surgery, University of North Carolina.

Figure 13.35. Courtesy of Dr. D. Nakayama, Department of Surgery, University of North Carolina.

Figure 13.36, D and E. Reprinted with permission from Nievelstein RAJ, Van Der Werff JFA, Verbeek FJ, Valk J, Verneij-Keers C. Normal and abnormal embryonic development of the anorectum in human embryos. *Teratology* 57:70–78, 1998.

Figure 13.37, A and B. Reprinted with permission from Nievelstein RAJ, Van Der Werff JFA, Verbeek FJ, Valk J, Verneij-Keers C. Normal and abnormal embryonic development of the anorectum in human embryos. *Teratology* 57:70–78, 1998.

Figure 14.3C. Courtesy of Dr. K. K. Sulik, Department of Cell and Developmental Biology, University of North Carolina.

Figure 14.8, A and B. Reprinted with permission from Stevenson RE, Hall JG, Goodman RM (eds). *Human Malformations and Related Anomalies.* New York: Oxford University Press, 1993.

Figure 14.9, D and E. Reprinted with permission from Stevenson RE, Hall JG, Goodman RM (eds). *Human Malformations and Related Anomalies.* New York: Oxford University Press, 1993.

Figure 14.10D. Courtesy of Dr. K. K. Sulik, Department of Cell and Developmental Biology, University of North Carolina.

Figure 14.11C. Reprinted with permission from Stevenson RE, Hall JG, Goodman RM (eds). *Human Malformations and Related Anomalies.* New York: Oxford University Press, 1993.

Figure 14.16, A and B. Reprinted with permission from Stevenson RE, Hall JG, Goodman RM (eds). *Human Malformations and Related Anomalies.* New York: Oxford University Press, 1993.

Figure 14.17, A–D. Courtesy of Dr. K. K. Sulik, Department of Cell and Developmental Biology, University of North Carolina.

Figure 14.25. Reprinted with permission from George FW, Wilson JD. Sex determination and differentiation. *In* Knobil E, et al. (eds): *The Physiology of Reproduction.* New York: Raven Press, 1988:3–26.

Figure 14.32C. Courtesy of Dr. K. K. Sulik, Department of Cell and Developmental Biology, University of North Carolina.

Figure 14.34, A–C. Courtesy of Dr. K. K. Sulik, Department of Cell and Developmental Biology, University of North Carolina.

Figure 14.35B. Courtesy of Dr. R. J. Gorlin, Department of Oral Pathology and Genetics, University of Minnesota.

Figure 14.40E. Courtesy of Dr. K. K. Sulik, Department of Cell and Developmental Biology, University of North Carolina.

Figure 15.1. Modified from Noden DM. Interactions and fates of avian craniofacial mesenchyme. *Development* 103:121–140, 1988, Company of Biologists, Ltd.

Figure 15.2A. Courtesy of Dr. K. K. Sulik, Department of Cell and Developmental Biology, University of North Carolina.

Figure 15.2B. Adapted from Noden DM. Interactions and fates of avian craniofacial mesenchyme. *Development* 103:121–140, 1988, Company of Biologists, Ltd.

Figure 15.5C. Courtesy of Dr. K. K. Sulik, Department of Cell and Developmental Biology, University of North Carolina.

Figure 15.6. Courtesy of Dr. K. K. Sulik, Department of Cell and Developmental Biology, University of North Carolina.

Figure 15.12. Reprinted with permission from Krumlauf R. Hox genes and pattern formation in the branchial region of the vertebrate head. *Trends in Genetics* 9:106–112, 1993.

Figure 15.13. Modified from Trainor PA, Krumlauf R. Hox genes, neural crest cells, and branchial arch patterning. *Current Opinion in Cell Biology* 13:698, 2001.

Figure 15.16A. Courtesy of Dr. J. Warkany. Reprinted with permission from Warkany J. *Congenital Malformations.* Chicago: Year Book Medical Publishers, 1971.

Figure 15.16, B–D. Courtesy of Dr. R. J. Gorlin, Department of Oral Pathology and Genetics, University of Minnesota.

Figure 15.17, C and D. Courtesy of Dr. K. K. Sulik, Department of Cell and Developmental Biology, University of North Carolina.

Figure 15.21C. Courtesy of Dr. K. K. Sulik, Department of Cell and Developmental Biology, University of North Carolina.

Figure 15.22C. Courtesy of Dr. K. K. Sulik, Department of Cell and Developmental Biology, University of North Carolina.

Figure 15.23C. Courtesy of Dr. K. K. Sulik, Department of Cell and Developmental Biology, University of North Carolina.

Figure 15.25, C and D. Courtesy of Dr. K. K. Sulik, Department of Cell Biology and Anatomy, University of North Carolina.

Figure 15.26, C and D. Courtesy of Dr. K. K. Sulik, Department of Cell and Developmental Biology, University of North Carolina.

Figure 15.27C. Courtesy of Dr. K. K. Sulik, Department of Cell and Developmental Biology, University of North Carolina.

Figure 15.29, A–C. Courtesy of Dr. M. Edgerton, Department of Plastic Surgery, University of Virginia.

Figure 15.29, D–F. Courtesy of Dr. R. J. Gorlin, Department of Oral Pathology and Genetics, University of Minnesota.

Figure 15.33. Reprinted with permission from Langman J, Woerdeman MW.

Atlas of Medical Anatomy. Philadelphia: WB Saunders, 1978.

Figure 16.1A. Courtesy of Dr. K. K. Sulik, Department of Cell and Developmental Biology, University of North Carolina.

Figure 16.2, D and E. Courtesy of Dr. K. K. Sulik, Department of Cell and Developmental Biology, University of North Carolina.

Figure 16.3, F and G. Courtesy of Dr. K. K. Sulik, Department of Cell and Developmental Biology, University of North Carolina.

Figure 16.8. Reprinted with permission from Moore KL. *Clinically Oriented Anatomy.* Baltimore: Williams & Wilkins, 1992:764.

Figure 16.10, E–G. Courtesy of Dr. K. K. Sulik, Department of Cell and Developmental Biology, University of North Carolina.

Figure 16.11, A–D. Courtesy of Dr. R. J. Gorlin, Department of Oral Pathology and Genetics, University of Minnesota.

Figure 17.1, D and E. Courtesy of Dr. K. K. Sulik, Department of Cell and Developmental Biology, University of North Carolina.

Figure 17.2, A–C. After Mann IC. *The Development of the Human Eye.* 3rd ed, British Medical Association. New York: Grune & Stratton, 1974.

Figure 17.3. Modified after Mann IC. *The Development of the Human Eye.* 3rd ed, British Medical Association. New York: Grune & Stratton, 1974.

Figure 17.4, A and B. Courtesy of Dr. K. K. Sulik, Department of Cell and Developmental Biology, University of North Carolina.

Figure 17.5. Modified after Mann IC. *The Development of the Human Eye.* 3rd ed, British Medical Association. New York: Grune & Stratton, 1974.

Figure 17.9. Modified from Ashery-Padan R, Gruss P. Pax6 lights up the way for eye development. *Current Opinion in Cell Biology* 13:706, 2001.

Figure 17.10. Modified from Ashery-Padan R, Gruss P. Pax6 lights up the way for eye development. *Current Opinion in Cell Biology* 13:706, 2001.

Figure 17.12. Reprinted with permission from Stevenson RE, Hall JG, Goodman RM (eds). *Human Malformations and Related Anomalies.* New York: Oxford University Press, 1993.

Figure 19.1, A and B, Modified after Ingalls.

Figure 19.1C. Courtesy of Dr. K. K. Sulik, Department of Cell and Developmental Biology, University of North Carolina.

Figure 19.2E. Courtesy of Dr. K. K. Sulik, Department of Cell and Developmental Biology, University of North Carolina.

Figure 19.3C. Courtesy of Dr. K. K. Sulik, Department of Cell and Developmental Biology, University of North Carolina.

Figure 19.4. Courtesy of Dr. K. K. Sulik, Department of Cell and Developmental Biology, University of North Carolina.

Figure 19.5. Courtesy of Dr. K. K. Sulik, Department of Cell and Developmental Biology, University of North Carolina.

Figure 19.6B. Courtesy of Dr. K. K. Sulik, Department of Cell and Developmental Biology, University of North Carolina.

Figure 19.8C. Courtesy of Dr. K. K. Sulik, Department of Cell and Developmental Biology, University of North Carolina.

Figure 19.14. Redrawn from Tanabe Y, Jessell TM. Diversity and pattern in the developing spinal cord. *Science* 274:1115, 1996.

Figure 19.16, A and B. Courtesy of Dr. M. J. Sellers, Division of Medical and Molecular Genetics, Guys Hospital, London.

Figure 19.20, C and D. Courtesy of Dr. K. K. Sulik, Department of Cell and Developmental Biology, University of North Carolina.

Figure 19.26D. Courtesy of Dr. K. K. Sulik, Department of Cell and Developmental Biology, University of North Carolina.

Figure 19.31. Reprinted with permission from Krumlauf R. Hox genes and pattern formation in the branchial region of the vertebrate head. *Trends in Genetics* 9:106–112, 1993.

Figure 19.32. Redrawn from Finkelstein R. Boncinelli E. From fly head to mammalian forebrain: the story of otd and Otx. *Trends in Genetics* 10:310–315, 1994.

Figure 19.33. Redrawn from Rubenstein JLR, Beachy PA. Patterning of the embryonic forebrain. *Current Opinion in Neurobiology* 8:18–26, 1998.

Figure 19.34. Redrawn from Lumsden A, Krumlauf R. Patterning the vertebrate axis. *Science* 112:1109–1114, 1996.

Figure 19.37. Courtesy of Dr. Nancy Chescheir, Department of Obstetrics and Gynecology, University of North Carolina.

Figure 19.38. Courtesy of Dr. J. Warkany. Reprinted with permission from Warkany J. *Congenital Malformations.* Chicago: Year Book Medical Publishers, 1971.

Figure 19.39. Courtesy of Dr. R. J. Gorlin, Department of Oral Pathology and Genetics, University of Minnesota.

Figure 19.40. Courtesy of Dr. R. J. Gorlin, Department of Oral Pathology and Genetics, University of Minnesota.

Figure 19.41. Redrawn from Cordes SP. Molecular genetics of cranial nerve development in mouse. *Nat Rev Neurosci* 2:611–615, 2001.

Index

Page numbers in *italics* denote figures; those followed by a t denote tables.